ROUTLEDGE LIBRAF
HUMAN GEOG

Volume 10

RURAL DEVELOPMENT

RURAL DEVELOPMENT
A Geographical Perspective

KEITH HOGGART AND HENRY BULLER

Routledge
Taylor & Francis Group

LONDON AND NEW YORK

First published in 1987 by Croom Helm Ltd

This edition first published in 2016
by Routledge
2 Park Square, Milton Park, Abingdon, Oxon OX14 4RN

and by Routledge
711 Third Avenue, New York, NY 10017

Routledge is an imprint of the Taylor & Francis Group, an informa business

British Library Cataloguing in Publication Data
A catalogue record for this book is available from the British Library

ISBN: 978-1-138-95340-6 (Set)
ISBN: 978-1-315-65887-2 (Set) (ebk)
ISBN: 978-1-138-96299-6 (Volume 10) (hbk)
ISBN: 978-1-315-65904-6 (Volume 10) (ebk)

Publisher's Note
The publisher has gone to great lengths to ensure the quality of this reprint but points out that some imperfections in the original copies may be apparent.

Disclaimer
The publisher has made every effort to trace copyright holders and would welcome correspondence from those they have been unable to trace.

RURAL DEVELOPMENT
A Geographical Perspective

Keith Hoggart and Henry Buller

CROOM HELM
London • New York • Sydney

© 1987 Keith Hoggart and Henry Buller
Croom Helm Ltd, Provident House, Burrell Row,
Beckenham, Kent, BR3 1AT
Croom Helm Australia, 44-50 Waterloo Road,
North Ryde, 2113, New South Wales

Published in the USA by
Croom Helm
in association with Methuen, Inc.
29 West 35th Street,
New York, NY 10001

British Library Cataloguing in Publication Data

Hoggart, Keith
  Rural development: a geographical
  perspective.
  1. Rural development
  I. Title II. Buller, Henry
  307'.14'091734   HN49.C6
  ISBN 0-7099-3756-3

Library of Congress Cataloging-in-Publication Data

ISBN 0-7099-3756-3

Printed and bound in Great Britain by Mackays of Chatham Ltd, Kent

CONTENTS

TABLES

FIGURES

PREFACE

In reading new texts in rural geography, one is often left with a strong sense of déjà vu. True, some emphasise village morphology and aspects of 'traditional' settlement geography more than others, but there is still a feeling of sameness. References are updated, new figures introduced and original diagrams drawn, yet the structure of material presentation is similar. The design is topic-oriented, not thematic. We have no objections to this format in principle, but for some time have seen the need for a more integrated approach to rural analysis. Many of the processes and causal relations identified when investigating specific topics (like housing, service provision and outmigration) are not only the same, but are mutually reinforcing. Their interactions also induce significant (subsidiary) by-products. It is in this spirit that this book has been written. We have attempted to provide an integrative, analytical approach to rural areas in advanced economies. Causation and the conseqences of societal change have been emphasised, in a framework which tries to draw out processes which operate at different geographical scales (and with varying intensities across space). In fact, the resulting text is neither as rounded nor as polished as we should have preferred. A good share of the blame for this lies in our own weaknesses. Our particular biases towards empirical analysis (rather than 'abstract' theorising) coincides with an absence of satisfactory theoretical models which are specifically for rural areas. In addition, our sense is that the social sciences are presently in a state of what might be termed 'calm turmoil'. There is a yawning gap between structuralist explanatory modes, which emphasise the dominant effects of socio-economic structures – and more behaviourally-oriented explanations. In caricature, these alternative approaches too often either accept structures as given and explain away behavioural divergences as 'contingent relations' or seek generalisation in uniformity of actions, with 'unexplained' variance put down to unmeasurable or yet to be identified variables. Certainly these extreme positions are rapidly losing their appeal. Yet a successful blend of structuralist and behaviouralist positions is some way off. Our contribution seeks to incorporate elements of what we have learnt from this debate, but it does not claim to have addressed this difficult issue head-on. Through the utilisation of a comparative, integrated approach, it does nevertheless seek to provide an alternative framework for examining the problems and potentials of rural areas in advanced economies.

This book has had an interesting and varied gestation. Conceived and started in the 'swamps' of suburban and East End London, most of its creation was against a backcloth of ghetto and gentrification in inner-city Philadelphia and the medieval Cathedral towers of Reims. Then, its latter stages saw drafts scurrying around a triangular path joining Paris, Oxford and central London. Chapters have been lost, then (thankfully) recovered in the trans-Atlantic mail. The whole manuscript saw almost two months of being 'impounded' by British Customs (when the shipping agent went bankrupt!). Typists have left for maternity and Sierra Leone. It has not been dull. On the way we have notched up our fair share of debts of gratitude. It is tempting to start these with the US Postal Service, for never actually losing a chapter. Given the number of our other items they 'mislaid', this must go down as a minor miracle. However, to praise such negative aspects of our experiences would inevitably provoke us into thanking the Thatcher Government for not completely demoralising or closing our educational institutions. To give any such praise would detract from the very real benefits we have received from other people. At a time when cut backs in personnel, funding and even good cheer have made their jobs doubly difficult, we have received immense help from support staff at King's College London and Oxford Polytechnic. They have added to their already enhanced workloads to help us. In particular, Clare Baynes and Clare Peppercorn undertook much administrative and organisational work on our behalf. Roma Beaumont and Gordon Reynell managed to piece together torn, crumpled, vilely drafted and poorly described maps and diagrams, to produce the figures for the text. Liz Howard very ably put up with poor writing, bad English, late submissions and innumerable second thoughts to produce the typed manuscript. Each of these has our sincere and deep thanks. Finally, we should really thank one another, if only because neither of us actually carried out the various threats we made concerning the other's well-being.

Chapter One

INTRODUCTION

Rural life and society is too often the subject of sentimental commentary. This is found in the writings of academics as much as non-academics. As W J Keith (1974:43) noted for prose on the English countryside:

> We tend unconsciously to think of the countryman as a quaint survivor of an earlier stage in the world's evolution - as Hodge, the rustic 'clown', the yokel with straw in his mouth. Once the point is made in these extreme terms, of course, its absurdity becomes immediately apparent, but it is important that we recognise this temptation to think in stereotypes. Any serious discussion of rural life and literature must begin by isolating and analysing our conventional attitudes to the subject.

It is not difficult to explain the quaintness of rural descriptions. In Britain in particular the general population has long held to romanticised images of rurality. The popularity of 'the cause' is seen in newspaper campaigns to 'save the countryside' against 'destruction' (e.g. The Observer Magazine 9 December 1984; see also Youngs 1985). In many nations this sentiment is evinced in the titles and arguments of books which 'attack' modern rural practices (e.g. Jim Hightower's (1973) Hard Tomatoes, Hard Times; Marion Shoard's (1980) The Theft of the Countryside; and Richard Body's (1983) Agriculture: The Triumph and the Shame). Advertisers likewise play on people's sentimentality to invoke associations between their product and 'the good life' of rural communities (Bouquet 1981; Goldman and Dickens 1983). That social researchers also appear prone to project romantic idealisations (Bernard 1973) is no doubt due to the comparatively slight theoretical advances of rural studies (and their seeming isolation from 'urban' theoretical insights). Certainly we are frequently told that rural analyses (in whatever discipline) are retarded in comparison with their urban counterparts (e.g. Bradley and Lowe 1984).

There are perhaps three main roots to the relative weakness of rural studies. First, there is the long and lingering adherence to notions of a rural-urban continuum. True, this is no longer accepted in its extreme form, for the vitality of critiques like Pahl's (1966) are well recognised. Yet the implication that rural areas are poorly integrated into, and slow to catch up with, dominant urban-centred activities is still present in much rural writing. Urban practices are seen to have 'penetrated' rural areas, whether in the economy (Berry 1970), in social values (Fischer 1975) or in political behaviour (Johnson 1972). The underlying implication is that rural areas are acted upon by urban forces; that resistance to change is endemic within the countryside (hence some analysts still distinguish rurality by residents' adherence to traditional values and attitudes - see Pacione 1984:3). Integral to this image is the ideal of the rural community. This concept provides a boundless vestibule for over-idealised commentary. The community, as Effrat (1974:2) observed, is like 'motherhood and apple pie' in that '... it is considered synonymous with virtue and desirability'. Indeed, even in its definition, there is a problem '... in separating the content of the conception from the value-laden imagery of warmth and camaraderie attached to it in many cases'. The purported existence of autonomous village communities, now breaking down under the impress of urban impulses, has induced nostalgic and prescriptive accounts of rural affairs (e.g. Ewart-Evans 1956). These start from an assumption of rural distinctiveness. Thus, in rural sociology, class and community have commonly been taken as polar opposites. Unlike their urban counterparts, rural social divisions have been ascribed primarily to status rather than class (Newby 1980). If this view is accepted, it is easy to find some justification for examining rural localities using distinctive methodological, theoretical and, to some extent, even conceptual frameworks.

Leading on from this first source of weakness, rural studies have over-emphasised issues connected with the consumption (or consequences) of socio-economic change. Apart from investigations of agricultural production per se, extremely little attention has been given to the production of socio-economic or other changes in rural areas. Thus, while studies of rural industrialisation have recently become popular, this owes much to the decline in urban manufacturing (e.g. Fothergill and Gudgin 1982; Keeble 1984); prior to this rural analysts were less concerned with the determinants of rural (industrial) production than they were with the consequences of that production - as seen through the narrow question of local multiplier effects (cf. Summers et al 1976).

Insights on processes of rural industrialisation, as opposed to their impacts on single localities, must still largely be culled from regional accounts of industrial change (e.g. Cobb 1982). Similar biases towards consumption-related issues are found for other activity spheres. With reference to service provision, for example, innumerable accounts exist of cuts in public transport routes, school closures, loss of village shops and the centralisation of health facilities. We know who loses most from these events and, to some extent, why they lose most. Comparatively little has been written about why such losses occur. Often in fact causation has been treated as if it were a 'natural' process; as an inevitable outcome of modernisation or economic advancement. At the same time, studies of agricultural production have been severely weakened by a mythological screen in which the family farm and a competitive, free-enterprise economy is pictured to be dominant in production relationships (Vogeler 1981). Even when production is emphasised, therefore, what drives and determines societal change in rural environments has been insufficiently grasped. Instead, research has concentrated upon the consequences of broader societal changes for rural residents.

A third weakness in rural investigations stems from the separation of agricultural concerns from 'other' issues (these are often referred to as 'rural development' issues; Copp 1972; Buttel 1982). In studies of agriculture there has been an emphasis on questions of production; these have largely been linked to issues of innovation diffusion, productive efficiency and farm structure. Yet with few exceptions (e.g. Heffernan 1972), how these affect rural localities in general has rarely been analysed. This bias is even evident in government policy, for a clear separation of agricultural (production) and other rural policy-making is prevalent in many nations (Buttel 1982). Recent years have seen a closing of the gap between these two; though this has occurred over a quite narrow range of subjects (these overwhelmingly being concerned with landscape issues - e.g. Bowers and Cheshire 1983). Predominantly rural studies remain sectionalised into either production oriented investigations of agricultural systems or more consumption related socio-economic descriptions of settlement systems. The twain meet with insufficient regularity.

When accompanied by the apparent willingness of rural researchers to treat their study sites as isolated entities (or at least as unique locales experiencing 'urbanisation'), the scene is set for an untheoretised, descriptive research agenda. This is precisely what we have in too many areas of rural research and particularly in rural geography (e.g. Clark 1982). Some writers appear to make a virtue out of their supposedly atheoretical stance. For example, in Ambrose's (1974) The Quiet Revolution: Social Change in a Sussex Village 1871-1971, he describes the book in these terms: 'It tries to depict the reality not in any particular way but just to depict it' (p.xiii). Ambrose goes further in refusing to use the word 'class' since people vary in infinitely complex ways and the '... usual attempt to order them into a limited number of categories is analytically imprecise, socially divisive and has outlived its political usefulness' (p.xiii). This is a peculiar commentary from someone who a year later published a book on urban areas (The Property Machine) with a clear class-theoretical, and even conspiratorial, overtone (Ambrose and Colenutt 1975). Class and theory are, implicitly at least, only of urban interest. Given the descriptive thrust of The Quiet Revolution, compared with the explanatory intentions of The Property Machine, we could also assume that studying causation is solely an urban issue as well. This is not to say that rural analysts have not recognised the need to investigate causation, nor have they failed to identify appropriate concepts. In a recent text on rural geography, for example, Michael Pacione (1984:7) informed us that:

> The final chapter of the book is concerned with the fundamental concept of power ... The revealed redistributive effects of rural decision-making processes underlines the conclusion that to comprehend contemporary rural society fully it is essential to understand the locus and exercise of power.

Why then, if political power is so fundamental to understanding contemporary rural society, does Pacione relegate its investigation to a final (almost tacked-on) chapter? Surely the very importance of power relations lies in the fact that they infiltrate every aspect of rural society. Hence, they must be integral to every chapter, not isolated into one.

Fundamental problems in rural studies arise from the weakness of a causal imperative in research frameworks and, where explanation is sought, the individualised nature of accounting frameworks (i.e. their poor links with broader social theory). This is not simply a problem in rural geography, for it exists in a variety of disciplinary approaches to the countryside (e.g. Newby 1983). An outcome of romanticised conceptions of rurality, seen particularly through notions of autonomous village communities, is a reluctance to theorise or generalise. The acceptance of localised communities as the societal core of rural areas has distracted analysts from the role of formal institutions and broader social forces. The idealisation of community as totality, requiring detailed investigation to elucidate its dimensions, has produced a mountain of unique accounts which seemingly defy either abstract theorising or generalisation. By contrast urbanists have been only too inclined to generalise. All too commonly we are told that we live in an urban society, where country areas can conveniently be compartmentalised into urban-centred regions and hence are 'urban' (e.g. Berry 1970; Simmons 1976). This is not a new phenomenon. What notably distinguished the earliest 'rural' community studies (the Lynds' Middletown) from their urban contemporaries (Park's Chicago investigations) was that while Park never ceased to generalise, the Lynds' consistently refused to do so (see Stein 1972). Urbanists have been prone to wrest any international, national, regional or local factor they happen upon and label it 'urban'. Rural investigators have too commonly shied away from all but the local. This division is of course a false one. There are no 'urban' forces, neither are there 'rural' ones. Social agents act, not geographical areas. Certainly, particular processes and patterns of behaviour are more common in urban than in rural areas, but they are neither 'urban' nor 'rural' per se. Social differentiation exists in both urban and rural environments. These are not distinct entities, but their dissimilar spatial structures can induce differences in social behaviour.

It is this that provides justification for presenting a book on rural development. Necessarily, because it is spatially selective in its areas of concern, this book is not directly concerned with the determination of universal processes or social structures. It is concerned with how these universal conditions produce dissimilar local manifestations. The perspective adopted does not envision rural localities as empty receptacles which act as passive recipients of mass society forces (their peculiar geographical properties perhaps then accounting for dissimilarities in the relative potency and hence the interaction of extra-local pressures). Instead, the manner in which local socio-economic systems positively and causally interact with more universal forces is explored. In essence this book seeks to elucidate the causal links between local and non-local processes which produce differences in development in rural areas.

This emphasis on development is perhaps not a common one in rural geographical circles. Textbooks on rural geography have largely been characterised by a concern with specific topics; in many instances investigated in an uncoordinated, isolated manner. Chapters on rural depopulation, counterurbanisation, national parks, public transport (or public service) problems and 'planning' are frequently present in these texts. These issues all have development implications, but authors have rarely teased these out in their work. Thus, Hugh Clout (1972:1) defined the subject

5

area of his <u>Rural Geography</u>:

> ... as the study of recent social, economic,
> land–use, and spatial changes that have taken
> place in less–densely populated areas which are
> commonly recognised by virtue of their visual
> components as 'countryside'.

The treatment of these changes is by individual topic, yet the
very emphasis on <u>change</u> predefines the subject matter as
developmental in orientation. If, as we contend here, development must be evaluated with reference to people (as individuals or social groups), rather than artifacts (services,
money, housing), then an issue–oriented approach must detract
from recognition of the correspondence and interweaving of
development processes. The intention in this contribution is
to focus on these inter–linkages.

Of course, to do this adequately we need to define the
concepts with which we will be dealing. Hence, the next
chapter is directed toward providing the reader with the
conceptual background to this book. In essence, our concern
in Chapter Two is with the interpretations of 'rural' and
'development' which will be employed in this text. This
conceptual review provides the framework for evaluating the
theoretical models of rural development presented in Chapter
Three. It should be stressed at this juncture that models of
rural development <u>per se</u> are notable by their absence. In
effect this third chapter is devoted to investigating theories
of societal organisation as a whole. These have implications
for all development processes and need assessing in
conjunction with models which account for variation across
localities. From here, the book takes an explicitly empirical
thrust. In organisation it adopts a spatial scaling sequence.
Chapter Four starts by concentrating on international
components of rural development. Its focus is on processes
and structures through which rural development is either
restricted or enhanced (but generally is broadly directed) by
the international interactions of governments and profit-
making institutions. A similar general theme runs throughout
Chapter Five. The main difference is that emphasis in the
fifth chapter is devoted to the national level of analysis.
Except that it is at a different scale, the aim in Chapter
Five is akin to that of Chapter Four; namely, to assess how
social structures and practices induce uneven spatial and
socio–economic distributions in rural development opportunities. The orientation of Chapter Six is somewhat different,
since its focus is on localities. These are social units for
which a corporate response from rural inhabitants is both more
feasible and more likely. Our interest here lies in the
manner in which rural residents in general seek to change

socio-economic conditions in their own locality. In addition to local pressures for change, Chapter Six also investigates how the broader societal forces examined in Chapters Four and Five conspire to produce inter-local disparities in development opportunities. Such inequities are of course integral to changes in societal conditions. They do not simply have a spatial dimension. In addition, there are differential impacts on social groups (distinguished by class, ethnicity, gender, power and status). Aspects of these social differences are integral to our analysis in all chapters. In Chapter Seven they form the core subject matter; here they are examined explicitly in terms of social processes within rural localities. Using the four geographical tiers of these chapters, it is intended to illustrate similarities and dissimilarities of causal processes at different spatial scales. This then leads us into our last chapter, in which the applicability of existing approaches to social research of rural development is examined.

Chapter Two

CONCEPTS

There are two parts to this chapter, each of which focuses on the meaning and components of a single concept. The two concepts are those which comprise the title of this book; namely, rural and development. In essence, the goal is to lay down a framework for the following chapters by outlining the interpretation of these concepts which is used in this text. No major or detailed review is intended. Rather, the aim is to indicate why particular conceptual approaches have been favoured over others.

ON 'RURAL'

The most strongly established and distinctive social research discipline which concentrates on rural (as opposed to agricultural) activities is undoubtedly rural sociology. Our quest for a definition of 'rural' begins with this discipline, because its central underpinning is that rural is a valuable analytical construct (with an empirical referent). It has to be acknowledged that rural sociologists have debated both the meaning and the usefulness of this concept virtually since the discipline was formally recognised. Certainly, many rural sociologists believe that 'rural' needs a more precise definition than has served in the past (e.g. Wilkinson 1985). Indeed the continuing absence of a generally accepted definition of 'rural' appears to have affected this discipline's self-confidence. Even its advocates have often criticised the discipline's poor theoretical standing and lack of analytical rigour (e.g. Newby 1980); both of which owe something to uncertainty over what the discipline's subject matter is (Bradley and Lowe 1984). While some see this as a fatal weakness, looseness of definition over a discipline's scope, content and even main themes is not problematical in its own right. In truth, any demarcation of disciplinary boundaries must be an artificial imposition. It can be justified because it promotes more detailed analysis of specific aspects of societies. Yet it carries the danger of promoting narrower and potentially misleading insights. To study, say, the economic aspects of transnational corporations without considering social, political and geographical dimensions inevitably leads to partial understanding. Similarly, much can be missed if researchers assume that economic, social and political processes are geographically undifferentiated. Geographical space is not a vacuum within which societal processes operate in an unhindered, uniform manner. Space constrains, structures and promotes opportunities; not simply because of simplistic geographical features (like distance), but also because current spatial arrangements are outcomes of past processes whose interactions generate (geographically) 'unique' constraints and opportunities (Urry 1981; Massey 1984). Perhaps the restraints of space per se have declined over time, but these are integral to past societal formations which continue to exert influence on current behaviour patterns.

If 'rural' is a valuable concept as a discriminator of alternative social formations (i.e. if it is considered worth-while distinguishing rural areas from others), then by impli-cation features of the rural environment should have causal effects on human behaviour. That universal agreement does not exist on what 'rural' actually is, is unfortunate but is consistent with the multidimensional character of many social science concepts (social class, for instance, embodies ambiguities over the positions of the petit bourgeoisie). In reality, differences in definitions of rural are not large. In the main, three substantive, but inter-related, meanings exist for this concept (Bealer et al 1965). These are the socio-cultural, the occupational and the ecological.

Socio-Cultural Definitions

Socio-cultural definitions of 'rural' rest on the assumption that behavioural and attitudinal differences exist between inhabitants of areas with low population densities (i.e. rural areas) and zones of high population density (i.e. urban areas). In this conception, rural is associated with an adherence to 'traditional' value-systems, which stress the merits of religious adherence, respect for 'elders', the importance of the family, a strong sense of community and suspicion of change in the socio-political status quo (the opposite supposedly applying to urban areas). Somewhat un-fortunately, this conceptualisation has taken on normative overtones in much non-academic and some academic writing (and has, on occasion, blurred the distinction between these two).

Even when researchers do not seem intent upon projecting a normative commentary on the desirability of small settlement living, sentimental language frequently pervades descriptions of rural life. Schaffer's (1976:61-2) anthropological investigation of the Irish community of Rosscarbery provides some portrayals of this practice:

> To live in Rosscarbery is to know the security of custom, and to know that few houses, barns or outbuildings are younger than 100 years, and that few families have been here for much less than that. It is to know your kin and your neighbour, and to know what to expect from each day. It is to know that when you walk out the door in the morning, to fetch the cows for the milking, that the heron will rise, flap twice, and glide further down the lake, and that the swans, with a furry string of cygnets trailing faithfully behind, will be swimming near the shore. It is to know that the days were longer in your youth and that piercing your ears will improve your eyesight and that if one of your cows dies you can avoid bad luck by throwing it in another man's field. To live in Rosscarbery is to be part of the community, to think you know your neighbour's business, and to be bound by what you know is right and held to it by the church and the potential judgement of others. It is to anticipate the fairs and carnivals and the tourists in summer and all the simple deviations from your daily routine. It is to give a favour, unhesitatingly knowing that it will be returned in kind. It is being whatever you are, without flourish, doing whatever you can, without derision. It is being with family, and sometimes with friends, on long winter nights, before the fire, with tea and cheese and homemade bread and local gossip and long conversations about politics and farming and the shape of days ahead. To live in Rosscarbery is sometimes good, and sometimes not. It is all this - and much more.

The words 'mythical', 'mystical' and 'ideological' all spring to mind as broad caricatures of this kind of commentary. It is the hoped-for-dream that many former urban residents are in search of when they move to rural areas (e.g. Forsythe 1980). The expectation is that the rural life will be quite different from urban living (Table 2.1). In some measure this expectation is not so far-fetched. In the United States, for example, those living in settlements of 2,500 persons or less are 93 per cent white, 80 per cent Protestant and 47 per cent

Table 2.1 Arizona Beliefs on the Attributes of Rural Areas

- Place in which to raise children

- Friendliness of people to each other

- Community spirit and pride

- Residents' voice in deciding community affairs

- Quality of religious life

- Residents' general satisfaction

- Respect for law and order

- Outdoor recreation opportunities

- Residents' general mental health

- Absence of illegal drug use

Note: Three quarters or more of a state-wide random sample
         of urban Arizona residents rated these factors as
         better in nonmetropolitan areas than in metropoli-
         tan areas.

Source: Blackwood and Carpenter (1978)

---

of adults have no high school diploma. The comparable
percentages for central cities with 250,000 plus residents are
68, 56, and 30 (Bryan 1981). We must nonetheless be careful
about transcribing such socio-demographic characteristics into
behavioural and attitudinal attributes.

    It is true that some people do exhibit value dispositions
representative of classical rural (and urban) ideals.   To
assume that these ideals are either dominant or spatially selective would nevertheless be a mistake.   For example, when
drawing on nationwide opinion surveys, Miller and Luloff
(1981) found that 88 per cent of their American respondents
could not be placed into a 'pure' (rural or urban) cultural
type.   Further, of those so placed, only 47 per cent lived in

a settlement type coincident with their dominant value orientation. This hardly forms an adequate basis for a geographically-bounded socio-cultural definition of 'rural'. Weakness exists not simply because comparable attitudes are maintained in dissimilar geographical environments, but also because a value-based definition of 'rural' requires that uniformity exists in rural residents' attitudes. Such unity does not exist. Flinn (1982), for one, has identified three distinct value dispositions which commonly exist in rural areas:

> small-town ideology - the belief that small towns promote a 'natural' life style and are the home of democracy;

> agrarianism - which holds that agricultural life is 'natural' and that the family farm is an ideal place to raise a family, important for democracy and the best mechanism for efficient, plentiful food production;

> ruralism - which portrays farmers as exploiters of nature and cherishes the countryside for the room it offers, better health and closer ties with nature.

Added to these are differences of social class and locale of residence, neither of which are coincident with the above categories (Flinn 1982).

There is abundant evidence from a variety of sources that casts doubt on the homogeneous socio-cultural character of rural areas. Davis (1973), for example, reports that open country dwellers in the Pisticci (southern Italy) are considered 'outsiders' by village dwellers, who characterise them as rough, liable to violent behaviour and not subject to the common rules of decency and cleanliness. Evidence from elsewhere confirms that this characterisation is not unique to this region or country (Burnett 1951; Pitt-Rivers 1960). Yet the reverse relationship also occurs, most especially in areas of large farms, where farmers often look down on small town dwellers and avoid their company (e.g. Oxley 1974; Bax 1976). Even within country towns and villages, highly stratified social structures have been identified, and these frequently have spatial manifestations (e.g. Ambrose 1974; Wild 1974). A further question mark over traditional images of rural areas is raised by the rapidity of population turnover that frequently exists. Population instability, resulting from an influx of 'outsiders' disturbing the 'tranquility' of 'traditional' rural communities, is not a recent phenomenon (although this impression is too readily left by commuter village and counterurbanisation studies). Population turnover has long been an enduring feature of rural environments. As an illustration, Wylie (1974:352) reported that of the 779 people who, in 1946, lived in Peyrane in the Vaucluse, only 275 were still there in 1959 (and only 137 of these were born in Peyrane; see also Gagnon 1976). The truth in Peyrane is that, with the exception of a small core of residents, population composition is dynamic. It is incorrect to assume that smaller localities are largely characterised by population stability or, if peripheral to major urban centres, by outmigration (e.g. Clout 1972). As Grafton (1982) has illustrated for English rural areas, those peripheral regions which (in the past) did suffer from net outmigration had similar levels of gross outmigration to those of metropolitan growth villages. Where they differed was in gross in-migration rates. Turbulence in rural population structures is common, not exceptional (see also Williams 1963; Wylie 1966; Brown 1967).

Without labouring the point, there is ample evidence that behavioural distinctiveness is not a clear-cut characteristic of rural locales. Certainly, there are differences in the social environments of rural and urban areas which induce dissimilar behavioural responses (e.g. Bechtel 1970; Newman and McCauley 1977), but these do not dominate societal behaviour. There are few social differences that merit a spe-cifically 'rural' designation (Dewey 1960). Even commonly assumed social discriminators, like psychological disorders (Mizruchi 1969), crime (Wilkinson 1984) and such maladies as cancer (Greenberg 1984) are in fact less of a distinguishing feature than a uniting element for city and remoter rural environments.

Our point here is not that rural areas do not contain in-habitants (and even localities) which fit a 'rural culture' ideal-type. Rather it is that (relative) population sparsity is not a primary causal mechanism which accounts for this. Social relationships in rural areas exhibit a wide variety of forms and can be subject to rapid change. As Pfeffer (1983) has ably shown, the same desire for maximum profits on the part of large rural landowners has, on account of peculiarities in timing, produced quite distinctive socio-economic environments in the United States. Social differences between the large landowner - wage labourer social structure of California, southern sharecropping systems and family farming communities in the Midwest are not the result of distinctive rural attributes. Rather, they emerged as peculiar geographical manifestations of the struggle between landowners and wage labourers. Yet, in a variety of circumstances, whether as an outcome of paternalistic landowner power relationships (Newby 1977) or due to the struggle for survival amongst homesteaders (Worster 1979), these different social forms have helped induce a sense of communal identity. Nonetheless, this kind of community spirit has been a fragile entity. Based essentially on a mutuality of poverty, in many instances it has broken down with increased wealth. A clear illustration of this is found in Ireland. Here, as emigrants abroad sent money back to their relatives, rural families came to depend less on their neighbours. Freed from the shackles of poverty, these outside funds allowed Irish families to pay for work assistance and entertainment. Once choice was (realistically) available, families chose to go their own way. Not surprisingly, as a consequence, the social integration of Irish rural communities has been severely weakened (see Brody 1973). Innovations like television have had a similar impact, for they have reduced dependence on neighbours for entertainment and have allowed people to pursue a more individualised social existence (e.g. Gallagher 1961; Brody 1973). Such social innovations demonstrate to us that rural residence per se does not lead to uniformity of social organisation. Yet these illustrations show this by focusing on change over time in single locales. What we also need to recognise is that marked intra-rural social differences exist at any one point in time. This is readily seen in comparing agricultural labour in the United States and England. The insecure, mobile and often brutal social existence which is imposed on much agricultural labour in the United States (Friedland and Nelkin 1971; Baker 1976) sharply contrasts with the tightly-knit occupational communities of agricultural workers in England (Newby 1977). It is not rurality per se; but the adaptations which emerge from struggles for power, wealth and status which induce these (and other) social forms (e.g. Blum 1971). A socio-cultural image of 'rural', while adding some useful insights in particular instances (e.g. Fischer 1975), is open to too many objections to constitute a valid conceptual base.

Occupational Definitions

As an alternative definition, social researchers have sought a basis for rural-urban distinctiveness in occupational terms. This approach rests on the dominance of primary industries, and particularly agriculture and forestry, in rural locales. This is a basis of division we do not accept. For one thing, this distinction has commonly been grounded in the notion that agriculturalists possess distinctive social values. Evidence from the past has provided some grounds for sustaining this belief (Glenn and Hill 1977). Yet the trend in farm structure is towards a bifurcation into large-scale, often corporate or partnership firms, and small-scale, increasingly part-time, farms (Vogeler 1981). In other words, it is now somewhat misleading to think in terms of the farm sector as a unitary whole. Class differences between farmers have grown (though they have long been significant in some areas) and are linked to significant attitudinal differences (e.g. Coughenour and Christenson 1983). In addition, the integration of farmers into non-farm labour markets (i.e. multiple job holding) is so highly advanced that treating farmers as a distinctive occupational group is unwarranted (Buttel 1982a). In any event, agriculture provides employment for too small a proportion of the rural work force to shoulder an occupational definition for 'rural'; and its share continues to fall.

Perhaps rather than occupation, it is the nature of the work environment that is critical. Certainly, this is the impression given in some analyses of tensions caused by the 'urbanisation' of the countryside. Vincent (1978), for example, characterises the tourist institutions in the Italian Alpine community of St Maurice as introducing capitalist relations into a non-capitalist family farm setting. This image is undoubtedly a misleading one if it is taken to imply that agriculture at anything other than subsistence level is not intricately linked with, and subject to the vagaries of, a market economy (e.g. Bernier 1976). Where this conception has a clear foundation in reality is in suggesting that rural areas have been characterised by small-scale family owned enterprises. Capitalist (employer - multiple employees) social relationships have been slow to penetrate into rural areas. Family farms, with no regular hired labour, are dominant in advanced economies; and this pattern of petty commodity enterprise has its counterpart in the small scale of commercial and industrial enterprises in rural areas. Within the same economic sector, perhaps attitudinal conservatism and a relative unwillingness to take risks (given a poor competitive position relative to larger enterprises) was the norm for smaller enterprises in peripheral locations over much of this century (e.g. Chadwick et al 1972). Yet scale of enterprise hardly constitutes a viable basis for separating out rural areas (recent research suggests that rates of innovation are as high or higher in small rural enterprises than in urban institutions; Pellenbarg and Kok 1985). In the past, enterprise size was closely linked to locational characteristics (albeit not in a dichotomous manner). Today, however, rural areas are increasingly housing large-scale activities. This especially applies to 'undesirable' facilities - like nuclear power stations - and, for profitmaking institutions, 'high risk' endeavours that seek the cushion of government financial support which goes along with locating in 'backward' areas (see Davies 1978). In addition, across the bulk of economic sectors, the trend in rural areas is towards larger sized units (as in agriculture - Vogeler

1981). Even here then, a work-based definition is inadequate.

Ecological Definitions

This leaves us with the third definition of 'rural'; that based on ecological characteristics. 'Rural' in this context refers to areas in which settlements are small, with substantial zones of open countryside between them. What exactly 'small', 'settlement', 'substantial' and 'open countryside' are is open to dispute. Whatever definitions are used are arbitrary. There are no sound theoretical reasons for demarcating any single measure of housing proximity (for a settlement), settlement size or distance between settlements as having critical classificatory properties. Green's (1971) argument in favour of a regionally based definition of rurality (distinguishing areas which are tightly integrated into the socio-economic affairs of nearby metropolitan centres from those that are not) has merit in this regard. It relies on a more extensive geographical categorisation than specifying individual settlement attributes. Yet it still rests on arbitrary decisions about what a metropolitan centre is (hence how many regions should be specified) and where regional boundaries should be drawn (so that while definitions of rural areas show broad similarity, they are quite different in detail; see Figure 2.1). With these problems in mind, it is more advisable to recognise that 'rural' is a 'chaotic' concept (Urry 1984). It lacks theoretical significance in its own right, but does signify a category of localities which might share properties of some causal importance. What distinguishes rural in our minds is the small size of settlements. Type of economic activity is not important. Durham's mining villages are as much rural as Newfoundland's fishing outports and hamlets on the Great Plains. They are different, but they are still rural. We claim no homogeneity amongst rural areas. What we do maintain is that small settlements raise particular potentials and problems for social, economic and political action which merit attention in their own right.

The pertinence of the preceding pages of this chapter might well be questioned given the simplistic approach to rurality that we eventually arrive at and adopt in this book. That the reader has been subjected to a roundabout dialogue is deliberate. There are significant differences in conceptualisations of 'rural'. These different beliefs not only affect the theoretical models which are deemed pertinent for explaining events in rural areas but also cast dissimilar shadows over the causal significance of 'rural' per se. It was therefore pertinent that we evaluated these alternative conceptions and made clear that, in our view, the only valid

Figure 2.1 Alternative conceptions of rural England and Wales

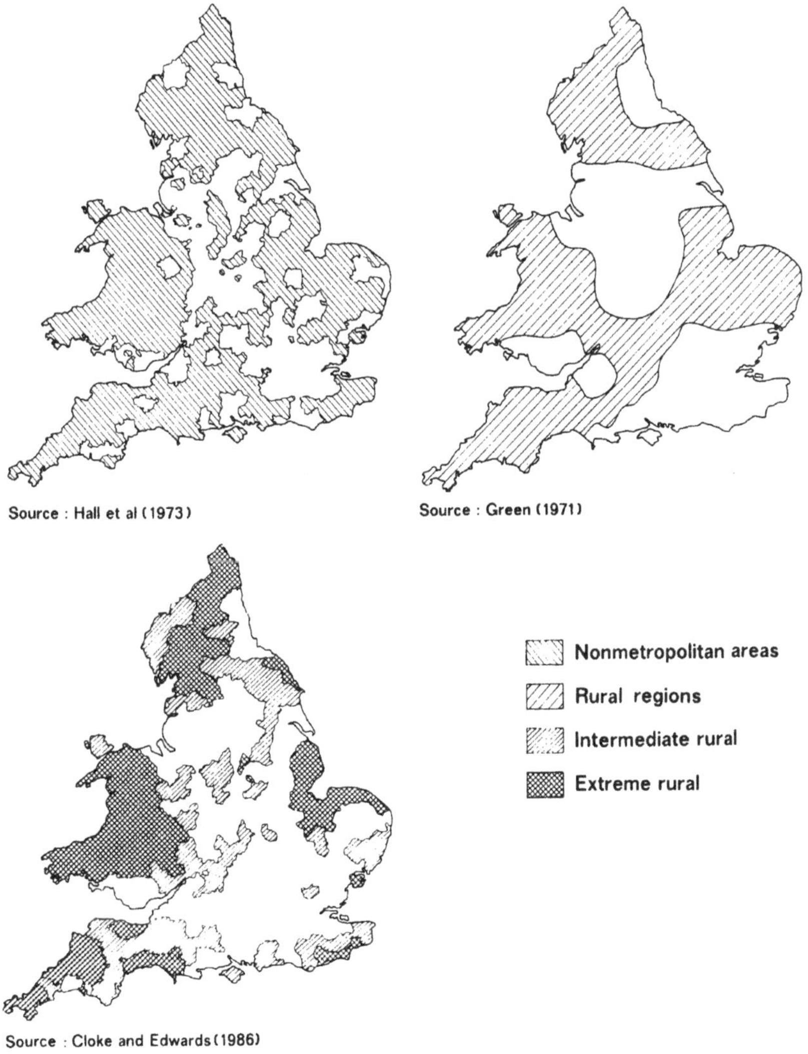

Source : Hall et al (1973)

Source : Green (1971)

Source : Cloke and Edwards (1986)

Nonmetropolitan areas

Rural regions

Intermediate rural

Extreme rural

means of distinguishing rural areas is settlement size. Further, while settlement size might well (in some circumstances) be causally significant, rural areas contain far too much variety for rurality per se to be afforded causal significance. 'Rural', like 'urban', is a generic term that covers a multitude of circumstances.

ON 'DEVELOPMENT'

Having established that 'rural' is a concept where interpretation is inevitably somewhat arbitrary, the reader should be prepared for a similar conclusion for the concept 'development'. As Welch (1984:4) put it:

> ... the term development has become little more than the lazy thinker's catch-all term, used to mean anything from broad, undefined change to quite specific events. As a purveyor of information, the term is virtually useless.

This is hardly an auspicious start for a book that carries 'development' as a primary element of its title. Yet it is difficult to disagree with Welch. When commentators speak or write about 'development' they frequently mean very different things. Land-use planning in Britain is often referred to as development control, yet the construction (or not) of housing conjures a quite different image from that of development as an innovative process leading to the structural transformation of society. It is in fact extremely difficult to reconcile the variety of definitions that exist for development, since at heart this concept is intrinsically value-laden. For some, the world-wide spread of Coca-Cola, Superman comics, the tie, and wage labour are all signs of development at work, for others they are decidedly signs of anti-development. Our goal in this section is not to propose a definitive model of development but to identify dimensions and problems in assessing development. What development actually is is a personal evaluation.

Even granted this, we can put forward a basic framework for understanding development. Disagreement on its meaning is not so great that no common features can be identified. At its most basic level, development is a normative concept which implies improvement. Stated more formally, development:

> ... covers the entire gamut of changes by which a social system, with optimal regard for the wishes of individuals and sub-systematic components of that system, moves from a condition of life

> widely perceived as unsatisfactory in some way
> toward some condition regarded as 'humanly'
> better. (Goulet 1971:333)

From this we can establish a number of principles. First, as
a wide variety of researchers have recognised (e.g. Copp 1972)
development is first and foremost about people. It is not
assessed with reference to electricity consumption, income,
formal education, employment levels or housing standards. All
these might well contribute to improvements in human well-
being, but they are merely means to an end, not the end
itself. Second, development must be assessed with reference
to more than single persons. This is perhaps most formally
proposed in neo-marxist analyses, wherein development is
evaluated with reference to the standing of social classes. On
a different plane, it can also be witnessed by comparing the
fortunes of, say, individual families with those of other
village residents. Consider, for example, the outcome when a
family migrates away from a rural settlement in order to
improve its lifestyle. This will probably lead to desirable
improvements for the family involved. But, can it constitute
development if that departure results in thresholds being
reached which prompt the closure of the village school, shop
or public transport service? Of course, it is unlikely that
the removal of a single family will have such an effect, but
the principle is a basic one (as seen in community responses
to stop emigration in times of economic hardship - e.g.
Worster 1979). Third, and obviously implicit within the last
two, there is the issue of value differences. There can be
little doubt that the dominant ideological value system in
western democracies is supportive of the idea of self-better-
ment. In the main this is interpreted with reference to
social mobility (which often has geographical overtones),
income increases and, through these, an improved level of
household consumption. Such values can directly clash with
other, more durable values. Thus, Emmett (1964:12) reports
that in the village of 'Llan' in Merionethshire

> ... most of them value Wales more than they value
> worldly achievement: they would rather see their
> sons shovelling manure or working in the factory
> than see them go to London. (see also Kenny 1961;
> Matthews 1976; Dunkle et al 1983)

To be 'un-touched by the spirit of the profit', as Morin
(1970) found many traders were in Plodemit, Brittany, is
frequently interpreted as representative of traditional value
systems characteristic of remoter rural areas. But is it
really this? An alternative explanation is provided in
the marxist idea that people's alienation from societal norms
is linked to the exploitation of workers under capitalist

relations of production and exchange. There is certainly evidence that 'capitalist' work relations are conceived as being alienating by many rural residents (see Vincent 1978; Bradley 1984). Viewed in this light, in rejecting the dominant (pro-growth) capitalist ideology, many rural inhabitants are actually expressing their preference for a more developed lifestyle (i.e. less alienated). From this perspective, 'traditional' values should not be condemned but should be praised because they encourage rejection of a dehumanising, anti-egalitarian model for society. As Goulet (1971:217) noted:

> A prosperous society whose members are manipula-
> ted by an impersonal system is not developed, but
> distorted. A society has 'anti-development' if
> its 'development' breeds new oppressions and
> structural servitudes.

This raises a fourth issue from Goulet's (1971:333) definition of development. This is that he refers to change and not growth. All too frequently development is assumed to possess economic growth consequences (as seen in income increases, new investment, more consumption). All Goulet stated is that conditions change. If the population sees it as an improvement that wealth is equalised (even if this involves a reduction in total wealth), this is development.*

In the above situation it is to be expected that for wealth to be equalised some must lose for others to gain. It is unlikely that the former will view the loss of their wealth as development. Hence we again return to the problem of value differences. This is certainly a tricky issue, for people can

---

* The reader should be aware of a weakness in this last sentence and in Goulet's (1971:333) definition. This is the implication that individuals can collectively decide on what development is (or what is 'optimally beneficial') in the abstract (or in an unbiased manner). This essentially liberal utilitarian view is flawed because people cannot evaluate alternatives 'outside' the values which have dom- inated their socialisation in society. In other words, existing social structures constrain and direct individuals' assessments of situations and their beliefs about desirable outcomes. Any aggregation of personal preferences into a so-called majority preference inevitably incorporates elements of dominant societal norms. These are strongly influenced by major power interests, which might comprise a minute proportion of the total population (for a theoretical discussion of this see Lukes 1974, with Gaventa 1980 provid- ing an empirical illustration of this point).

see the same occurrence as development in one circumstance but view it as the reverse in another. This is well illustrated in Cottrell's (1951) paper, 'Death by dieselisation'. In this study of Caliente (Nevada), Cottrell reported that community residents professed support for 'The American Way', with its emphasis on economic growth and technological progress. Yet, when the economic life of their town was threatened by such progress (the change to diesel trains from coal-fired engines meant this railroad station was no longer needed as a fuelling and watering point), residents criticised the railway company for not taking loyalty and community morale into account in its closure decision. If it happens elsewhere it is development, if in your own backyard? This is perhaps a somewhat unique example, but the same principle is at stake for village school closures, the decline of public transport, the location of nuclear power plants, oil wells and strip mining sites, and change in landscapes to provide forestry, water or tourist resources. It is not an isolated problem. It manifests itself in a variety of different guises and emerges as a continuous process.

With so many potential bases for disagreement, Welch's (1984) contention that on its own the term development is virtually meaningless has much merit. Further, from the preceding commentary it should be clear that it is impossible to reach universal agreement on what development is. Nevertheless, we can attempt to impose some order on the 'chaos' that appears to exist. Perhaps the most widely recognised model which attempts to do this is that of Abraham Maslow (1970). It must be acknowledged from the outset that Maslow's schema is linked to the individual psyche (particularly the development of personality) rather than broader socio-economic development processes. Yet, while subject to question marks over both its rigidity and personal focus (e.g. Macpherson 1977), it has been adopted with credit by a number of researchers to analyse development issues (e.g. Manzer 1974).

The basic framework Maslow (1970) presented is a hierarchy of needs. In essence, he argued that people's 'needs' can be categorised by their relative importance (it is perhaps pertinent to view 'need' in this context as something that people are least prepared to sacrifice). Certain needs are more important than others and people are less willing to forego their satisfaction. In order of need of priority, Maslow suggested the hierarchy:

> physiological needs, like sleep and food - these
> are needs that must be satisfied in order to
> preserve life.

safety needs, like freedom from anxiety, a sense
of security and stability - viewed at its
simplest level, if food is obtainable for sustenance, but is irregular in supply, then the
stability of a regular food supply is a development. Suggestive of the significance of safety
needs are some researchers' observations that it
is making a living rather than making a profit
that is critical for many rural businesses (e.g.
Morin 1970; Dunkle et al 1983).

belongingness and love needs have seldom been
analysed, but there are numerous indications of
their importance. Fitzgerald (1977), for
example, has pointed out that studies of children
raised in institutional settings, where
physiological needs are well-provided for but
emotional needs are (comparatively) ignored,
illustrate that a deprivation of 'belongingness'
feelings can cause both emotional and physical
damage. In rural settings the pertinence of
these needs is well illustrated in reports that
the self-estimation of villagers rests as much on
a sense of 'belonging' as on self-achievement
(e.g. Emmett 1964; Strathern 1982). It is also
seen in the willingness of some rural residents
to forego improved public services and higher
incomes (resulting from outmigration) in order to
preserve the social vitality of their community
(e.g. Matthews 1976; Richling 1985).

esteem needs incorporate desires for achievement,
adequacy or mastery, along with an accompanying
positive reputation. In essence this refers to
the need for self-confidence (i.e. for an
individual's belief in his or her own worth and
capabilities). An important dimension of such
needs is that they can lead to very similar
behaviour patterns for social groups which
actually have quite different goals. Hence, the
fact that lower-income groups want to increase
their income and live in more comfort does not
mean that they wish to copy middle class or
upper class behaviour patterns (Duhl 1963).

self-actualisation, described by Maslow (1970:
46) as, 'What a man [sic] can be, he must be. He
must be true to his own nature'. This is the
most difficult of Maslow's categories. It
effectively signifies that there is no single
model for 'improvement'; that people should

pursue dissimilar higher order goals according to their own preference structures.

aesthetic needs, which Maslow believed we know less about than other needs and which some reviewers of Maslow's hierarchy do not include in their scheme (e.g. Fitzgerald 1977; Macpherson 1977). This recognises that a 'craving for beauty' is an integral part of people's life and experiences.

That aesthetic needs were placed lowest in the hierarchy indicates that, in Maslow's view, aesthetic needs are only important once other needs are satisfied at a critical level (i.e. aesthetic considerations are of little consequence when people have insufficient to eat and fear for their lives). That this assumption is conditioned by national cultures should be clear. A comparison of the United States and Britain, for example, clearly reveals that aesthetic considerations (most especially regarding landscape) are much more strongly developed in Britain. This applies even though the basic need of employment (which has strongest impact on physiological and safety needs) is probably much less satisfactorily met in Britain than in the United States.

There are a number of points that need to be raised about this hierarchy of needs. First, Maslow saw room for overlap of needs. In other words, one need does not have to be completely satisfied before the next has any significance. The hierarchy suggests that new and 'lower priority' needs only receive attention and articulation when more basic and essential motives are satisfied to some 'critical' level. Second, the hierarchy is obviously value-laden and is culturally specific. Third, the scheme does not distinguish societal and personal needs or draw attention to potential conflicts between them. It might, for example, be in the interests of some social groups to promote the satisfaction of aesthetic needs even though this will hurt other groups who are a long way from having their more basic needs met (thus middle class, ex-urbanite pressures for a 'pretty' landscape in commuter belt zones can directly restrict employment, housing and service facilities for working class residents; e.g. Connell 1978). This raises a fourth point, which is partly a component of the third, but does raise more fundamental issues. This is that Maslow's hierarchy does not make allowance for the distribution of societal benefits (it is of course not intentionally designed for society specific evaluations). Most people would accept that to have more of something that is beneficial is a development, as it is to possess the same amount of that entity but of higher quality. These are quantity and quality components which inevitably form part of any assessment of development (Maslow's hierarchy focusing on the quality aspect). In addition, however, there is the distributional component. The question raised here concerns the extent to which development can be said to exist when societal benefits are unevenly allocated.

Clearly, any answer to this question will depend upon value judgements and the particular distribution pattern under consideration. Even those who believe that inequality generally arises because 'the best' rise to the top of the pile, might be somewhat perturbed by a social system in which millionaires coexist beside a large body of starving people (most especially if the regime itself is totalitarian). Society is however composed of people with markedly different ideological stances. Some people find little discomfort in their support of oppressive regimes like present-day Chile and South Africa or Stalinist Russia. In part this emerges from pure self-interest or emotion-charged biases (like racism). To an important degree, it arises on account of dissimilar views on what aspects of inequality are most important. In the United States, for example, there is no intention to promote equality of economic returns. Instead, the principal egalitarian goal (hypothetically at least) has been to promote equality of political opportunity. Quite the reverse applies in the Soviet Union. Little emphasis is given to equality amongst individual people in their political rights, but more attention is given to ensuring that major inequalities in wealth do not arise. Different conceptions of what is important obviously lead to dissimilar interpretations of how egalitarian a society is and in what ways changes will be developmental. For our purposes, distributional issues will be assessed in terms of Max Weber's (1948) three main dimensions of social stratification. These are:

> social class – which distinguishes those who own means of production (land, capital, machines, etc.) and earn a living off the labour of others, from those who must sell their labour to survive;

> social power – which refers to the ability to obtain desired ends, by personal (or group) action, by the acts of others, or by so structuring feasible behaviour options (for others) that preferred goals are attained even if no action takes place;

> social status – which can effectively be equated with people's prestige in other people's eyes.

Improvements in a social group's standing along any of these dimensions is generally considered a development. However, while the gains of one group cannot be directly equated with losses elsewhere (i.e. this is no zero-sum situation), it is still true that gains for one are often associated with losses for another. Hence value judgements about desirable distributions of societal rewards are critical to evaluations of development. This is not integral to Maslow's schema.

Even with these limitations, Maslow's hierarchy has much to recommend it as a framework for helping evaluate development issues. Conditioned by the inevitable value judgements which underlie all assessments of 'development', the hierarchy does provide cues for evaluating the developmental characteristics of socio-economic changes. For instance, in a conflict between employed rural residents who oppose a new mining venture on aesthetic grounds and unemployed residents who will gain jobs from the scheme, Maslow offers a clear framework for assessing the relative importance of these needs. Of course such situations are complex, since development issues commonly involve conflicts of a local versus national nature, as well as contrasting benefits which are short-term and those that are long-term in duration. These introduce further dimensions into any evaluation of development which re-emphasise the arbitrary and value-laden nature of this concept. Even so, Maslow has provided one of the few schemes which provides guidance for evaluating conflicting claims over (a wide variety of) developmental issues. For this, it merits attention as a general guideline for evaluation.

What this framework does not provide, however, is a dynamic component. It suggests that development is a state of being. To move from a situation defined as 'worse' to one recognised as 'better' is seen as development, but the actual process of movement is not conceptualised as possessing developmental properties. This is probably the most widely held interpretation of 'development', but it does ignore the deeper implications of this concept. As Ball (1974:1) has written:

> Development is a process - a state of becoming.
> As such it involves change. However, development
> is not just the situation at the beginning nor at
> the end of change. It is instead the ongoing
> evolutionary transformation that modifies what
> exists at the beginning to what exists at a later
> point in time.

Put in a more specifically rural context, Copp (1972:519) argued the same theme:

> I would propose that rural development is a pro-
> cess, through collective efforts, aimed at impro-
> ving the well-being and self-realisation of

> people living outside urbanised areas.   In other
> words, I believe that the ultimate target of
> rural development is people.   It is not infra-
> structure, it is not factories, it is not better
> education, or housing or even communities.   These
> are only means; if they leave the spirit of man
> [sic] untouched, they are but sound and fury.

There is a deep significance in acknowledging   the process
nature of development.   In its recognition there is a   clear
belief that development must not simply change people's physical living conditions, but also their control over these
conditions. Placed in this context, it is a questionable 'development' if the existence of an improved living condition is
dependent upon the goodwill of 'outsiders' (Schneider et al
1972); for this represents merely an increase in dependency.
Development as a process on the other hand requires improved
socio-political understanding and activity which fosters
autonomous control (Bertrand 1972).   'Outside' aid   can
certainly help in this process, but if, on the withdrawal of
such aid, people are incapable of sustaining the improvements
that have happened then what occurred was not development but
a short-term improvement in living conditions.

A humorous presentation of this in action can be found in
Jamie Uys's 1984 film 'The Gods Must Be Crazy' (Twentieth
Century Fox).   This film portrays how the dropping of a Coca-
Cola bottle from a private aircraft initially led to a marked
improvement in living conditions for a nomadic band in
Namibia.   The glass of the bottle was harder than other
materials, so in using the bottle like a hammer it softened
food easier. By blowing into it it made music, as it did when
hit with a stick.   A variety of improvements resulted in
living conditions for members of the band. However, being the
only bottle available, band members were soon in conflict over
who would use it and for what.   If it could soften food, it
could also soften heads.   Previously unknown violence erupted
within the band, leading to the conclusion that the gods were
crazy to drop this disruptive item into their midst. Perhaps
the band leader's search for the end of the world - from which
to drop the Coke bottle - is symbolic of modern society,
perhaps not.   Certainly, the fact that the nomads were
mentally unprepared for this new possession, signifies that
development is a learning process. Of course, it must also be
acknowledged that people can be prepared for substantial
living improvements even though the 'possessions' they need to
make this good are denied. The poor resources of southern
blacks during the 1950s and 1960s US civil rights movement
provides one instance of this; here, dependence on the good
will of whites undoubtedly acted as a restraint on progress
(Nelson 1979).   Ideally, for development to take place, both

26

mental and material conditions need to be improved in a mutually reinforcing process.

The conceptualisation of development as a process brings to the forefront the question of whether development is a short-term or a long-term phenomenon. From a process perspective, development evaluations require longer time frameworks than most researchers appear willing to grant. This is because development per se must touch the spirit of people and not simply the physical fabric of their surroundings. Unfortunately, practitioners (or policy-makers) with developmental intentions seem particularly short-sighted in their desire to derive an immediate impact from their development efforts (Copp 1972). This often leads to them being patronising towards (or even critical of) those who resist 'modernisation' pressures. Arguably, however, a more appropriate interpretation for rural residents' opposition to socio-economic innovations is to see this as an attempt to maintain local control in the face of 'anti-developmental' pressures to increase dependency (e.g. Schneider et al 1972; Matthews 1976; Wallman 1977). How acceptable this interpretation is depends on the context in which events are occurring and the political ideology of the evaluator. In the abstract, maintaining local control might appear a desirable goal. Yet to accept this view at face value requires one's approval of steps taken by recent in-migrants to the countryside to use their superior power resources to restrict future in-migration and residential growth within 'their' new-found locality. As various researchers have observed, such behaviour is less pro-developmental in orientation than it is self-interested. Further, given the manner in which growth restriction discriminates against potential new in-migrants and less wealthy existing residents, there are good reasons for calling such local practices 'anti-developmental' (Connell 1978; Buller and Lowe 1982).

PERSPECTIVE

The objective of this chapter has not been to present the reader with a detailed specification of what 'rural' and 'development' are. Instead, the intention has been to identify the looseness of these concepts. For 'rural' we have adopted an ecological (or geographical) definition since we believe that this is the only valid basis on which rural areas can be distinguished (without raising the spectre of 'rural' possessing theoretical significance at least). Even then, at the margins, there are no clear-cut indicators of what is or is not 'rural'. Nonetheless, even recognising that the term 'rural' disguises a wide variety of socio-economic and political conditions, it is believed that the spatial structures of such areas impose restraints and offer possibilities for action that justify their being analysed in their own right. As for 'development', the main point we wish to make on this concept is the value-laden nature of its assessment. In this circumstance it is tempting to adopt an explicit, single evaluative framework and ignore other perspectives. Since it is inevitable that our own values are woven into the fabric of this text, substantial bias in our commentary will undoubtedly exist already. We accept that this is inevitable and certainly do not wish to down-play its importance. Yet we would have difficulty presenting a unitary framework for development at anything other than a most abstract and superficial level. There are simply too many issues of great complexity to enable us to do this. What weight, for example, should be given to the interests of 'minority' rural populations (who are perhaps negatively affected by changes which benefit the majority?). Then there is the problem of assessing whether a short-term benefit will turn into a long-term disbenefit (or vice versa). These issues alone make it difficult to evaluate whether specific changes are pro-developmental or anti-developmental. Between the two writers of this text there is certainly disagreement and uncertainty in this regard with respect to a number of critical issues. Hence, our intention in this book is not to preach about what rural development is. Rather the aim is to analyse the determinants and consequences of socio-economic change in rural areas. The reader must decide whether the trends and processes specified are developmental or not and in what ways.

Chapter Three

THEORETICAL FRAMEWORK

It might seem that the looseness of our definitions of development and rural precludes establishing a theoretical base for understanding rural development. This would be so if we were looking for a peculiarly rural development model. But rural, as we have demonstrated, is a concept which lacks explanatory power. Explanations of development in rural areas must therefore be tied to processes and structural arrangements which exist in society at large. Consequently, the aim in this chapter is not to present a development theory for rural areas alone, but to examine theoretical relationships between society-wide socio-economic forces and development processes in rural locales. 'Rural' in this context is treated as one sub-set of a number of locality-types. Each of these is both acted upon by wider societal forces and, through the combined effects of individual and multi-locality processes, helps mould these broader pressures. Although locality-specific effects are weaker than societal ones, there is a mutuality of effect which provides a convenient organisational format for this chapter. To start, we concentrate on theories pertaining to societal organisation. Our argument here leads us to support explanations which draw most heavily on neo-marxist and weberian interpretations. These then provide an over-arching framework within which locality-specific forces can be analysed and interpreted. It is through this appreciation of society-wide and locality-specific processes that our investigation of development will be conducted.

If we accept that 'development' is heavily value-laden, then it follows that no single theoretical perspective is likely to emerge. This point has been well argued by Peter Taylor (1983). He pointed out that there are at heart only two kinds of theory - traditional and critical. The first of these effectively supports the status quo, albeit at times in the guise of reformism (as in the 'welfare approach' to human geography). The second seeks to expose the mechanisms of societal organisation in order to promote radical change (i.e.

as part of a process of 'overthrowing' the existing order). With such divergent basic aims, it comes as no surprise that a vast chasm separates the methodologies, analytical questions and causal propositions of these two theoretical stances. It must be emphasised that these divergences are not restricted to development models, but infiltrate every aspect of social research. In order to understand the relevance of these divisions to development theory, it is therefore pertinent to appreciate the broader foundations of this basic divide.

DIVISIONS IN SOCIAL THEORY

Theories of societal stability and change overwhelmingly trace their roots to one of two 'grand theories'. Although referred to under different titles by various writers, these are most commonly known as integration theory and conflict theory. As outlined by Dahrendorf (1959), the basic premises of integration theory hold that: (1) every society is comprised of a relatively persistent, stable structure of elements (e.g. families, communities, institutions); (2) every society has a well integrated structure of elements; (3) every element has a function which it contributes to the maintenance of society (for families, the socialisation of children into the conventions of society is one such function); and (4) every society is based on a consensus of values amongst its members. Following this line of reasoning through, the unequal (geographical, social or other) distribution of societal benefits is jointly explained and justified on the grounds that: (1) tasks are crucial to the continued existence of any society to a greater or lesser extent; (2) the skills or capabilities needed to perform the most crucial tasks are limited in supply; (3) personal sacrifices must be made by individual people to ensure that a capacity exists for crucial task performance (e.g. lack of income during periods of training, and/ or localities living with externalities like pollution, social disruption and changing work patterns in order to benefit from industrialisation); (4) to induce acceptance of these sacrifices, 'capable people' must be offered access to scarce and desired rewards; and (5) the resulting unequal distribution of societal benefits becomes institutionalised because it is positively functional (in other words, it is beneficial for the continued existence of society as a whole). Expressed in short-hand form, the integrationist theme is contained in the Davis-Moore thesis that social inequality is an unconsciously evolved device whereby societies determine that their functions of greatest importance are performed by those persons (or in those geographical areas) with the best capabilities (for a review see Lopreato and Hazelrigg 1972 or Blowers et al 1976).

Conflict theory rests on quite different assumptions. Here it is believed that society is subject to change at every point. What generates this pressure for change is dissensus. This produces conflict over the distribution of societal benefits. In its broadest setting, conflict theory does not attribute the origins of dissension to one social group. Instead it conceptualises every element of society as making some contribution to disintegration and change. What binds society together, producing order from potential chaos, is coercion. In order to maintain their privileged standing, the 'haves' (i.e. elites) must coerce 'have-nots' (i.e. non-elites) into accepting their lowly positions.

Clearly there is little common ground between integration and conflict theories. Fundamentally, integration theory holds that the basic factor which maintains social order and social integration is value consensus, whereas force and constraint are hypothesised as the primary mechanisms in conflict theory. In terms of political action, the integrationist message is that we should accept the status quo (or at most should seek to finely tune social practices for its more 'efficient' operation). By contrast conflict theorists contend that the exploitative nature of societal organisation justifies its being overthrown. Despite their seeming incompatability, some theorists have argued that both perspectives are needed for comprehensive understanding. Ralph Dahrendorf (1959), for example, has presented the view that these conceptualisations are complementary because society is Janus-headed (its two faces can be accounted for by different theories). As he described it: 'We cannot conceive of society unless we realise the dialectics of stability and change, integration and conflict, function and motive force, consensus and coercion' (1959:163). There is merit in Dahrendorf's claim that both perspectives need recognition. There is also no doubt that their basic incompatibility means that they cannot be equal partners in a unified theoretical model (despite the efforts of Lenski 1966). Ultimately, one or other of these perspectives must be taken as the primary framework for an explanatory structure. In this book the framework adopted favours conflict theory. It is nonetheless accepted that elements of integration theory are worthy of acceptance. There is, for example, broad acceptance for the continuation of numerous social inequities (like higher wages for more skilled workers). Even when general agreement exists, however, when viewed in an historical context this is frequently seen to be a product of power relationships (for specific rural examples, see Newby 1977; Gaventa 1980). After all, social consensus is aided by socialisation, primary organs of which are controlled by specific social groups:

Members of the underclass are continually exposed

> to the influence of dominant values by way of the
> educational system, newspapers, radio and tele-
> vision and the like. By virtue of the powerful
> institutional backing they receive these values
> are not readily negated by those lacking other
> sources of knowledge and information. (Parkin
> 1971:92)

As Westergaard and Resler (1975:403) pointed out:

> ... beneath the surface of industrial peace and
> the 'consensus' tapped by opinion surveys lie
> these untapped and ill-articulated resentments of
> widespread, indeed routine, popular distrust, of
> a common sense of grievance, a belief that the
> dice are loaded against ordinary workers, which
> involves at the very least a rudimentary
> diagnosis of power and a practical conception of
> conflicts of interest ...

In rural areas this underlying disquiet has been articulated
in agricultural protest movements which have periodically
arisen in a wide variety of spatio-temporal settings (Burbank
1971; Brym 1978; Jenkins 1982; Smith 1983). These are not
unique events of singular cause and distinct objective. They
are usually populist in character, being in line with mainstream support for material progress, but rebelling against
its prevailing character and distributional consequences
(Kitching 1982). These outbreaks do not signify that agriculturalists in general are openly pushing for major changes
in societal organisation. Yet by the frequency of their
occurrence (in a variety of spatio-temporal settings), they do
nonetheless reveal that an underlying sense of disquiet does
exist. Further, they show that this unease can be fairly
readily provoked into overt political action. Although on-
going, open conflict is not the norm, conflict of interest
certainly is. Once we remove the requirement that value
consensus exists, then integration theory clearly loses its
causal properties. If consensus does not exist, then the
distribution of societal benefits is not accepted by all. If
acceptance is not present, there must be some means of
ensuring its continuance; otherwise there would be constant
flux. Since integration theory assumes an equilibrium
position, 'force' must be required to counteract pressure for
change. Integration theory thereby collapses under the
weight of its own assumptions (so long as value divergences
exist over societal organisation). Examples of the repression
of 'alternative' values – as in the outlawing of Poland's
Solidarity Movement, Ku Klux Klan strikes against black
Americans or even the recent British Government's dismantling
and rate-capping of Labour local authorities – are all icing

on the cake. They provide specific instances of value dissensus being squashed by the coercive arm of societal elites.

CONFLICT THEORIES

Almost universally, social researchers acknowledge Durkheim, Marx and Weber as the three most eminent social science theorists (Giddens 1971). Their publications underpin modern social theory and provide the basic thrust for theoretical positions on rural development (e.g. Buttel and Flinn 1977). Marx is generally taken to be the archetypal conflict theorist, although, as Lopreato and Hazelrigg (1972:74) have pointed out, 'Max Weber must be viewed as moving in the Marxian orbit'. It is no surprise then to find some notable similarities in these two theorists' ideas. Yet there are also substantial disagreements between them (e.g. see Harloe 1977). These are reflected in the substantial disputes which exist between marxist and weberian advocates (e.g. Saunders 1982). The underpinnings of the approach adopted in this book are readily seen in the differences between marxian and weberian traditions.

In very simplistic terms, Marx saw capitalist society as being divided into two principal strata - the bourgeoisie and the proletariat. The key distinction between these was that the bourgeoisie owned means of production (like factories or land) whereas the proletariat did not. In order to live, the proletariat had to sell its labour to the bourgeoisie (i.e. assuming no gifts or inherited funds, people cannot produce or purchase food or shelter if they do not own means of production, so they must sell their only resource - their labour power - for money). From this vantage point, social change (and hence development) is driven by conflict between bourgeoisie and proletariat. Capitalists (the bourgeoisie) induce conflict by pressuring for lower wage costs and higher product prices (hence squeezing workers at both ends), in order to maximise their profits. Workers (the proletariat) fuel the conflict by resisting these pressures and pushing for high payments for their labour. The bourgeois quest for maximum profits is conditioned by competition between capitalists themselves (hence it is inaccurate to think of the capitalist class as a unified social group). Failure to extract as high a level of profit as competing capitalists results in a weakened competitive position (e.g. insufficient funds for future investment or an inability to compete in a price cutting war). This heightens the chances of bankruptcy, and so raises the probability that a capitalist will be pushed into the proletariat. Changes in the precise personnel of the capitalist class are largely unimportant. What is critical is the continued existence of capitalist relations of production. It is these that make members of the capitalist class the elites of society. Through their economic power, they also acquire political power and social standing.

There is some controversy amongst sociologists as to whether Max Weber completed Marx's view of the distribution of societal rewards, offered a different conception or took on the role of devil's advocate with respect to Marx's writings (Littlejohn 1972; Lopreato and Hazelrigg 1972). Where Weber's views coincided with Marx was in his treatment of social stratification as a phenomenon closely allied to the distribution of, and struggle for, power. Where he added to Marx's conception of social stratification was in his articulation of its three dimensions - class (ownership), power (control) and status (prestige). Where they fundamentally disagreed was in the importance they attached to each dimension. For Weber, these dimensions could operate somewhat independently (the British Queen has considerably more prestige than the chairperson of IBM, and so probably has Mother Teresa, but their ordering in class and power terms are quite different). Class differences are seen to have fundamental effects on power and status, but for Weber it was power which really determined the stratification system. For Weber this was becoming ever more apparent in the growth of bureaucratic organisations. Bureaucracies, he foresaw, were the concrete administrative form of a rationality of action that was increasingly penetrating all spheres of advanced capitalist societies. Over time the imperatives of such organisational forms would weaken class considerations. The accuracy of precisely this suggestion has now engaged the attention of social analysts for some time. There are those like Dahrendorf (1959) who have contended that as control in large corporations has become separated from ownership (i.e. as family-owned private companies have declined in economic importance and giant corporations with shares owned by thousands of people have come to dominate advanced economies), private property ownership has lost its explanatory value. As Galbraith (1967) argued, large corporations are likely to pursue organisational goals of growth and security rather than profit-maximising goals in this situation. This view is not exactly that of Weber, who recognised that the bureaucratisation of company management could actually reinforce capitalist practices (since the maintenance of an institution depended on its continuing to operate within the constraints of capitalist rationality; see Scott 1982). Nevertheless, Weber's conceptualisation of the multidimensional nature of social stratification usually provides the framework within which the rewards and causes of social organisation are understood.

An obvious restriction on the applicability of marxist theory is its focus on capitalist societies alone. This makes marxist theorising narrower in scope (or perhaps applicability is more accurate) than the writings of other theories of social organisation; albeit this narrower scope does allow for a richer understanding of one social system. Even so, there are writers in the marxist tradition who acknowledge some utility for multidimensional images of society. Thus, Parkin (1971) has contended that a class, power, status framework might be appropriate for the United States, which is highly differentiated by race, ethnicity, religious affiliation and regional differences, but that the dominance of class cleavages in most European nations makes a multidimensional view inappropriate. Clearly, however, the utility of class analysis is much reduced in, say, the USSR. Yet the economic, political and social structures of Eastern Bloc nations are elite dominated (Djilas 1957; Nove 1982). Moreover, in many instances, their economies are characterised by trends of urbanisation and neglect of peripheral areas (Demko and Fuchs 1984), which marxist theorising too commonly interprets as the outcome of capitalist production relationships. Similarly, marxist conceptualisations of capitalist systems leading to Third World nations suffering under exploitative (trade, diplomatic and financial) ties with advanced capitalist nations, has its counterpart in the communist bloc. Perhaps the widespread 'looting' of East European nations immediately after 1945 can be ignored (on the possible grounds that it can be justified by what happened to the Soviet Union in the Second World War). But more recent practices, like the Soviet manipulation of the commercial value of Afghanistan's exports to the superpower's advantage (including re-selling some goods to other East European nations at higher prices; Fullerton 1984), are too commonly portrayed as products purely of capitalist production relations. Explanations for such practices need to take more into account than class relationships; power differentials must be considered. Emphasising the critical role of power relationships per se, Nove (1982:603) observed that '... in [the United States of] America there is a power elite and a class structure, while in the USSR the power elite is the class structure, or rather its apex'. From the above points, we wish to make no claims about either errors in marxist theory or the value of a universally applicable theory. Yet the existence of similar spatio-economic trends in two quite dissimilar social systems (even granted the USSR's actions are restricted in some measure by a dominant capitalist world-system) does favour a model of social organisation which incorporates more than class conflict as the primary mechanism of social change. Max Weber actually offers little in this direction, since death interrupted his work and he never presented a comprehensive theory of social organisation.

Some insights can nonetheless be derived from the work of Vilfredo Pareto. Pareto's writings are not widely acclaimed amongst social researchers. This is partly on account of their distorted use as intellectual backing for Mussolini's Italian fascists and partly for their assumed anti-egalitarian overtones. In essence, Pareto concurred with the marxian view that society was divided into two basic camps (elites and non-elites); although the boundary between them is treated with much less rigour by Pareto than Marx. Yet whereas Marx saw social change stemming largely through class struggle, for Pareto it was competition between elites which was most significant. For Marx class struggle would lead to an egalitarian society. For Pareto the idea of a classless society was wishful thinking:

> ... the demand for equality is nothing but a disguised manner of demanding a privilege ... People agitate for equality to get equality in general, and then go on to make countless numbers of distinctions to deny it in the particular. Equality is to belong to all - but is granted only to the few. (Pareto 1935:para 1222)

Lest the reader thinks Pareto's views are appropriate only for earlier ages, it is pertinent to note the comments of one of the world's most eminent political scientists about the USA of today:

> The revolutionaries of contemporary America do not seek to redistribute privilege from those who have it to those who do not. These radicals wish to arrange a transfer of power from those elites who exercise it to another elite, namely themselves, who do not. (Wildavsky 1971:29)

Changes in elite structures do occur: for '... the history of man [sic] is the history of the continuous replacement of certain elites: as one ascends, another descends' (Pareto 1901:36). But while, 'History is the graveyard of aristocracies' (Pareto 1935:para 2053), in the end it is still dominated by (new) 'aristocracies'. For Pareto, the overthrow of capitalism will not produce equality, merely a new elite (cf. the USSR). Although this view has not endeared Pareto to those advocating radical social change, for its cautionary tone and particular insights it is important.

Pareto's writings are more 'comprehensive' than Marx's, in that they are applicable in contexts where elites and non-elites are distinguished by more than class relationships. Yet they have not received the refinement that followers of Marx's work have provided and, in their original form, they lack the richness of Marx's explanatory structures. This is particularly so in Pareto's treatment of non-elites. In effect Pareto treated non-elites as a residual class. His concern was not social change but social stability (or at least the persistence of inequality no matter what constitutes the primary principles of social division). As such his theorising was only tangentially within the conflict mould.

Pareto does not offer a convincing alternative to either Marx or Weber, but does add insights which help further our understanding. Different theoretical positions cannot be clubbed together - taking an idea from here and another from there - in a simplistic manner. A uniting theoretical posture is required which binds the end-product of such cribbing. This involves either the formulation of a new (more integrative) theoretical model (as Lenski 1966 sought to present) or acceptance of the primacy of one existing theoretical model (with emphasis placed on the added depth which other models bring). The latter of these two represents the approach favoured here. Too often analysts have seemed so determined to propagate the primacy of one favoured theory, that they have effectively blinded themselves to the insights other perspectives offer (marxist critiques of weberianism appear particularly prone to this). Our commentary is informed by a belief that the capitalist drive for material benefits (i.e. capital accumulation) and its concomitant search for domination of the proletariat (i.e. class conflict) so structure social relations in capitalist societies that they must provide the primary explanatory accounting framework. However, it seems a disservice to the spirit and intent of Marx's writings to over-emphasise the determining effects of social structure and of materialist considerations on human behaviour. Geographers have too commonly been guilty of this kind of economic reductionism. In recent years, however, developments in sociology with an explicit geographical focus (e.g. Urry 1981), have helped emphasise weaknesses in the translation of structuralist ideas into specific behavioural outcomes. At the same time, they have offered improvements in our understanding of the underlying constraints and general thrust of materialist explanations (e.g. Clark 1985; Dickens et al 1985). To date, however, developments in this direction have not progressed sufficiently for the insights of theorists like Pareto and Weber to be ignored.

Pareto's insights emphasise the competitive nature of elite relations. For geographers in particular, such inter-elite conflict adds to our appreciation of geographically-based social alliances. Thus, conflicts over (say) automobile exports from Japan into Europe and North America are not simply instances of competition between capitalists. Workers' jobs are at stake and national economies (hence political futures) are threatened. Class relations are intrinsic in such conflicts, but their basis is not understandable in terms of traditional capital-labour antagonisms. Competition is between capital-labour alliances in different elite nations, not simply between capitalists.* The composition of elites and non-elites varies in more complex ways than are usually portrayed in marxist idealisations. Similarly, weberianism informs us that the causes of social action extend beyond social class concerns to include (non-material) power and status considerations. We cannot, for example, readily understand the poor standing of industrial capital in Britain, unless we recognise how the dominance of a status system based on the values of a landed aristocracy denigrated industrial affairs (Tylecote 1982). The very structure of the British capitalist class, with its comparatively weak links between banking and industry (Moran 1981), and the peculiar emphasis on overseas loans instead of home investments (Edelstein 1982), are both linked to this phenomenon. Further, the traditions established in the nineteenth century have persisted and go some way towards accounting for the twentieth century decline in Britain's international economic position (Tylecote 1982). Similarly, at a more local scale, it is difficult to explain the emergence of the Tennessee Valley Authority without incorporating the personal visions (and fortuitous rise in political position) of President Franklin Roosevelt and Senator George Norris (McDonald and Muldowny 1982). In neither of these cases did fundamental changes in social structure occur, but differential patterns of employment, income and home circumstances have resulted. Neither of these social conditions has fundamentally redirected national social change but, for the people directly involved, they have had substantial developmental implications. Since status and (non-class) power relations do have identifiable implications for such developmental processes, there is merit in going beyond the class-bound explanatory framework in marxist writings.

For our purposes, the applicability of this argument is intensified by the limited scope of the subject matter of this book. An arbitrary division has been introduced by focusing on rural areas per se rather than on capitalist society as a whole. We are not attempting to explain the broad path of change in capitalist society, but its manifestations in a restricted range of geographical environments. Of necessity, this entails focusing more on the interaction of local and

---

* From this perspective, workers in advanced nations who co-operate with capitalists in promoting exploitative trade relations with Third World nations are (in this instance at least) part of the elite.

societal forces than on the determination of society-wide processes. In addition, an outcome of our narrow geographical focus is that attention is directed towards the consequences of pressures for change as much as with their determinants. In combination, these features orient this book towards analysis at the level of local population groups. Only a few years ago this would probably have directed attention away from structural impositions on rural behaviour patterns (as evinced in the topics covered and formatting of earlier rural geography texts). In recent years, however, there have been impressive advances in showing how society-wide structural arrangements interact with specific, localised social forms to produce spatial divergences in socio-economic change (e.g. Urry 1981; Massey 1984).* Such developments offer the potential for more substantive theoretical insights; ones that position the behavioural overtones of weberian interpretations within the context of structural forces. Such a move is an important step in integrating class, power and status considerations into causal frameworks. Just as some marxist researchers have over-relied on social class to explain human behaviour, weberian ideas have seemingly been grasped a bit too firmly by some social researchers because they lay less stress on class relations. As with Lopreato and Hazelrigg (1972:92), we suspect that a somewhat uncritical recourse to Weber betrays more of an ideological aversion to marxism than a pronouncement of merit for weberianism. Certainly the willingness of some researchers to proclaim the death knell of class as an explanatory variable, promoting status or political power in its place, is misplaced. A clear demonstration of this can be found in British electoral studies. Ragin (1977), for example, has shown that the supposed usurping of class-based voting by status-based regionalist parties in Britain has been misconstrued by inappropriate research designs. Similarly, Warde (1986) has revealed that local class structure has an increasing effect on constituency voting trends, even though the relationship between personal class position and ballot casting has weakened. Both these researchers do nonetheless recognise that, for voting at least, status does have a part to play. That this finding has broader applicability is suggested by Runciman's (1968:38) assertion that:

> ... the rival claims of class over power and power over class [or status over or under either] serve to cancel each other out. If there is the least plausibility in both of them, then the

---

* These insights have been more frequent in recent years but have in fact been with us for some time, as investigations like Littlejohn's (1963) <u>Westrigg</u> show.

claim by either to universal priority becomes to
that degree impossible to maintain.

This is a sentiment that has guided our appreciation of much
of the material presented in this text. But a proviso is
needed in interpreting our support for this position. This is
that differences in explanatory power must be recognised
between forces which structure dominant social relations and
those that lead to specific social practices.   As will be
evident in the chapters that follow, social class is conceptualised as having the most potent consequences for this
structuring process (albeit status and non-class based power
relations are important),   whereas these three bases of
division (and others, like gender) have more complex and less
easily disentangled effects on social practices.   Of course,
since social structures are determined by behaviour patterns,
what is actually being stated here is that the direction of
impact of class relations is less contradictory than that of
non-material power or status.

DEVELOPMENT AND LOCATION

Our search is for explanations for spatial uneveness in
development trends.   The most comprehensive, established body
of work in this field is furnished by regional development
theory (see Cooke 1983 for a review of the main categories of
regional development models).   In fact,   however, while
theorising is most advanced at the regional scale, there are
common strands in approach and emphasis at all spatial scales.
This might suggest that the critical differences are not found
in the geographical scale of theoretical models, but in their
explanatory structures.   Certainly, there are substantial
dissimilaritites in this regard.   Some analytical models are
explicity concerned with developmental processes, while others
focus on their resulting patterns.   Both equilibrium and dis-
equilibrium models are available.   Some emphasise production,
others exchange. If not explicitly, then at least implicitly,
each of class, power and status have primary roles in one or
other explanatory model.   Yet despite the wide range of
theoretical compositions available, similarities and consis-
tencies exist amongst them.   Not surprisingly, these models
fall into the main camps which divide social   science
theorising.   On the one side there is the 'blame the victim'
school, whose adherents portray low levels of development as
the fault of those who experience it. Taking the opposite
position are those who conceptualise underdevelopment as the
outcome of exploitation (or perhaps as a result of a
'deliberate strategy' is more accurate). Within both of these
schools there is a significant (but not universal) tendency to

portray development patterns and processes as being spatially
defined.   For some it is one area that exploits another area,
while others present underdevelopment as the inability of an
area to provide an environment conducive to development.   As
Urry (1981) has pointed out, such a conception is indicative
of a spatial fetishism, for only social agents act. Viewed in
this manner, one is led to question the explanatory utility of
geographically   grounded   development   models.   Only if
geographical units (localities, regions and nations) carry
theoretically meaningful markers - in that they act as organisational frameworks around which social action is motivated
and oriented - are they valid theoretical constructs.

Social   researchers have for many years argued over
appropriate geographical scales for their primary concepts and
theories.   These disputes not only range over issues of scale
per se, but also focus on whether significant concepts have a
geographical grounding.   Thus, for the concept 'community',
many analysts signify their bias by attaching the designation
'local', while others see communities ranging up to the nation
and existing in a non-place realm (Bernard 1973; Effrat 1974).
Our belief, in line with Stacey (1969) is that assumptions
about local communities are over-drawn, over-sentimental and
untheoretised.   Nevertheless,   the   popularity   of   this
conception is based on social reality, insofar as local social
systems and social processes do exist (some with 'community'
attributes, some without).   What we need to establish is how
these are distinguished and why they are important.

One perspective we have much sympathy with is that of
Dickens et al (1985:18):

A local social process worthy of the name must
refer to something active and specific to localities, although not necessarily unique to one
locality,   rather   than   local   deviations   to
national level processes.   Similarly, a national
process worthy of the name must not merely be the
average of a mass of local ones; rather it should
apply to all areas in practice.

In this light, a locality is a socially defined unit which is
distinguished by active and specific local differences in
causal processes.   Herein lies the theoretical import   -
locality is an arena for separate causal processes.   A
significant statement in this regard is David   Harvey's
(1985:125-64) commentary   on the significance of   urban
politics.   From a marxist perspective, capital accumulation
and   class struggle are dominant social processes; ones that
transcend any particular locality boundaries.   Grafted onto
this framework, Harvey's contribution lies in explaining how

the dominant forces of capitalist societies induce locality-specific social responses (in which causation takes on a broader foundation than social class relations). This arises because it is at the local level that competition (capital-labour and between capitalists) takes on a more personalised form and loses its more explicit class foundation. In addition, differences in rates of capital accumulation (across industrial sectors) and in technological advance lead to dissimilar requirements for public and private sectors. This induces further political clashes, (say) on account of dissimilarities in interest between those who want better public services and will pay for them and those who seek as low a tax rate as possible. The underlying causal process is the same for all localities here, but because the sectoral mix of local areas differs, because rates of technological change vary geographically and sectorally, and as the intensity of competition varies across localities (being tied to the former two factors), the political activity generated is truly local in character. On the question of how such localities should be identified, the reader is cautioned not to look for clearly separated and distinct areas that can be plotted on a map. Further, it is not necessary to search for areas with 'community' type attributes. As Crenson (1983) has argued, there is in fact considerable evidence that clearly defined neighbourhoods may emerge even if local residents do not exhibit noticeable tendencies towards neighbourly activities and sentiments. Localities can be defined by either sentiment or by the conflicts that bind residents together. Frequently, however, local areas find the roots of their identity in their relations with outside groups and institutions (Suttles 1972; Crenson 1983).

This 'response to outsiders' factor is also a key element in maintaining the role of nation-states as a causally significant geographical division. The sentiment associated with nationalism is clear; it sees articulation in sports events, elections, media coverage, indeed in a huge array of factors in people's everyday lives. As Anderson (1983) has convincingly argued, such sentiments are in no sense 'natural'. The nation-state is an imagined community. It cements everyday lives around a capitalist economy which carries within it all the required seeds of an anarchic society (since competition is such an intrinsically vital component of capitalism). Nationalism is an enforced organisational form: one that enables capitalist social relations to be imposed with authority (i.e. through the rule of law); one that sets up an institutional arrangement to provide services which otherwise would not be available (like defence), since competition restricts inter-capitalist co-operation; and one that provides a shield behind which national capitalists can defend themselves against competition from 'outsiders' (e.g. setting up trade barriers). Nationalistic sentiment provides a further blessing, for it offers the ideological camouflage, 'in the national interest', behind which grossly inequitous and socially divisive public policies can be screened. How effective nationalism is as a screen is nonetheless variable. Nation-states generate their own unique histories of social class struggle which induce distinctive social forms and practices. Hence, the nation, both with regard to its external relations and as a result of internal social processes, provides a key organisational structure which carries theoretically significant markers. This is not so for regions, since they not only lack a strong institutional structure but also lack a coherent 'corporate' identity.

This does not mean that region cannot be a useful descriptive tool. Nor should it be taken to imply that regional identities are not important. We accept that, even without the aid of ethnic differences, regional identities do exist (e.g. Townsend and Taylor 1975); albeit they are of variable intensity. Further, there is no doubt that a community of development experiences can exist on a region-wide basis and this can prompt a relatively unified socio-political response to underdevelopment. Thus, even conservative politicians in such remote rural areas as Newfoundland and Appalachia have called on dependency theories to explain the poor standing of their regions (Overton 1979; Claval 1983). Most usually, however, without the backing of ethnic, occupational, or some long-term and clear-cut administrative distinctiveness, such regionally-based political responses are rare (e.g. McDonald 1979). Although each of us can probably identify regions with social coherence relevant for developmental processes, a longer list of 'regions' that lack social unity can be put forward (similarly, geographically-bounded local 'communities' vary in their social integration). While this raises questions about the usefulness of a spatial delineation of development process, it does not mean that space does not have significant development implications. In the case of North American Indians, for example, the reservation system represented a deliberate use of spatial distinctiveness to restrict and control the development potential of native people (e.g. Adams 1975; Foraie and Dear 1978). At times the reverse situation has occurred. Specific social groups - like the Hutterites in Canada, for example - have enclosed themselves within specific geographical confines to exclude 'developments' their leaders believe are undesirable (Flint 1975; Wallman 1977). More commonly, however, where no strong institutional structure exists (as nations provide) geographical boundaries are highly permeable. Regions and localities alike contain persons and social groups with widely differing lifestyles and development prospects. Though a single geographical area might be widely accepted to be 'deprived', the characteristics of this deprivation may be typical of only a small proportion of its population (Cullingford and Openshaw 1982). Area of residence provides a communality of development experience, but it cannot account for the totality of that experience. Further, its causal role is insufficient to merit areal aggregations being accorded a pre-eminent position.

This can be readily demonstrated by elaborating on the Newfoundland and Appalachian situations mentioned above. In these two areas, socio-political leaders have recently proclaimed that exploitation by core metropolitan centres has caused their region's underdevelopment. This argument has a strong local appeal (especially for politicians), since 'blame' for a region's poor socio-economic standing is laid 'elsewhere'. No 'inadequacy' in the local population and no lack of effort on the part of the local leaders is allowed for. 'Outsiders' are to blame. The similarity of political rhetoric and argument in these two regions is notable. The similarity in the causes of their relative poverty is much less marked. Appalachia has an economic history characterised by marked levels of absentee-ownership (Claval 1983). For many Appalachian localities, the dominance of coalmining brought rule by 'the company', which controlled local shops, schools, police, housing and, of course, employment. The continuity of this power relationship over decades was linked to the lack of alternative opportunity structures, while the wealth of the area (in the form of profits from mining) was being drained from the region by absentee mine owners (Gaventa 1980). Superficially, a similar situation existed in New-foundland. Here local merchants dominated rural settlements, with fishing providing the productive base. As Alexander (1974) has made clear, however, at this point the similarity ends. The 'exploitation by metropole' thesis is less convincing in Newfoundland's case on account of the unwillingness of the local elites to mobilise themselves. Newfoundland might well have suffered from a lack of new investment but this was less on account of expatriation of profits, than it was due to local leader's unwillingness to invest in their own area. As an added dimension of this relationship, in areas like Appalachia and the North East of England (Carney et al 1975), elites actively discouraged investment in other economic spheres in order to preserve their hegemony over the local economy. Irrespective of the character of economic linkages which cross regional boundaries, therefore, a causal accounting for regional (or local) underdevelopment needs to make allowance for the effects of 'local' elite domination.* At this time, the accounting frameworks of geographically-based theoretical models are little suited for incorporating such considerations.

The essential point is that a sense of 'us' versus 'them' can be directed toward a local rather than an 'outside' body. Socio-economic change (or the persistence of stability) can be induced by 'outside' or 'inside' elites. Even for localities, which have lost autonomy and are increasingly influenced by national and international forces of change (Warren 1978), local agents still have significant effects on local conditions. This is illustrated in Elizabeth Bird's (1982) investigation of Luss in Dunbarton. Following the defeat of Bonnie Prince Charlie in 1746, this community was created as a new settlement. Its residents had to adapt from their previous clan-based social system - with its characteristic links of mutual aid and protection between chief and clan - to the paternalistic 'Lord-of-the-Manor' system favoured by the new anglophile laird. Thus, a basic change in social structure resulted (under force of arms) from a non-local event (the Jacobean uprising). That local elites controlled the specific manifestations of this upheaval is readily identified by the dissimilar policies which lairds pursued (the most notorious of which was the eviction of Highland residents, and their transhipment out of Britain, in order to make way for more profitable sheep farming - e.g. Prebble 1963). For Luss, a second major transformation has occurred in recent decades due to a 'local' event - namely, the death of the last 'traditional' laird in 1948. His son (the new laird) has (perhaps unintentionally) introduced significant changes to the area by breaking down its previous 'enclosed' character. Workers have been hired from outside the locality, houses have been rented to non-estate workers (because of higher rents), managers have been placed on farms in place of tenants (as profit from farming is taxed at a lower rate than the unearned income of rent). To understand this change from paternalistic to profit-maximising estate management practices, it is not sufficient to focus on the values of each laird. Their goals cannot be divorced from the broader socio-economic environment. Thus, the death of the old laird in 1948 brought heavy penalties in death duties. Even had the new laird wished to continue the traditional pattern of pater-

---

* The reference to elite domination rather than class conflict here is deliberate. Classes are defined with regard to ownership of the means of production. In Newfoundland, fishermen were generally self-employed. The exploitative relationship they experienced under local merchants was not that of any employee (or labour), but occurred in the exchange of produce (fish) for income. This kind of 'exploitative' relationship is very characteristic of rural economies, for petty producers who are self-employed occupy a substantial share of the labour market (family farmers constitute a key component, for instance).

nalistic relations, the legal responsibility for paying death duties would have necessitated either changes to enhance income generation or a reduction in the laird's standard of living (or both). The choice, then, boils down to that of elite or communal 'loss'. Elite dominance not surprisingly resulted in 'adverse' changes being imposed on the non-elite.

Two points should be taken from this example. First, there is the question of adaptation. Notions of mass society and, quite distinctly, an emphasis on structuralist explanation (at least implicitly) help promote the image that localities have little freedom of action from extra-local forces (or that local autonomy is only present for minor items). Bird's (1982) Luss example illustrates how local agents can adapt to extra-local pressures. Hence national and international change should not be expected to induce uniform local responses. In Luss, local - extra-local interactions were mediated predominantly through one local agent (the laird); this situation is characteristic of landed estate or peasant social systems, just as it is of localities dominated by a small, unified elite (e.g. Silverman 1965). However, in localities with a looser integration of social groups, more complex local - extra-local interactions occur. As one illustration, Bennett (1969) found distinctive relationships with the 'outside world' amongst agrarianists on the Canadian Prairies: Hutterites remained aloof and dealt with outside agencies only at arm's length; ranchers resembled the Hutterites in disliking intimate relations with 'the outside', yet (although they saw themselves as free-enterprisers) when 'necessary' they did not hesitate to influence government policy in their own favour; meanwhile, farmers were aggressive participants in higher-level politics, for their socio-economic condition was heavily dependent on government initiatives. These functional (economic sector) and cultural divisions within localities are significant for they are linked to dissimilar 'messages' being projected to 'outside' agencies concerning local residents' preferred courses of action. Perhaps most commonly, this kind of divide runs along social class lines. Characteristic of such a division is the resistance of ranchers in the US West to new energy projects (Freudenberg 1982) or large-scale farm operators' opposition to manufacturing growth in East Anglia (Rose et al 1976). In both instances working class residents would have benefitted from new job opportunities with higher wages. The potential consequences this would have for local salary payments was at the heart of local landowners' aversion to such endeavours. Alongside so clear a separation of local interests, however, convergence of local residents' preferences commonly exists. Granted such convergence can result from lower class attempts to 'stay on the right side' of local elites or from a failure to appreciate that particular courses of action will not help their cause (e.g. Gaventa 1980), yet it can also arise from a genuine communality of interest. This frequently occurs when localities are threatened by 'outside' forces (as seen in opposition to new nuclear power plants). Even in these cases, however, intra-local conflict is commonly present (as seen in the contrasting positions taken by those who welcomed the new jobs which a third London airport would create for the Stansted area and those who complained about population growth, air pollution and the introduction of 'undesirable elements' into the region). While extra-local forces do impinge on localities in major ways, their precise impact is conditioned by local adaptations. These are conditioned by internal local power relations.

A second point raised by the Luss example is the manner in which social agents use localities (or regions or nations). The kind of socio-economic change taking place within a locality is heavily conditioned by spatio-temporal circumstances. To give a very simple example, many houses at some 70 kilometres from Paris have experienced a series of quite different roles within this century. Around 1900 they were occupied by farmers who worked in adjacent fields. As agricultural restructuring occurred they were abandoned, to be taken up in more recent decades by second home owners as weekend retreats. Since the mid-1970s, however, improved transportation links and an increased preference for country living amongst urbanites has led to these buildings becoming primary residences for daily commuters to Paris (Clout 1984). As Massey (1984:117-8) put it:

> ... local areas rarely bear the marks of only one
> form of economic structure. They are products of
> long and varied histories. Different economic
> actitivities and forms of social organisation
> have come and gone, established their dominance,
> lingered on, and later died away. Viewed more
> analytically, and concentrating for the moment on
> the economic, the structure of local economies
> can be seen as a product of the combination of
> 'layers', of the successive imposition over the
> years of new rounds of investment, new forms of
> activity ... So if a local economy can be analy-
> sed as the historical product of the combination
> of layers of activity, those layers also repre-
> sent in turn the succession of roles the local
> economy has played within wider national and
> international spatial structures.

In essence, past economic activity enhances the potential for specific kinds of future activity and restricts the emergence of alternatives. On account of the greater resources at their disposal, it is usually changes in the behaviour of socio-economic elites which lead to the activation of this potential (albeit behavioural change may well have been 'forced' or at least encouraged by pressures from non-elites). Thus, US manufacturers have increasingly sought low-cost production sites for mass produced, technologically unsophisticated items (like textiles) in order to ward-off foreign competition (and circumvent demands for higher wages in the northeastern manufacturing belt). This has led to a substantial industrialisation push in rural areas of the southern states (Cobb 1982). Yet the industrial emergence of the South cannot be explained adequately without understanding the area's social history. In Kirkpatrick Sale's (1975) terms, the southern 'cowboy-elite' has enforced an ideological environment based on the three Rs of rightism, racism and repression which has produced depressed wages, anti-union legislation and minimal tax payments ideally suited to maintaining low production costs (Nicholls 1969). The fears of the 1950s and 1960s about the South's inability to industrialise have been transformed into concern over the old industrial heartland's inability to maintain jobs. Over these time periods, underlying social conditions in the South have barely changed. What has altered is the competitive situation of US manufacturers. This has encouraged them to seek alternative production locations with more favourable environments for profit-making. Note that this locational change has been much less pronounced for high technology products. Northern states still provide more of the facilities capitalists require for these enterprises. Local socio-economic environments interact with the preferences of elites to produce specific spatio-temporal patterns. The process is dynamic; a nation's core at one time could conceivably become its backwater at another.

In bringing about such a reorganisation in local (or regional) fortunes, local and extra-local elites can compete, cooperate or compromise. Local leaders might promote extra-local developments in preference to local ones if this advances their own position. At a national level this is evident in manufacturers' relocations of manufacturing plants from advanced nations to cheaper production sites in Third World countries (Fröbell et al 1980). At a regional level, a pertinent example would be the horse-trading that took place between Italy's northern bourgeoisie and the southern landed classes, whereby southern elites were guaranteed a continued dominance of local social systems (with military aid where necessary) in return for supporting the nation's unification and hence enlarged markets for northern manufacturers (Graziano 1978). Similarly, within individual rural localities, the desire for an unchanged or 'pleasant' living environment has induced some elites to restrict improvements in their locality, even when this would not challenge elite domination (e.g. Kendall 1963; Wild 1974). The specific examples of elite restrictions on 'local' development given in this paragraph are of differing character. The restrictions on development opportunities identified by Kendall (1963) and Wild (1974) offer instances in which personal preferences (in part prestige inspired) were primary determinants. For manufacturer relocations and southern Italian landowners, by contrast, elite behaviour owed much to the actual or threatened actions of non-elites (i.e. to demands for higher wages in existing manufacturing sites and for land reform). From a conflict theory perspective, elites and non-elites wage a constant struggle over their respective standings in socio-economic and socio-political hierarchies. The characteristics of such struggles are not uniform. Frequently, non-elites appear to accept their subservient standing, with elites having free rein to direct societal development. As Harry Cleaver (1979) argued at length, and John Gaventa (1980) demonstrated in a particular empirical context, this does not mean that societal structures are different in 'non-conflictual' social settings. Predominantly, what we see in these social environments is a pattern of elite domination so intense that non-elites have been conditioned (over a long period of time, through socialisation into specific norms of behaviour) into an accepting, non-questioning role in the social structure (see also Newby 1977). The points that need emphasising here are that both elites and non-elites must be analysed for comprehensive understanding, that the character of their social relationships varies and, consequently, that the absence of overt conflict does not invalidate a conflict theory perspective.

In terms of its geographical content, a point that should be drawn out is the manner in which elite - non-elite relations reveal dissimilar social patterns at different geographical scales. Smaller geographical units (like single villages or towns) are less likely to be dominated by inter-elite competition. At this scale it is easier for a small group to dominate. This permits a clearer articulation of (elite) personal idiosyncracies into developmental processes. In addition, these geographical units are of a magnitude that promotes specialisation of socio-economic activities. This can foster non-elite solidarity (as in mining villages), but it also reduces the scope of elite - non-elite conflicts. Thus, dormitory (or commuter) villages are distinctively residential areas for which capital-labour problems in manufacturing do not appear to be relevant (hence any non-elite solidarity built around the locality is not carried into the employment sphere; inducing inter-local competition in elite - non-elite conflicts is thereby less likely). On account of their idiosyncratic character, localities will be subjected to (and reflect) societal pressures at higher spatial levels to different degrees (e.g. a locality composed of people from a variety of ethnic or racial groups is less likely to see its social conflicts dominated by locality-specific social class considerations). Futhermore, the smaller the geographical scale, the more likely that residents find their mutual interests cut across class lines (as in their common experiences resulting from playing in the same cricket or baseball side, supporting the same football team, opposing proposals to close the village school or drinking in the same pub). As Taylor (1980) made clear, it is at this geographical scale that people experience and evaluate the society in which they live. Hence the determinants of their behaviour patterns are inevitably infused with more complex considerations than social class.

At the next step up in the spatial hierarchy – that of the nation-state – class relations are more evident. The significance of nations as elements in development processes rests on their providing the key organisational framework within which public and private institutions act. Most readily, this is seen in the presence of the national governments of countries. In addition, however, it is at the national level that both capitalists (in business confederations and individual corporations) and workers (in trade unions) are most coherently organised. Further, while there are regional separatist movements in some nations which challenge the general applicability of the phenomenon, nations also provide the most fundamental reference point for communal identity, cultural distinctiveness and political mobilisation. This is the geographical scale at which world-wide social processes are transposed into specific areal form. Here, the ideological overtones of nationalism promote the strongest impulses which both supersede locality-specific interests and also promote 'unique' responses to social, economic and political forces of a transnational nature. Hence, it is at the national level that social class (or other social units of mobilisation like ethnicity) finds its most coherent expression. It is also at this level that the ties of social class break down in response to calls for special treatment on nationalistic grounds. At the national level there is a strong impulse for socio-economic advancement within a geographical area per se (as opposed to advancement for a particular social group). In particular, this is made clear in responses to 'threats' from other nations. However, increasingly the nation-state is declining as a locus for decisions with major developmental implications (as seen in the growth of transnational corporations or in organisations like NATO or the European Economic Community). In order to understand how this process affects rural development, we need to appreciate causal processes behind supra-national organisations and the international actions of state governments. This, and the consequences of these agents' actions for rural areas, constitutes the starting point of our examination of rural development and the subject for the next chapter.

PERSPECTIVE

The explanatory approach used in this text emphasises the importance of two spatial scales in determining development processes - the national and the local. These are the loci around which individuals and social groups possess a corporate (or communal) identity. Hence, they tend to be the focal points around which development efforts are organised on a geographical basis. Certainly, development processes are not usually articulated with geographical units as their primary considerations; it is thereby generally incorrect to conceptualise development as having geographically specific outcomes (rather, it has people specific outcomes). This does not weaken the importance of local and national geographical arenas, for these are the units about which broader change processes are given unique expression. Of course, they are not the only geographical scales at which development processes operate. Consequently, it is critical to understand how corporate bodies at these geographical scales behave toward one another and are influenced by (or more accurately adapt to) broader forces (i.e. their 'external' relations), as well as the manner in which 'internal' divisions operate at these scales. These divisions are basically of two kinds. There are those between elites and those between elites and non-elites. Technically, there are also factions amongst non-elites, but the slight impact these have on development (except insofar as they weaken non-elite alliances in conflicts with elites) has left them with a small role in this text. Our use of the expressions 'elite' and 'non-elite' in the above contexts is certainly lacking in theoretical rigour. It would be much more satisfying if we could tie our titles to specific social divisions such as social class, race/ethnicity or gender. However, significant bases of social division do not fall exclusively into one of these categories. In part this arises because of the limited geographical scope of this text. Being interested solely in rural areas does introduce complications. It orients our work more towards the consequences of development processes than would be the case if our concern was at a broader geographical scale. Our belief is that significant social divides are more plentiful when the consequences of socio-economic change are examined than they are when their determinants are investigated. Even for the determinants of development, however, there is no single classification of populations which captures the primary categories which produce development patterns. As we will argue in the following chapters, a variety of social divisions have critical effects and their importance varies with the geographical scale of analysis.

Chapter Four

THE INTERNATIONAL CONTEXT OF
RURAL DEVELOPMENT

Textbooks on rural areas have often neglected the international dimension of their topic. Although cross-national comparisons have been made, the consequences of inter-national activities for rural locales are usually omitted. This is perhaps not surprising, given that there are no convincing criteria for separating international events with rural implications from other international occurrences. In truth, no such boundaries exist. Yet, it is undeniable that rural development paths within individual states, and their development implications for single localities, are affected by forces at a transnational or even global scale. From a theoretical perspective this is most forcefully emphasised in Wallerstein's world-system model (e.g. Taylor 1980). This stresses that a single, world capitalist market exists, wherein processes of capital accumulation primarily operate on a world-wide, rather than intra-national, basis. Certainly, this perspective does not assume that states are equally powerful; some, more than others, can distort markets to favour their own nationals. Hence, within a world-system framework, the international standing of nations is signified in their classification as core, semi-periphery or periphery. As an explanatory device, this categorisation has potentially unfortunate overtones of geographical determinism (Agnew 1982). Nevertheless, it merits attention since it emphasises that international standings do affect development processes at a local level.

A world-economy perspective actually goes further than this for it draws attention to the dissimilar benefits that the same development processes have in nations of different international standing. Even though this text only deals with advanced capitalist nations, this is an important point to make, since development models are still frequently portrayed as being universally applicable. As François Perroux (1983:27) put it:

> The most widely-held general economic theory of
> the past century has built upon the experience of
> the developed countries, in response to
> clandestine pressure from their ruling classes,
> by English-speaking authors writing from England,
> whose prosperity depends on external trade and on
> finance. This general theory, in many respects
> implicitly normative, serves the interests of the
> country in which it originated both by its
> premises and by its construction; if applied
> uncritically in the developing countries [or
> semi-developed countries], it would be
> detrimental to them since the 'market' to which
> they would in fact be subjected is one in which
> they participate against a background of deep-
> rooted, universal and lasting inequality.

Stated boldly, models of development are conditioned by power
relations. This is unequivocally demonstrated in the United
States. While this nation is now a primary advocate of free
trade between more advanced and less advanced nations, in the
nineteenth century its leaders tenaciously adhered to import
restrictions in order to protect its fledgling industries from
British exports (Hoffman 1982). The modern day equivalents of
this are seen most visibly in calls for a New International
Economic Order and in Third World opposition to the pro-
western biases in International Monetary Fund (IMF) and
General Agreement on Tariffs and Trade (GATT) procedures (Hart
1983; Corbridge 1984; Griffith-Jones 1984).

Even in rejecting dominant development models, critics
often remain conditioned by their prevailing ideas. Thus, the
world-system model, and its associated but less generalised
dependency theory, largely presents a polemical opposition to
dominant diffusionist ideas rather than an autonomous
theoretical position. In a sense, the world-economy
perspective turns diffusionism on its head; rather than the
diffusionist idea of core capitalist development carrying the
periphery in its wake, dependency theorists stress that core
development is built upon peripheral underdevelopment (see
Browett 1980 for an elaboration of this argument). In
essence, the world is divided into core and peripheral areas
(with intermediate or semi-peripheral locations), which are
imbued with causal properties (Agnew 1982). At the extremes
the contrast is between raw material exports from Third World
nations and manufacturing sales from advanced economies; terms
of trade then produce the unequal benefits which channel
income gains toward advanced nations. Yet, both the
geographical conceptualisation and the causal processes of
this unequal exchange are too simplified (Delacroix 1979;
Browett 1980). In the first case, many advanced economies

Figure 4.1 Net US trade balance in agricultural and non-agricultural products 1960—85

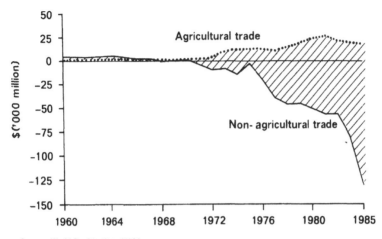

Source · World Food Institute (1985)

---

have long relied on staple exports for their economic well-being. The cases of Australia, Canada, Denmark and New Zealand come immediately to mind, but even the industrial giants have primary product dependencies (the United States, for example, currently accounts for 50 per cent of the world trade in coarse grains and, as Figure 4.1 shows, continues to rely heavily upon agricultural exports for foreign currency earnings – Frundt 1975; Tarrant 1981). In the second case, not all primary product exports are linked to supplier weakness (the failure of the Union of Banana Exporting Countries stands in marked contrast to the success of the Organisation of Petroleum Exporting Countries). Further, there is no inevitability concerning whether exporters of primary products are price-takers or price-makers. There is a dynamism in both production and exchange relations which can rapidly alter developmental potentials (as in the increase in OPEC oil prices from $1.70 a barrel to $10.75 a barrel between 1970 and 1974). This dynamism is not captured in static concepts like core, semi-periphery and periphery. To better understand this dynamic element, the processes through which international development impulses are generated needs elucidation.

Our references to the world-system model should not be taken as confirmation that this perspective provides the kind of encompassing explanatory framework required for this task. World-system theory contains some deeply embedded flaws that make its explanatory prowess suspect (e.g. Browett 1980; Agnew 1982; Chirot and Hall 1982). Further, while the world-system perspective offers a 'leftish' account for world development patterns, explanations grounded elsewhere in the political spectrum also exist. Thus, while world-system theory links national economic standings to determining processes focused around core nations, in Michael Beenstock's (1983) transition theory it is the actions of peripheral nations (or some of them at least) which have led to recent realignments in world economic patterns. Since this text is specifically concerned with advanced rural economies, we have not devoted time to exploring the alternative strengths and weaknesses of such models. Our concern is to emphasise that theorists have increasingly recognised that intra-national development is strongly conditioned by transnational economic, social and political processes. No theoretical position has so far traced these global processes downwards in order to explain development patterns in rural areas (the world-system recognition that different processes operated at world, national and local geographical scales is some distance from this, even though it is the closest to it conceptually). Frankly, on account of the arbitrary nature of 'rural' as a concept, we do not believe that this goal is feasible. What is nevertheless both possible and highly desirable is to illustrate how development within the rural localities of advanced economies is conditioned by international occurrences.

What we wish to demonstrate in this chapter is that the global scale imposes itself on the national and local scales in two main ways. First, it does so by structuring development potentials. In other words, it helps establish prevailing social practices - like the norms of social behaviour, technological standing and national military alliances - which promote particular behaviour patterns and discourage others. Such structures are not unchanging, for they are dependent upon social agents continuing to reproduce them (i.e. adhere to their conditions). However, changes in such structural arrangements imply significant alterations in a social order. They are consequently difficult to change once established. Hence, the determination of primary structural arrangements goes a long way toward directing future paths of development (see Lukes 1974). Alongside the structural context of development processes, our second kind of international effect - specific behavioural acts - are more easily appreciated and interpreted. Grain embargoes, a fall in petroleum prices and restrictions on immigration can all have immediate consequences for rural development. They are easily seen in operation and their consequences are readily felt. However, the literature on rural areas has as yet given them insufficient weight.

The organisation of this chapter follows the separation of international effects into structural and behavioural components. The absence of specific theoretical links between development at a world scale and in rural localities has meant that specific examples have been called upon to illustrate the central dimensions of international processes. In examining structural contexts, for example, particular attention has been given to the settling of the US West, the power of transnational corporations and the rise of European Community institutions. The first of these has been included to show how the processes associated with United States expansion (into what was then non-US territory) produced specific rural environments which persist to this day. The second is presented to demonstrate the limits to national government actions and how many rural economic institutions are but small parts of multinational enterprises. Our examination of European Community institutions seeks to draw out the manner in which international government cooperative ventures produce their own peculiar structural impositions, which (while somewhat different to market forces) lead to significant inequities in development potential. In the second section, on the behavioural manifestations of international events, we investigate the nature of western trade with the Soviet bloc, foreign aid to Third World nations and immigration policies. These are used to show how national foreign policies fluctuate over time. Our ultimate intention is to illustrate that these oscillations produce short-term changes in rural fortunes. By selecting specific examples, we hope to demonstrate how this occurs when the rural impacts of government foreign policies are both intentional and incidental. In addition, our aim is to show that international effects, as much as national and local ones, have significant implications for the social distribution of developmental benefits.

THE STRUCTURAL CONTEXT

In a report published in 1975, the Economic Council of Canada estimated how incomes would change across economic sectors if Canadian barriers on foreign imports were removed. Symbolic of the relative weakness of rural interests in the policy-making process, rurally-dominant economic sectors received scant import-protection compared with their urban counterparts. The smallest estimated fall in percentage value-added for any manufacturing sector once import controls were removed was 2.22 for transport equipment (the highest, for petroleum and coal products, was 44.40). By contrast, the figure for agriculture was 0.52, while each of forestry (0.66), fishing and trapping (2.59) and mining, quarrying and oil drilling (0.35) would have seen rises in value-added were tariffs removed. To understand how such structural imbalances emerged, it is necessary to appreciate the character of capitalist development in Canada. The insulation of Canadian industry from the vagaries of world market competition, and the concomitant acceptance of agriculture's openness to such competition, has not arisen by chance. High rates of manufacturing protection were introduced as part of Prime Minister John A. Macdonald's National Policy of 1879. This was aimed at uniting the nation and promoting its economic well-being. Imprinted on the character of this policy, however, there was the hand of British financial interests which oriented investments toward staple exports (Naylor 1975). With nineteenth century financiers funnelling British portfolio investment into staples exports, while warming to a profit bonanza from railway expansion which was underwritten by government (e.g. Chodos 1973), little investment capital was left for industrial endeavours. The defeat of the Confederacy in the US Civil War removed opposition to industrial import tariffs in that nation and helped promote counter Canadian tariff walls. These reactive measures were essential since, without 'artificial' inducements, the porous border which separated these two North American states provided scant resistance to investors moving south (since confederation in 1867, the outflow of Canadian emigrants has amounted to three-quarters of its immigrants confirming that 'Canada was truly a demographic railway station'; Gertler and Crowley 1977:53). The movement of US manufacturers into Canada (to enlarge the US sales market) was heralded with joy by Canadian financial and commercial elites of the day. Their profit-making horizons still largely ignored industrial production (except as a portfolio investment). By the time populist agricultural movements began to challenge the privileged position of industry, the patterns of today were entrenched. Financiers were reaping bounties from US manufacturers and the decision-making centre of the economy had shifted out of Canada into the United States. As Rotstein (1976:109) described it, what was:

> ...so effectively blotted out was the problem of power, particularly the locus of economic power. Thus, when power began to shift out of the Canadian economy to Chicago, New York and Washington by virtue of the highly visible and active hand of the multinational corporation, and

> by virtue of our dependence on Washington's legal
> and economic policies, we were barely conscious
> that anything of importance had happened. All we
> could do was ask whether the 'market' was
> functioning properly and whether profits
> (somewhere, anywhere) were being maximised.

Such a process could hardly take place without resentment
being stirred in peripheral places (e.g. Wood 1924; Overton
1979) and amongst 'marginalised' peoples (e.g. Adams 1975;
McRoberts 1979). For the bulk of Canada's urban-
industrialised population, and most especially for its
economic elite (Grant 1965; Clement 1977), however:

> It would be a mistake to evoke the image of
> Canada as a seething colony struggling to break
> loose. Canada bears rather the signs of a
> successful lobotomy to which it has voluntarily
> assented. The routine of daily existence is
> comfortable, decent and sane. (Rotstein 1976:98)

Thus, the emergence of structural inferiority for Canada's
rural economic activities resulted from elite accommodation
and encouragement. The same certainly could not be said for
Britain.

A critical difference between Britain per se and its
former colonies lies in the latter not having an existing
agrarian elite to resist the impulses of industrial and
commercial entrepeneurs (though royal land grants resulted in
a similar social structure in the US South, with the US Civil
War representing the ultimate struggle between industrial and
agrarian elites). For British colonies in general, propertied
interests in the 'home land' were principally concerned with
heightening returns on investments (primarily loans).
Patterns of rural settlement and production in colonial lands
were conditioned by the tariffs, credit, railways, technical
innovations and land rehabilitation programmes that local
industrial and commercial elites promoted. Yet these were
closely tied to the surpluses which British agrarian
capitalists (the landed aristocracy) filtered into the 'new
territories' (Thomas 1972; Edelstein 1982). As Denoon
(1983:124) made clear: 'Neither Kruger nor Rhodes, Balmacedo
nor Pellegrini could stray far from the class interests of
their real constituents'. For British landed elites, interest
in the new lands focused upon profit-making per se. In
Britain itself, however, maximising returns on investment was
a lesser consideration than how the profit was made.
Ingrained in the aristocratic mentality was a rejection of
industrial enterprise (e.g. Tylecote 1982). This inevitably
resulted in inter-sectoral elite competition. Quite simply,

the basis of profit-making in the industrial and agrarian sectors was (and still is in many respects) incompatible. For agrarian capitalists, higher food prices boosted profits. For industrialists, it intensified worker pressure for higher wages. In tracing the dialectical relationship between these two, we can outline the changing distribution of power in British society.

To mediate between landed aristocrats and Britain's rising industrial elites, the Corn Laws were passed in 1815; originally with the intention of guaranteeing supplies at fair prices for both consumer and producer (Crosby 1977). By the early nineteenth century, however, the strength of agricultural interests was made apparent in grain prices which favoured growers. The growing muscle of industrial capital was nonetheless made evident when the Corn Laws were repealed in 1846. After this, with food prices rising throughout the 1850s and 1860s, with railways helping to distribute crops to more distant markets, and with mechanical innovations reducing production costs, the demand for protectionist measures lost momentum. By the end of the century, when crop prices fell significantly, there was insufficient agrarian strength to overturn the now dominant manufacturers' preference for laissez-faire trade policies. Although the roots of this industrial power usurpation can be traced back to the English Civil War (Hill 1961), the transference of primacy from agriculture to industrial capital was effectively a 'bloodless coup'. The power struggle between agrarian and industrial elites in the United States, by contrast, was a much bloodier affair. Its origins can be directly traced to that nation's 'foreign' affairs.

Settling the US West

Every nation develops and embellishes myths about its own heritage which eventually symbolise actual or projected attributes of national character. That these are myths is inherent in their nature, since they emphasise the positive aspects of events and (at best) only tangentially gloss over the key implications of the events themselves. Thus, for Britons, descriptions of the Charge of the Light Brigade are festooned with hyperbole. This was a 'glorious' episode in British history, wherein bravery, discipline and an innate sense of mission were visibly portrayed. That this episode owed much to ingrained incompetence built upon a rigid class system which explicitly scorned ability in favour of 'breeding' is rarely drawn out. The Charge of the Light Brigade finds a place in British self-portrayal. Yet the mass slaughters at Ypres, Empire resuscitating measures like the

1956 Suez invasions and the continuing flight of capital from Britain (averaging £1.7 thousand million every three months in portfolio investment alone during 1983), which are tinged with the same mentality, are less willingly put forward as symbols of national character. It is by delving behind the public facade that we come to understand causality. This is particularly so for the settling of the western United States.

The conventional image of western expansion has been ably portrayed in films like 'How the West Was Won'. Rugged individuals who matched courage with conviction and risk-taking with compassion personify the myth-makers' western settlers. With government land grants providing material sustenance to supplement the personal freedom immigrants had attained by escaping persecution in Europe, and by the enterprise of railroad and other investors, impetus was provided for these new areas to blossom. As in fictional accounts, like James Michener's Centennial, barring the sour taste of a few 'bad eggs' and the heightened tension of brave acts, the whole process comes across as manifest destiny. Behind the smokescreen, there lies a far more complex, and indeed more interesting, story to western expansionism.

Implicit within these fictional accounts is the assumption that the West was 'naturally' American. This is far from the case. As Gareth Stedman Jones's (1972:217) short but cogent review makes clear: '...the whole natural history of United States imperialism was one vast process of territorial seizure and occupation'. That process is clearly seen in the progress of the nation's westward expansion (e.g. Figure 4.2). It demonstrates too that the definition of the 'West' in itself varied over time. With France owning the Louisiana territories, Mexico controlling Texas and California, Spain holding Florida and with British colonies to the north, the fledgling USA had no pre-ordained rights or expectations of continental control. Such expectations nonetheless soon emerged. In 1812, for example, the US sought to conquer Canada, with some Congressmen pressing for accompanying invasions of Cuba and Florida (and calls for the elimination of all native Indians on the continental mainland). War with Mexico followed in 1845, the result being the acquisition of California and Texas. The motivation behind these steps was equally present in attempts to buy the Virgin Islands in the 1860s (with the possibility of purchasing Cuba, Greenland, Iceland and Puerto Rico also being explored), and in the war with Spain in 1898 (from which the US obtained its first overseas colonies). In essence, the aim was to expand the market for US capitalists. Possession of internal markets was critical for US industrialists (in particular), since the earlier dawning of Britain's industrial revolution, the size of the British market and the dominance

Figure 4.2 Dates of admittance to the USA

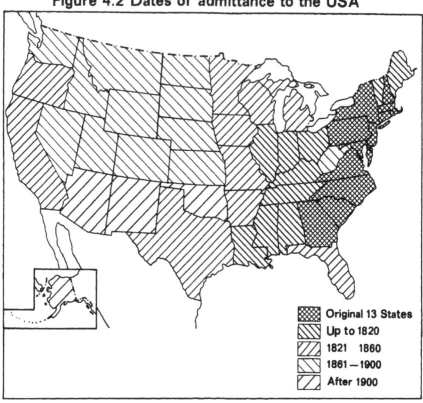

Legend:
- Original 13 States
- Up to 1820
- 1821 1860
- 1861 — 1900
- After 1900

of Britain's state apparatus in the international sphere, all placed US manufacturers at a disadvantage in external markets. Hence, the crux of US foreign policy was to convert foreign relations into internal ones; to annex adjoining lands into a 'greater' United States.

A component of this expansionary thrust was the successful colonisation of the West by agricultural commodity producers. This provided the joint benefits of raw materials for industrial producers and markets for manufactured goods. The 'hopes' pinned on the West did nonetheless meet with fierce opposition from southern planters (who feared that the growth of 'free' labour areas would result in pressures to end slavery). Various compromises were made to stave off conflict (as in agreements over whether new states joining the Union should allow slavery or only have free-labour work environments), but alternative preferences for western expansion also clashed with differences in goals for trade policy. While southern planters desired free trade to facilitate the uninhibited export of cotton to Britain (while importing goods cheaply in return), northern manufacturers wanted import restrictions to protect their fledgling enterprises. Vested interests were too entrenched for compromise solutions. The US Civil War (1861-5) was the result. That this war is presented to the general public (as in school texts) as a conflict over the end of slavery tells much about ideological manipulation by dominant power interests. As Abraham Lincoln made clear during the war, in a letter to New York Times editor, Horace Greeley (22 August 1862), slavery was a side-issue:

> If there be those who would not save the Union unless they could at the same time save slavery. I do not agree with them. If there be those who would not save the Union unless they could at the same time destroy slavery. I do not agree with them. My paramount object in this struggle is to serve the Union and is not either to save or to destroy slavery. If I could save the Union without freeing any slave I would do it, and if I could save it by freeing all slaves, I would do it; and if I could save it by freeing some and leaving others alone, I would also do that. What I do about slavery and the coloured race, I do because I believe it helps to save the Union, and whatever I forbear, I forbear because I do not believe it would help save the Union. I shall do less whenever I shall believe what I am doing hurts the cause, and I shall do more whenever I shall believe doing more will help the cause. I shall try to correct errors when shown to be errors; and I shall adopt so fast they shall appear to be true views. I have stated my purpose according to my view of official duty; and I intend no modification of my oft-expressed personal wish that all men everywhere could be free. (Couloumbis and Wolfe 1982:88)

That the emancipation of slaves was proclaimed at all probably owes more to the realisation that Britain would not enter the war on the side of its southern trading partner if the North projected its goals in moral terms. Certainly, as events after the war were to show, there was no intention of allowing emancipated slaves the equality of opportunity which the

American constitution purportedly gave them (e.g. Greenberg 1981; Trelease 1981; Mann 1984).

For the development of the West, the Civil War had more immediate outcomes. Freed from the constraints of southern Congressmen, northern industrialists (through their control of the war-time Congress) passed a variety of measures that had previously been blocked. One of the most significant of these was the Homestead Act (1862). This Act gave homesteaders ownership of 160 acres of unsettled land, provided they lived on and improved it and paid a minimal registration fee. There can be no doubt whatsoever that this Act aided settlement expansion in the West. But its frequent linkage to Jeffersons's idealistic beliefs about family farming enhancing democracy, or its presentation as an idealistic and visionary step to develop the West, is both contrived and misleading. The United States was not governed by idealists but by hard-nosed capitalists. The federal government had consistently made substantial profits on its land sales (thus Louisiana, Florida and the Gadsden purchases, the acquisition of the Oregon territory and the annexation of Texas cost around four and a half cents an acre, but the land was sold, under a 1785 ordinance, for over a dollar an acre - Vogeler 1981). As these land sales helped reduce taxation their replacement with a 'give away' programme was not immediately welcomed. Only when it was believed that land sales had ceased to be a source of government revenue was the Act approved (Vogeler 1981). This is not to say that 'gifts' were not a component of western land deals. Mark Twain's dictum that there is no inherently criminal class in America except Congress must surely owe its roots somewhat to western land acquisitions. Congress willingly turned a blind eye when its 'special friends' were obtaining land. For the Ohio Company, for example, Congress suspended its 1785 ordinance so that 1,000,000 acres could be sold for around 10 cents an acre (with the Yazoo land fraud resulting in most of present-day Alabama and Mississippi being sold for less than two cents an acre by the Georgia legislature - Barnes 1978). With requirements for titles and deeds often being ignored, Congress stood by while speculators allowed settlers to improve land, only to be subsequently evicted from it (Frundt 1975). Speculators were favoured by liberal credit terms for those purchasing at least 10 square miles, though similar credit was denied homesteaders. Hardly surprising, in this light, that Barnes (1978:138) felt compelled to state that:

> The history of the giveaway of America's public lands - hundreds of millions of acres over a century and a half - constitutes one of the largest ongoing scandals in the annals of modern man [sic].

The giveaway itself was to the already rich. 'Free' land under the Homestead Act amounted to just under 80 million acres, which contrasts with the 128 million acres given to railroads (with an additional 2 million to wagon and canal companies), 140 million acres to the states (largely passed on to large companies), 100 million acres of federal land sales and sales of 100-25 million acres of Indian lands (Vogeler 1981).

All this tells us about is the causal framework and character of land settlement. We need to go beyond this to appreciate impacts on rural development. Clearly major changes occurred in population structure and land tenure, but of greater importance in appreciating spatial differences in present-day rural conditions, is the effect family farm homesteading had on rural social structures. To fully comprehend the causal mechanisms involved here, it is illuminating to compare the settling of the Great Plains with that of California and the South (Pfeffer 1983). Plantations were an institutionalised component of the southern landholding structure on account of royal land grants in colonial times. Hence, inequality of property ownership originated in the establishment of the southern agrarian economy. In California, a similar structural inequality in landholding was instituted (McWilliams 1939). Here, however, the basis of that inequality was an 1846 Congressional decision to acknowledge vague and imperfectly registered Spanish land grants. The significance of these, and the monopoly power they provided for their owners, is suggested by the size of the holdings themselves. In 1871, for instance, 16 families in California each owned at least 84 square miles of the state (McWilliams 1939), with 25 landowners still possessing 61 per cent of all privately owned land within the state in 1973 (Vogeler 1981).

In a sense, the occupation of the South is a special case since it was instituted by the colonial power. The patterns in California and the Great Plains are, however, markedly different (though they both stemmed from a national government that was consistent in its promotion of expansionary policies). The differences which emerged in these regional land ownership structures cannot be tied to the presence of southern planters in Congress, since California was slave-free and so constituted a 'threat' to the southern slavery system. More realistically, they can be explained by the different opportunity structures which were afforded to landowners in different locations. It is clear that the maintenance of a plantation-style agrarian structure depended upon the availability of a cheap, captive labour force. In the South this was available through slavery. In California, Mexicans and imported Chinese coolies provided the necessary labour inputs. On the Great Plains, however, no such labour force existed. As we have seen, the federal government was more than happy to sell large tracts of land to those who could pay. Some of these landowners did seek to develop plantation structures. With an initial 33 per cent return on capital, for instance, much British investment flowed into Great Plains cattle ranching. However, herds were soon decimated by severe winters and drought, so large-scale investors left the Plains and the region came to be dominated by smaller enterprises (e.g. Worster 1979). This process was aided from the outset by the difficulty of attracting industrial and other urban workers to homesteading areas. Given a sufficient labour pool, the speculators could feasibly have established a Plains landed aristocracy similar to that of California and the South. However, without the carrot of landownership - which undermined the speculators' position - settlers could not be enticed into the area in sufficient numbers.

Of critical importance is the fact that these early social structures have changed little over time. In the South, the Civil War induced few changes in land ownership structure. Certainly, former slaves were 'free', but through the intimidation of the Ku Klux Klan (Nelson 1979) and restrictive labour legislation (Greenberg 1981; Mann 1984), the actual position of blacks changed little. The symbolic freedom that was enjoyed made continuation of the old-style plantation system extremely unlikely. Through crop sharing (Pfeffer 1983; Mann 1984), neo-plantations (Aiken 1971) and restrictions on black landownership (Nelson 1979), however, a social structure built around substantial wealth inequalities was perpetuated. In California the process was different - for here cheap labour was obtained by exploiting a succession of different immigrant groups (Chinese, Japanese, 'Dust Bowl' evacuees, Filipinos and Mexicans - see Baker 1976) - but the end-product was the same for rural areas. In sharp contrast, the Great Plains system of family farming - attained more as a default option than a goal (albeit some allowance must be made for the federal government wishing to reward Union soldiers by giving them homesteads after the Civil War) - has come to epitomise North American expectations about desirable rural living. More recent fears about the decline of family farms seemingly ignore the fact that family farms have been a minor or subservient organisational form in many parts of the US for a long time. Nonetheless, the existence of dissimilar organisational structures is significant both because these can be seen to stem from similar elite aims and because they condition future development options.

Important as these points are, the reader might possibly be puzzled over our treatment of settling the US West in a chapter on international effects on rural development. Its inclusion is based on two points. First, we do not accept that the US had any manifest destiny to incorporate these areas within its boundaries; hence US involvement in the West was for a long time in the realm of 'foreign affairs'. Second, the pattern of expansion revealed in the West has long characterised US foreign policy; that some of this expansion was in areas now within the USA should not stop us from recognising the strength of the underlying expansionary urge. This, of course, was not something that was restricted to the United States. Between 1870 and 1900, for example, Britain added four and three-quarter million square miles of territory to its Empire, with France aquiring an additional three and a half million and Germany, one million square miles (Healy 1970). The US was neither militarily strong enough, nor sufficiently motivated (with parts of continental North America still to be settled) to be at the forefront of this particular 'rush for the spoils'. The basic pressures for more markets and cheaper (or guaranteed) raw material supplies were nonetheless still present – they became manifest in sending troops or aiding military involvements in Samoa (1878), Morocco (1880), Korea (1882), Congo (1883), Hawaii (1887-93), Venezuela and Brazil (1893-9), Cuba and the Philippines (1898-9) and Haiti (1915) (see Healy 1970, 1976; Frundt 1975). These pressures continue to provide a major stimulus for US foreign involvements. Today, however, the vehicles through which economic control is secured are somewhat different. Territorial possession provides an expensive, potentially troublesome and organisationally complex means of guaranteeing markets (relative to returns at least). More satisfying in its lesser demands on government intervention, its flexibility and its healthy profit-loss balances, is corporate control. This is granted to the United States by the operation of its transnational companies.

Transnational Corporations

From his evaluation of US agricultural policy-making, Frundt (1975:2) concluded that:

> Agricultural policy is an instance of state policy in general which operates to aid and legitimate control by large corporations over the forces and relations of production.

The basis of Frundt's charge was the manner in which corporations dominated the farming sector. This charge is valid. It is not simply that agricultural processors and merchants dominate farming enterprises, but that domination is by a few, large corporations. To give one example, the

Federal Trade Commission has found that four out of 1,200 US canning companies (Campbells, Heinz, Del Monte and Libby) make 80 per cent of canning profits (Vogeler 1981). For the European reader, the pertinence of this point in a section on transnational corporations will be clear; for these same 'agricultural' companies also dominate the shelves of European supermarkets. Not that these corporations are agricultural per se. Frequently they are multinational conglomerates for which agricultural endeavours comprise a small component of their enterprise. Thus, Unilever owns BOCM Silcock (Britain's biggest feed manufacturer), Bachelor Foods, Bird's Eye, Walls, East Sussex Farms Ltd, Dale Turkeys, Matteson's Hams, plus Flora and Blue Ribbon margarine (Norton-Taylor 1982); Shell is an important carrot producer in California, and has been active in buying US seed companies like North American Plant Breeders, Agripo Inc, Midwest Seed Growers, Tekseed Hybrid Co, and Rudy-Patrick (Congressional Quarterly Inc 1984); Uniroyal, Boeing, Getty Oil, Kaiser Industries and insurance companies like John Hancock Mutual and Connecticut General are all 'farmers' in the United States (Cortz 1978); and the energy company Tenneco is one of the largest vertically integrated agricultural companies.* The picture in Britain is a little different. As a Conservative member of Britain's Parliament concluded, the main beneficiaries of government agricultural policy have not been farmers, but large companies like ICI, Shell, BP and Fisons: 'In a campaign of remarkable success, they have been at the centre of the argument for protectionism and subsidies, of scorn for "cheap food" and the call for agricultural expansion "in the national interest"' (Body 1983:introduction).

The gains which accrue to corporations from agricultural enterprises are frequently not from farming alone. Irrespective of how loudly fears are expressed over the purchase of agricultural land by non-farmers (e.g. Cortz 1978; Lapping and Clemenson 1984) or over the 'destruction' of family farming by vertically integrated corporations (e.g. Flinn 1982), these are still relatively small elements of the farming sector (the percentage of US farms other than family or partnership owned was only 0.87 in 1982).** As Mann and Dickinson (1978, 1980; Mann 1984) have reported, agricultural production contains impediments to capitalist development

---

* Vertical integration refers to companies in allied product lines which are under the same ownership. This effectively means handling a variety of stages in the production and sales process under one company. The Dominion Stores supermarket chain in Canada, for example, owned farms that produced goods which were processed in that company's own factories and were then sold within that supermarket chain.

which reduce its attractiveness for direct corporate involvement. These restrictions include: (1) production cycles taking a long time, so tying up capital, reducing profit rates and restricting responsiveness to price fluctuations; (2) a seasonal work pattern which makes for inefficient utilisation of machinery, management and labour; and (3) uncontrollable factors like weather, disease and pests. The disjuncture of agricultural production and mainstream manufacturing enterprises is such that initial corporate interventions into this sector have often been followed by some spectacular financial disasters (see, for example, Cortz's 1978 description of how the introduction of industrial economies of scale ideals by S.S. Pierce, a Boston distributor of food products, turned a recently acquired California subsidiary, Pic'n Pac Foods Inc, from a profitable endeavour into insolvency in only a few years).

In truth, corporate involvement in agricultural production is often not primarily aimed toward profiting from corporate farm production. The prolonged inflationary spiral of the 1970s, when farmland appreciation outran other investments, saw a rapid increase in non-farm holdings of agricultural land. This did not represent a wider involvement in agriculture, but a shift in fluid resources to a point of high returns (Munton 1977; Lyson 1984). As higher profit margins arose in other investment sectors, the flow of funds into agricultural land was reversed. Not that all such investment was removed. Through tax breaks, such as preferential rates for farm use, deferred taxation, income tax credits and tax rebates (Lapping and Clemenson 1984), corporations have relied on agricultural involvements to reduce their overall tax payments. This is no new phenomenon. In 1965, for example, of the 119 US farmers reporting income of more than one million dollars, 103 wrote off farm losses against other income (Bible 1978); in the same year, only

---

** The reader should be aware of two points here. First, Rodefeld (1974) found that corporate ownership in the US was greatly understated in official documents (by up to 216 per cent in Wisconsin). Second, in some nations, like France, non-farm ownership of land is encouraged by the state to reduce farmers' capital investment needs (Bryant and Russwurm 1979). In fact, through contracts for farm produce, ownership of agricultural processing and domination of marketing and sales sectors, corporations do not need to own land or farms to control their operations. This is an issue to which we will return in later chapters, since it has direct relevance for recent attempts to promote a rural social stratification schema based upon land ownership (e.g. Stinchcombe 1961).

9,244 out of 17,578 US corporations reporting farming as their principal business recorded a profit for tax purposes. Agricultural interventions are just one component of a general trend in tax avoidance that results in both US (Table 4.1) and British (Figure 4.3) corporations making comparatively slight

Table 4.1 Effective Tax Rates by Industry for 275 Major US Companies 1981-4 ($ millions)

Industry	Profit	Tax	Rate(%)
Airlines	441.7	(33.1)	-7.5
Financial	4,826.2	(139.2)	-2.9
Railroads	13,308.4	236.0	1.8
Telecommunications	41,035.8	927.5	2.3
Paper and Forest Products	7,158.8	258.3	3.6
Aerospace	13,354.0	587.8	4.4
Chemicals	13,810.8	827.7	6.0
Utilities	53,956.1	3,788.6	7.0
Electrical, Electronics	21,735.3	2,539.5	11.7
Conglomerates	5,427.5	736.7	13.6
Services, Trade	12,720.6	1,994.2	15.6
Rubber	2,417.4	462.2	19.1
Oil and Gas, Coal, Mining	77,615.7	14,955.4	19.3
Building Materials, Glass	3,442.7	677.6	19.7
Drugs, Hospital Supplies	16,228.7	3,289.3	20.3
Food and Beverages	22,686.3	4,770.5	21.0
Automotive	5,952.4	3,805.9	23.9
Computers, Office Equipment	27,709.1	6,742.9	24.3
Miscellaneous Manufacturing	7,752.4	1,927.3	24.9
Publishing and Broadcasting	6,735.3	1,691.5	25.1
Construction	2,435.3	708.9	29.1
Leisure, Personal Care	7,799.5	2,513.8	32.2
Textiles	2,390.2	801.0	33.5
Tobacco	11,300.2	4,096.1	36.2
All Industries	400,546.1	60,265.7	15.0

Source: McIntyre and Wilhelm (1985)

Figure 4.3 British government revenue sources 1983 / 4 — 1984 / 5

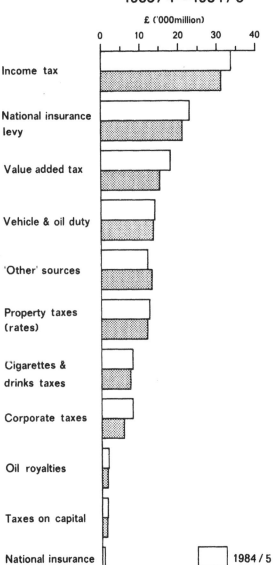

Source : The Guardian 14 March 1984

tax contributions.   For agriculture per se, this is a significant issue, though its importance is commonly more symbolic than economic.

Corporate involvement in agricultural endeavours occurs across the full spectrum of farm-related activities. Just two US companies handle 50 per cent of the world's grain shipments, for example, while the proportion of feedstuffs and fertilizers supplied to farmers by companies of national and international scope has been estimated at 70 and 95 per cent respectively (Gale 1977). As the Canadian Royal Commission on Farm Machinery (1969) reported, two companies (John Deere and International Harvester) are so dominant internationally that they effectively determine tractor prices on an inter-continental basis. Geographical differences in farm production costs arise because of this dominance. It was this factor that prompted the establishment of this Royal Commission. The Commission itself ended up confirming that tractor prices bore no relationship to production costs. As an illustration, John Deere's model 710 tractor, which was manufactured in West Germany, had a net wholesale price to dealers that was $C1,200 lower in Britain than in Canada, even though there was a $C450 tariff for importation into Britain, whereas it entered Canada tax free. In this example we see one dimension of the world-wide organisational framework which transnational corporations bring to economic relations (as is clearly seen in the locus of production and trade patterns associated with Ford's J-car; see Figure 4.4). Although such global relationships are not commonly linked to agricultural endeavours, it is difficult to disentangle agricultural from non-agricultural concerns. The estimates of Baker (1976), of one in seven US jobs arising in the 'food industry', and of Schulman (1981), with one-sixth GNP, one-fifth employment, and one-quarter US exports in the hands of agribusiness,* are bettered by Carlsson et al's (1981) suggestion that one-third of US workers are in agribusiness. Yet, by their magnitude, these estimates reveal the difficulty of distinguishing agricultural and non-agricultural activities. Even more impracticable is distinguishing rural from non-rural endeavours. This means that we cannot directly identify the impact of transnational corporations on rural areas. Instead, we must understand the organisation and operations of transnationals and, from this, tease out their rural

---

* Agribusiness refers to all organisational structures associated with supplying inputs to farming (fuel, seeds, chemicals, equipment), the farm production process, the marketing and processing of farm products, the distribution of these products (nationally and internationally) and their marketing and sales (Carlsson et al 1981).

Figure 4.4 Ford's J-car assembly lines

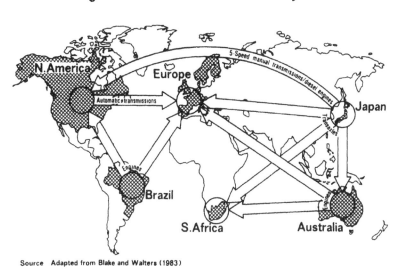

Source   Adapted from Blake and Walters (1983)

---

implications.

One of the most important contributions to our understanding of transnational corporate behaviour is Stephen Hymer's (1972) model of transnational corporate organisation. This proposes that transnationals are arranged in a hierarchical manner. Hymer suggested three levels, though the principles behind his model are sufficiently general that other levels could exist. In brief, the apex to the organisation is the international head office (Figure 4.5). This contains the major decision-making elements and it is around this locus that the most technically advanced activities take place. This is the focus of senior management and the most significant specialist services (like research and development, legal services, and accounting services). Secondary to this core are regional or functional head offices. These are either oriented to serve a particular market across the range of corporate activities or they act as mini international headquarters for a more restricted activity area. Companies vary in the structures they adopt (Taylor and Thrift 1982) and in some instances, there is a mixing of these two organisational principles. For geographically structured

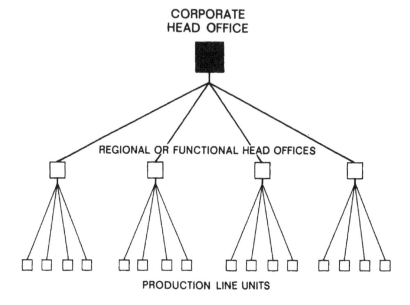

Figure 4.5 Hymer's model of corporate
organisational structures

CORPORATE
HEAD OFFICE

REGIONAL OR FUNCTIONAL HEAD OFFICES

PRODUCTION LINE UNITS

Source : Ideas from Hymer (1972)

organisations, these secondary offices commonly contain
services that are nationally or regionally important. Thus, a
US corporation like the Ford Motor Company will employ legal
experts and accountants at its main British office on account
of differences in legal, technical and taste conditions in the
US and UK, (to allow some sensitivity to a host country's
labour and marketing situations, these offices must be given
some decision-making autonomy). They are nonetheless very
clearly second level offices. Thus, in a Canadian
manufacturing sector which is around 60 per cent foreign
controlled, just 20 per cent of subsidiaries are estimated to
have substantial decision-making autonomy (Clement 1977:98); a
further 60 per cent have high operational decentralisation but
final authority for decisions rests elsewhere. On a broader
front, just 6.9 per cent of US transnational company research
and development is undertaken outside the parent nation and 75

per cent of that appears to be restricted to adapting corporate products to host countries' legal or market requirements (Lall 1979).

Yet, while their decision-making capabilities are restricted, these secondary units still possess more organisational power than production or line units. These third tier units have little decision-making autonomy. In Wales, only 22 per cent of foreign subsidiaries undertake their own marketing, while 60 per cent of Scottish transnational affiliates have the geographical area in which they can sell commodities restricted (Watts 1981). Despite their lowly corporate position, these units are generally the biggest employers of labour. Governments, which fear the electoral consequences of high unemployment, are not surprisingly actively involved in enticing these plants into their country (or province/state or even locality). Evidence clearly shows that these government inducements can be critical in making locational decisions (e.g. Forsyth 1972; Thoman 1973). Since these incentives can be extremely generous, they act as strong attraction forces. Thus, to induce Volkswagen to build its new car plant in Pennsylvania, the state government gave a $40 million loan (with interest at 1.75 per cent a year for 20 years, then 8 per cent for the last 10 years), a $30 million bond issue for railway development, $4 million to train workers, and reduced property taxes for the first 15 years of operation (Fry 1980). Similarly, the Ontario Government gave the Ford Motor Company $C40 million (and the federal government added $C28 million) to set up an engine plant within that province. In general, in fact, transnationals bring surprisingly little money into a host nation. By drawing on government grants and loans from local sources, they are able to expand while drawing little capital from their parent nation (Taylor and Thrift 1981; Yeats 1981). For host nations, these financial contributions provide some leverage in plant locations. In Britain and a number of other European countries, the transnational orientation towards wealthier regions has, in good measure through financial inducement, altered in recent decades so that new plant placements have tended to favour poorer regions (Thoman 1973; Smith 1982; Dicken 1983).

That finance alone is not a sufficient inducement is signified by the continuing dominance of core areas as plant locations for transnational companies in Canada (which is perhaps a special case as it adjoins the United States), Australia and France (Taylor and Thrift 1981; Blackbourn 1982). More so for high technology plants and research divisions than for line units, but still seemingly dependent on the magnitude of the investment (and new jobs) proposed, governments allow their priorities to be distorted in return for new transnational investment. Thus, IBM threatened to establish a new plant in France if the British Government, which was promoting employment growth outside south east England, did not allow it to locate in Havant, Hampshire; the Ford Motor Company was allowed to break the 1964-70 Labour Government's wages and prices guidelines in return for locating its engine plant at Bridgend; while Michelin induced Nova Scotia to change its labour relations code to restrict trade union activity when it located a plant there (see McDonald 1977; Blackbourn 1982).

We do not need to go to the Third World for evidence of transnational companies distorting national development policies for their own ends (Gonick 1970; McDonald 1977). This is an important point to emphasise. It is largely units with the lowest socio-economic benefits, that lie in the poorest organisational power positions, which nations are invited to fight over. These units are the most unstable element in transnational corporate structures. A primary reason for this is made clear in Raymond Vernon's product cycle hypothesis (Norton and Rees 1979; Vernon 1979). The reader should be advised that this hypothesis is no longer taken as the 'best' explanation for the location behaviour of transnational corporations (though see Markusen 1985 for a modern, neo-marxist theory with significant similarities in expected developmental outcomes). Today, much greater emphasis is placed on the oligopolistic character of transnational behaviour; whereby large firms follow each other in order to maintain sales shares in the world market (Vernon 1979; Taylor and Thrift 1982). For most of the post-1945 era, however, the predictive capacity of the product cycle hypothesis far outshone that of other models. Essentially, this hypothesis holds that companies which operate in a global market are obliged to play their innovative gambles for high stakes. Not surprisingly, they prefer to keep research and development activities (and early product market testing) close to their decision-making core (the international headquarters). This allows face-to-face contact between key research, legal and financial experts. As a new product becomes established and its sales increase, competitors will be encouraged to capture part of the burgeoning market by offering an alternative (but closely related) product. Initially, sales outside the core market draw supplies from core plants, but as sales increase companies are encouraged to locate plants in 'foreign' markets. Intense competition for sales can even intensify cost-cutting pressures to the extent that when the product is well established, its production becomes routine, and it no longer requires tight supervision from core departments, the complete production process is transferred out of (high waged) core areas into low cost sites. In its early stages, this process was precisely that which characterised the movement of textile plants out of the US manufacturing belt into the South. Today, it is commonly associated with relocation beyond the core nation. Hence, we find a flow of textile plants from the advanced nations into the Third World (Fröbell et al 1980), just as Third World sites have become more important in car manufacturing (McDonald 1977; Figure 4.3) and companies like Heinz have found richer pickings for their canning operations in poorer European nations (e.g. Eccles and Fuller 1970).

The important contextual condition that must be taken into account in evaluating these patterns is the dynamic of capital accumulation (Markusen 1985). It is pertinent to restate the fact, '...long ago emphasised by Marx... [that capitalism] is by nature a form or method of economic change and not only never is but never can be stationary' (Schumpeter 1943:82). As Massey (1984:49) points out:

> One of the main characteristics of the developing geography of capitalist societies so far has been precisely the development of new areas and the abandonment of old. Even if we accept that the existence of production in a region does not in itself make that region a centre of accumulation, the whole history of capitalism indicates that what we need is not some single concept, such as centralisation, but a whole range of concepts which allow the analysis of complicated, and changing, structures of geographical inequality.

At the lowest rung of the production process we encounter the least innovative product lines; those whose life cycles are nearest to their terminal point. With large government grants (or cost subsidies) the life span required for these plants to recoup company investments is considerably shortened. Hardly surprising, then, that these plants have been reported to have more unstable life durations than proprietor-owned plants (Tornedon 1975; Clark 1976; Erickson 1980; Anderson and Barkley 1982); albeit instability declines the larger the plant size (not unexpectedly given this has clear implications for the total investment figure). It is certainly the case that transnational company plants offer a significant capacity for the creation of employment; as Forsyth's (1982) example from Scotland demonstrates. Such conclusions are nonetheless time-bounded. Transnational line investments mark time in particular locations on their way to points of higher profitability (in 1966, for example, RCA's television plant moved from Indiana to Memphis to save costs but had relocated to Taiwan by 1971 for the same reason; Markusen 1985). Corporations constantly restructure to maintain and enhance their profitability (e.g. Massey and Meegan 1979). The name of the game for line units is locational instability.

For head office sites, by contrast, a greater element of security is present.   This security is in no sense absolute. The corporate world is habitually the scene of major takeovers and mergers (Leigh and North 1978; Stephens 1982). These can induce some short-term employment enhancement (Green and Cromley 1982), but aquisition is frequently linked to ultimate job losses (Leigh and North 1978). Certainly there is a clear trend toward the loss of senior decision-making functions from peripheral areas, as major corporations take over more vibrant 'local' companies (e.g. Smith 1982).   For the very largest corporations, though, there is considerable (middle-term) stability of location.   This applies not simply at the national level, but also in the precise geographical areas within nations that house major corporate headquarters (e.g. Goddard and Smith 1978; Cohen 1979).   This is a key point. The United States is the parent country of head offices accounting for more than half the world's total (foreign) transnational investment. It benefits substantially from this in two regards.   First, US corporations truncate top management and technically advanced sections from host nation's company structures (Britton 1980; Taylor and Thrift 1981); that is, they keep the highest paid and most advanced corporate sections within the United States (and by utilising market power to acquire new subsidies, perpetuate this pattern).   Second, through procedures like transfer pricing* and more direct repatriation of profits, the wealth of the United States is substantially enhanced to the detriment of other nations.   In 1982, for example, · on account of transnational company actions, Latin America had a net capital

---

* Transfer pricing refers to the adjustment of prices (i.e. artificial 'rigging') so that products cross national boundaries, usually from one section of a corporation to another, at the price which maximises company profits, rather than reflecting 'true' production costs.   Hence, a manufacturing plant could technically operate at a loss in order to avoid high tax rates in one country; its products then being sold at high prices in a country with low tax rates, giving inflated net profits.   The   potential importance of transfer pricing is indicated by a 1972 Bolivian Government study which found over-pricing on pharmaceutical goods of 155 per cent, with figures for electrical and rubber goods and chemical imports standing at 44 per cent and 25 per cent respectively (Yeats 1981 reported that over-pricing of up to 3,000 per cent occurred on individual items, and listed a variety of studies on other nations which confirm the magnitude and indicate the universality of this practice).

inflow of $19 thousand million but lost $34 thousand million in profit and interest payments (Griffith-Jones 1984; similarly Perry 1971 reported a net outflow of $2.6 thousand million from Canada to the United States over the 1960s). As an outcome of world-wide activities, nations housing the decision centres of transnational control benefit from immense infusions of funds. This significantly enhances national wealth and, via multiplier effects, has very obvious implications for the well-being of rural residents in core nations. Of course, as recent concern in the United States over foreign ownership has shown (e.g. McConnell 1980), no nation is wholly 'core'. All experience the enhanced insecurity (and employment benefits) of line units and all advanced nations are loci of some transnational head offices. In degree, however, there are clear differences in standing. The developmental experiences of rural areas are constrained, prompted and promoted by the standing of their homelands in transnational production processes (and the stability of their position within that mode of capital accumulation).

The magnitude and scope of transnational corporations is such that states are significantly constrained by their actions. As General Motors's chairperson, Alfred P. Sloan, explained to stockholders on the eve of Germany's 1939 invasion of Poland, his corporation was 'too big' to be worried by 'petty international squabbles' (Solomon 1980:79). Reminiscent of oil companies and corporations like Union Carbide defying an international ban on trade with Ian Smith's Rhodesia (Astrow 1983), major corporations like Ford, ITT and General Motors manufactured and traded with both Axis and Allied powers in the Second World War; they would reap golden profits no matter who won.* The ability of transnational corporations to act in this manner is partly derived from their size. Ranking sales as equivalent to gross national products, General Motors would be the 25th largest 'country', with Exxon (27th), Shell (34th), Ford (35th), Mobil (39th), Texaco (44th) BP (48th) and National Iranian Oil (50th) all falling within the top 50 (Couloumbis and Wolfe 1982). Size alone is not a sufficient explanation. Critical to their

---

* O'Sullivan (1986:86) provides further indications of the power of transnational corporations, or perhaps more accurately of the high priority national governments give to preserving capitalist production relations, by reference to some Second World War practices. These include the acceptance of cartel arrangements between DuPont and IG Farben which restricted the production of US bomber nose windows and the preservation of patent agreements between Bendix, British Zenith and Siemens which constricted British production of aircraft carburettors.

influence is the greater flexibility of transnationals and the (relative) ease with which they outmanoeuvre national governments. This flexibility comes not simply from the absence of electoral stringencies, but also arises from their adaptability in manipulating and directing capital resources. At present, for example, 61 per cent of US manufacturing imports from other OECD nations are between parent companies and their subsidiaries (Helleiner and Laverge 1979). In addition, transnational company resources are easily transferred from one country to another. United States Tariff Commission figures for 1971 show this, for in that year transnational corporations had $268 thousand million in liquid assets compared with foreign exchange reserves of just $87 thousand million in the major industrial nations (Lauter and Dickie 1975). Perhaps not surprisingly in these circumstances, governments have become alarmed over the growing power of transnational companies. Especially since these corporations are drawn disproportionately from the United States (where they at least act in alliance with the federal government, even if they cannot really be said to owe allegiance to it), the enhanced role of transnationals implies losses of both economic and political autonomy (e.g. Kavanagh 1974; McDonald 1977).

Our discussion of transnational corporations has been designed to bring out both the manner in which national governments are constrained in their behavioural choices (i.e. their options are structured) and how transnational locational decisions have clear spatial effects. The precise character of these effects is nevertheless various. Bornschier and Chase-Dunn (1985) have made clear that when multinational manufacturers are locating plants intended to serve world markets they principally search out low cost sites, and offer little enhancement for 'local' income levels. When transnationals are producing for the markets of host nations, however, they tend to be higher technology enterprises in which wages constitute a small percentage of total costs; hence they often bring positive benefits for average income levels. It is inappropriate therefore to impose a blanket condemnation on transnational companies for their impacts on poorer regions. In rural areas the benefits from transnational activity are divided spatially. Enterprises are distinguished by their technological and intra-organisational power positions. Transnational production is increasingly international in scope (of the Harvard Multinational Enterprise Project's 180 investigated corporations, for example, 32 per cent had switched to global production structures by 1980; Taylor and Thrift 1986). This enhances the importance of access to international airports, to national government offices and to major financial institutions. Commuter belt rural areas might well obtain some trickle-down effects from this, for these

high technology and senior decision-making establishments are
predominantly city oriented. However, for more peripheral
rural areas, the transnational corporate urge is to utilise
government subsidies, seek low cost sites (and low local
taxes) and keep wages down (thereby resisting unionisation),
in order to enhance the world-wide competitiveness of their
product. Peripheral rural areas have gained jobs in this
process (Figure 4.6), just as Third World nations have
(Fröbell et al 1980; Markusen 1985; Taylor and Thrift 1986).
For both, the jobs provided are unstable. Yet the same point
can be made about the impacts of government programmes in
peripheral rural areas. In Western Europe at least,
government programmes themselves have taken on an increasingly
multinational character in recent years.

The European Community

The formation of the European Economic Community was linked to
fears that Franco-German conflict might once again revive if
unambiguous efforts were not made to intermesh their socio-
economic (and political) fortunes (Parker 1983). On a number
of fronts - industrial, agricultural, energy, defence,
monetary and political - integration was seen as desirable
(even if only as a long-term goal). In particular economic
sectors like iron and steel production, but also more
generally to restructure industrial enterprises and merge
European companies in order to meet the US corporate challenge
(Owen 1983; Swann 1984), European Community institutions have
taken on an important role in structuring the economic affairs
of member states. Significantly, however, it has been through
the Common Agricultural Policy (CAP) that the cause of
European integration has been most advanced. As Hill
(1984:118) reported:

> It is a monument to the determination of
> politicians to work together for a united
> Community. Because the CAP is the first major
> common policy involved in the integration
> process, it has become the symbol of cooperation.
> In this spirit of determined cooperation the
> politicians have nurtured the CAP in defiance of
> economic logic and the long-term interests of the
> Community.

The prominence of agriculture in this process is relatively
easy to understand. By the time European Community
discussions were taking place in the 1950s, governments were
already key agents in determining both agricultural production
levels and farm incomes (e.g. Wright 1955). Agriculture was,

Figure 4.6 Manufacturing employment change in
England and Wales 1971 — 78

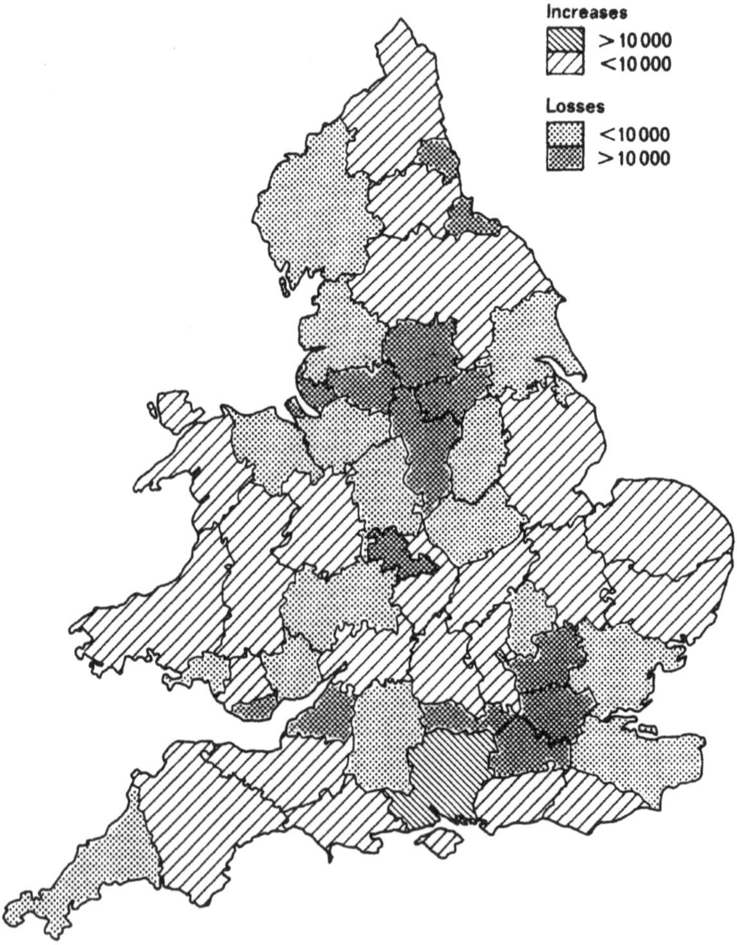

Increases

▨ > 10 000

▧ < 10 000

Losses

▦ < 10 000

▩ > 10 000

Source : Adapted from Keeble (1984)

Figure 4.7 Countries of the European Community

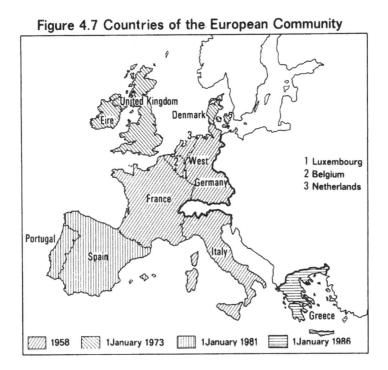

furthermore, a 'troublesome' economic sector, insofar as it was declining in employment terms and was in need of structural reform. Yet the importance of agricultural employment both in the original Community Six (Figure 4.7) and in most later members (Table 4.2) made farmers a group of electoral importance. This had long been the case in France (Wright 1955; Berger 1972), but even in nations with smaller farm populations, the strategic position of the farm vote in affecting party returns gave this occupational group a disproportionate political influence (e.g. Averyt 1977; Andrlik 1981).

The international context of rural development

Table 4.2 Agricultural Labour Forces in the European Community

	Entry Year	Percent Economically Active in Agriculture	
		1970	1984
Original Six Members*			
Belgium		4.9	2.5
France		13.7	7.1
Italy		18.8	8.9
Netherlands		8.1	4.5
West Germany		7.5	3.1
Later Entrants	**(Entry Year)**		
Denmark	(1973)	11.1	5.8
Ireland	(1973)	26.5	18.8
United Kingdom	(1973)	2.8	1.8
Greece	(1981)	46.0	34.2
Portugal	(1986)	33.3	23.6
Spain	(1986)	26.0	14.4

* Luxembourg not included

Source: United Nations, Food and Agriculture Organisation, 1985, Production Yearbook 1985, Rome.

The electoral implications of the CAP have been critical to the policy's development. Underlying the basic economic rationale of the Community as a whole has been the desire for free trade between member states. For industry this goal has found accord with the joint notion that economic efficiency and national economic power can be enhanced through larger organisations. For agriculture, the idea of an open market - which would have disassembled the peculiar national production subsidies which each nation's farmers had come to rely upon and even adapted to - was not an attractive proposition. If

84

agriculture was to be integral to the unification process, its
practitioners had to be 'bought'. This was achieved within
the Treaty of Rome, which established the Community, in
recognising five main goals for the CAP: (1) increasing farm
productivity; (2) providing a decent standard of living for
farmers; (3) ensuring market stability; (4) guaranteeing
security of food supplies; and (5) providing reasonable food
prices for consumers (Averyt 1977). The promise of the CAP,
then, was for higher, more stable rural incomes, with cheaper,
more home-grown products for consumers.

How the CAP has fared in meeting these objectives is open
to some dispute. As the President of the British National
Farmers' Union, Richard Butler, made clear, there has been
much on the farmers' side to commend in the programme (e.g.
The EEC farm policy that is not as black as it is painted, The
Times 3 March 1981:12). Amongst academic assessors, however,
a less favourable conclusion has been common:

> The failure of the CAP is comprehensive. It has
> totally failed to improve the relative position
> of the agricultural population even on average.
> Within this it has benefited rich farmers and
> regions rather than the poor, largely at the
> expense of consumers, the poor bearing the major
> burden because they are both more numerous and
> spend a higher proportion of income on food.
> Major transfers of income between countries,
> caused by common financing, have also to a
> considerable extent been inequitable. Harsh
> economic effects have been imposed on third
> countries [especially in the Third World],
> generating ill-will and the potential seed of
> destructive trade wars. The national incomes of
> community countries have been reduced by an
> unknown but significant extent. The CAP has even
> failed to be a common policy and has added to
> rather than reduced distortions to competition
> within the community. (Hill 1984:117-8)

The range and intensity of weaknesses signified by this
critique indicate that more than inefficient actions, corrupt
practices or inopportune timing are involved. The weaknesses
specified are such that structural handicaps must be critical.
This is undoubtedly the case.

There are certainly numerous grounds on which the CAP can
be criticised (for detail see Hill 1984; Bowler 1985; Duchêne
et al 1985), but in a rural development context three
particular issues are important: first, how have agricultural
returns changed compared to consumer incomes; second, have

benefits (particularly for rural residents) been equivalent in each member nation; and, third, has the distribution of gains and losses been random or systematic amongst farmers themselves. In answering these questions we must not fall into the mistaken belief that the CAP (or other European Community institutions) operate as autonomous programmes, in which the decisions of Brussels or Strasbourg are handed down with authority. For one thing, a long period was allowed before the agricultural programmes of member nations had to meet on a common path (unified product prices only came into effect in 1967 for instance). Before this stage was reached, the idea of the European Community as an autonomously powerful supra-national political unit was effectively dead. De Gaulle's insistence that unanimous voting was required for major Community decisions, added to a 1965 six-month French boycott of Community affairs (to express displeasure over a proposal to enhance the power of Community organisations), left no doubt that the CAP would not be able to plan European agriculture in a unified manner. Politicians were not going to transfer their loyalties from national to European institutions (not yet anyway). The CAP was to be a compromise package that lacked direction or long-term aims. 'There is a strong sense that each country is still inclined to point out how its neighbours might mend their ways the better to retain its own' (Duchêne et al 1985:132). When proposals to induce a European direction to agricultural production have been forwarded (like the Mansholt Plan of 1968 which proposed financial aid and price reductions in order to promote larger consolidated farms), national governments have frequently ignored them (George 1985). It is not simply that electoral considerations have proved too strong here, for clashes of values are also involved. Whereas the Danish and French see small farms as cornerstones of rural vitality, others, such as Britain, continue to promote even larger farm enterprises.

To be fair, the Treaty of Rome set the CAP nearly incompatible goals. This particularly applies to the provision that farm incomes increase while 'reasonable' food prices be maintained. The inherent incompatibility of these aims is one root cause of public disenchantment with the programme. Numerous academic and journalistic reports inform us that for most of the CAP's history, the scales have been weighted heavily in favour of higher farm incomes and food prices.* Estimates vary but claims that the CAP increases food prices to 20 per cent or more above world levels are not

---

* Body (1983) reported that the mean average level of farm income support from 1953 until Britain's entry into the CAP was 61.6 per cent. Since then it has been over 100 per cent of total net farm incomes.

uncommon (e.g. To cap it all, only those rich members get the cream, The Guardian 1 December 1983:22). With estimates of subsidies per farmer running at around £10,000 a year (Bowers and Cheshire 1983), there clearly appears to be an imbalance in CAP support towards agriculture. This, however, is a conclusion that must be treated with caution. Since 1950, for example, the prices farmers have been paid for their products have consistently fallen in real terms (in Ireland the fall has been the smallest but even here stands at 1.0 per cent per year; Duchêne et al 1985). In addition, the benefits farmers have received have been very unstable. This was seen in the British Government's differential pricing of the 'green pound' during the last decade. At all times, the transfer of food products between European Community nations is governed by a special foreign exchange rate (the so-called green currencies); it is through calculations based on these green currencies that farm product prices are assumed to be equivalent in all member states. The green currencies mirror foreign exchange rates but are not subject to their daily fluctuations. Instead they are determined by inter-governmental agreements. Since nations have veto power over the level at which green currency exchanges are set, national governments have been able to manipulate food prices (for imports and exports) in pursuit of their domestic policy aims. Between 1974 and 1979, when faced with record inflation levels, the British Government did not alter the green pound in line with the value of sterling. As a result, it became overvalued by as much as 30 per cent (Bowers and Cheshire 1983; Hill 1984). This reduced the price of food imports, thereby weakening Britain's inflationary surge, reducing Britain's farm incomes and decreasing the willingness of British farmers to make new investments. In this case, potential electoral losses were insignificant. Not only was Britain governed by the Labour Party (for whom farmers traditionally provide few votes), but it also stands apart in having a very small share of its total population engaged in agriculture. Not surprisingly, perhaps, starting with Jim Callaghan's Labour administration, and forcefully carried on by Margaret Thatcher's Conservative governments, it has been Britain that has led the assault on high (and higher) farm product prices within the Community.

Attacks on high farm prices have been popular in Britain. However, what has made British governments adopt their antagonistic stance is not merely vote catching. The method of calculating European Community contributions, coupled with Britain's small agricultural work force (discriminating against it receiving large CAP payments) has led to Britain being the largest net contributor to Community funds, despite being one of its poorer members. In fact, with the notable exception of Ireland, the tendency has been for richer

Table 4.3   Gains and Losses from the Common Agricultural
            Policy

	Percent Change in Gross National Product from CAP Payments	Gross National Product Per Capita (EEC = 100)
Denmark	2.27	140.2
West Germany	-0.17	134.9
Belgium/Luxembourg	-0.47	120.0
France	0.32	115.8
Netherlands	0.93	115.2
Britain	-0.59	77.7
Italy	-0.46	61.6
Ireland	8.80	47.6

Source: Hill (1984)

Community nations to receive income supplements from the CAP while poorer nations have paid for these economic enhancements (Table 4.3). The significance of this point comes from the fact that the CAP has taken around 90 per cent of the Community's budget for most of its life. Not that this situation is inevitable. Britain has been offered increased Community payments via the Regional Development Fund and through social programmes on a number of occasions. The rejection of these offers has owed much to the Thatcher Government's goal of reducing government commitments to all such 'welfare' programmes. What must also be borne in mind is that while nations like Britain might lose out on an agricultural front, they could benefit in the industrial field. Unfortunately this has not been the case (relative to longer-term Community members at least). The United Kingdom's entry into the Community did not produce the economic bonanza many people expected. Britain joined the Community just when economic growth impulses deteriorated. Continental producers had a decided advantage as they had already made the necessary adjustments for changing from a national to a Community-wide market (Owen 1983). The end-product is that Britain in particular has reaped few rewards from Community membership (though this is a relative statement only). Compared with

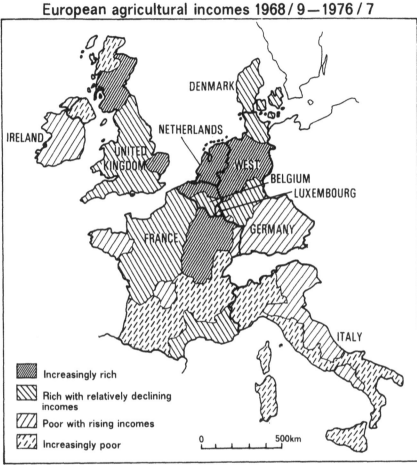

Figure 4.8 Typology of regional trends in European agricultural incomes 1968/9—1976/7

Source : Bowler (1985)

other members, Britain in general (and hence its rural areas indirectly) has been relatively disadvantaged in development terms since joining the Community (though this statement should not be read as a commentary on what might have happened had Britain not joined the Community).

Just as Community procedures have tended to favour richer

member states over poorer areas, so have richer agricultural areas within states (and across them) benefited at the expense of poorer areas (Figure 4.8). In the main, the reason for this is that CAP payments have discriminatory effects amongst farmers themselves. Some of these effects are product oriented. Thus CAP subsidies have changed oilseed rape from an unprofitable to a highly profitable crop for British farmers (Bowers and Cheshire 1983). Similarly, Britain's entry into the Community bought huge income increases for cereal farmers, much smaller increases for sheep farmers, and, on account of feed cost rises, reduced funds for pig and poultry farmers (Josling and Hamway 1976). Thus, based upon farm size and farm income in particular, the broad effect of the CAP has been regressive, exacerbating already considerable differences in farm income levels between sectors of European agriculture and implicitly favouring owners of large (cereal producing) holdings. The geographical effect of these biases is evident. Bremen in West Germany, for example, had, in 1977, eight times the income per labour unit of the west of Ireland and sixteen times that of Molise in Italy (Cuddy 1982:206). Josling and Hamway's (1976) conclusion that the wealthiest quartile of British farmers would receive a net gain of nine per cent in total farm incomes due to CAP payments is likewise representative of the broadly accepted conclusion that the CAP has redistributed farm incomes towards the already wealthy (Bowers and Cheshire 1983; Hill 1984; though, as we shall see in the next chapters, this is a common criticism of government agricultural policies).

Inevitably, because agricultural enterprises are closely bound by structural features, the net effect, both of the CAP's product oriented discrimination and pricing policy and the relationship of these to the a priori location of agricultural wealth, has been the exacerbation of regional economic disparities. This has earned it an 'anti-regional' tag (Martins and Mawson 1982:191), one that was fully appreciated by Britain during its negotiations over entry in 1972. In regional terms the beneficiaries of the CAP have been the richer agricultural areas of Europe (Cuddy 1982; Clout 1984). The losers have been the mountainous and peripheral regions already handicapped by geography and terrain (Bowler 1985). Nevertheless, specific measures do exist for addressing imbalances in farm incomes and the viability of farm holdings. Directive 268/75, to name one, provides funds on a regionally selective basis to maintain incomes in areas with natural handicaps (Reichenbach 1980). In effect, this policy seeks to stabilise population levels in areas of outmigration and encourages farmers to become 'gardeners' of areas of natural beauty. Another measure with redistributive intentions was the guidance section of the European Agricultural Guidance and Guarantee Fund (EAGGF).

This provides funds for restructuring agricultural enterprises to help make them more 'viable' units. Unfortunately, predominantly it has been farmers from the wealthier agricultural regions that have sought and obtained assistance under this guidance section (Vanhove and Klaassen 1980:421). The failure of the EAGGF in this regard, and the general regional effects of CAP as a whole, lie behind motivations to establish specific ameliorative policies at the Community level. Yet the fact that these ameliorative policies account for only around 2.6 per cent of the total Community budget (Cuddy 1982:207) gives some indication of their relative position in the hierarchy of broad Community goals.

Significantly, from the outset the European Community has resisted specific regional development policies (other than in the singular case of southern Italy) preferring instead to make regional aid ancilliaries to other policy instruments (thus the European Coal and Steel Community (ECSC), formed in 1952, or the European Investment Bank (EIB), formed in 1959, have both addressed regional development issues, the former through the retraining of workers in declining industrial regions, the latter through infrastructural investment). One reason for this recalcitrance has been a conflict over the means of achieving spatially selective development aid (the end, continued economic expansion, has not been questioned). The conflict here is between the ideals of a free-trade economy which are enshrined in the Treaty of Rome, and the notion of positive discrimination, to which a strong regional policy would need to be addressed. In the confident days of Community formation, balanced across-the-board growth was regarded as attainable. By the late 1960s, however, it was clear that while some regions and economic sectors were expanding, others were static or in decline. The relative inability of the ECSC, EIB and Social Fund to redress this widening imbalance led to demands for a more unified approach to regional development across the Community. In part these demands arose out of a growing recognition of the divisive effects of existing Community policies (most notably, the CAP), but, in part, they also reflected a concern that the long-term goals of the Community were likely to be frustrated by large-scale regional disparities. Certainly, concern for the future viability of the proposed Economic and Monetary Union (which, it was acknowledged, would exacerbate existing imbalances further, and reduce the abilities of member governments to directly assist their poorer regions) led the Council of Ministers to propose the establishment of a specific regional policy in March 1971. By that time, individual member states had operated national regional policies for some years. However, as these national policies ran counter to the Community's free trade goals, the Council was keen to coordinate and restrict their impact (earlier attempts by the Community to control them through notification procedures having failed).

A further note of discord arose with the entry of the United Kingdom into the Community in 1973. In return for payment of disproportionately large contributions to Community funds, the UK, whose entry added to the number of Community 'peripheral regions', demanded redress in the form of regional aid. Subsequently, Britain has been accused of using Community aid both to reduce (rather than supplement) its own regional assistance budget and to underwrite its welfare system (Arbuthnott and Edwards 1979:117; Martins and Mawson 1982). At the time of its entry into the Community, however, Britain was a key promoter of the establishment of the European Regional Development Fund (ERDF).

The present work is not the place for a detailed assessment of the activities and effectiveness of the ERDF. The reader is directed to texts which are already in existence for this (e.g. Pinder 1983). Suffice to say that in its eleven years' existence, the ERDF has had to face the problems of a small budget (compared with that of the CAP whose effects it is largely redressing), competition amongst member states as to who gets what, problems over the definition of what is a 'less favoured area',* weaknesses in the coordination of national regional policies and, more recently, the enlargement of the Community to include nations with a large proportion of workers in low-productivity agriculture. Domestic political considerations have also prevailed. In order not to offend 'urban' voters by openly subsidising small farmers at a time of high food prices, the designation of 'less favoured areas' has been limited (Figure 4.9). A major reassessment of the ERDF and Community regional policy in general, which was undertaken in 1981, threw light on these and other operational problems. In the first five years of its existence, the ERDF concentrated overwhelmingly upon assisting infrastructural provision and the establishment of production units in Italy, the UK, and Ireland (and latterly Greece) where some 80 per cent of the 14,524 projects were located (Klein 1982). Yet all these projects were linked to the domestic regional policies of member states. This called into question the need for a supra-national development fund, other than as a 'topping up' mechanism.

---

* Kiljunen (1980:215), for example, points out that the assisted areas of West Germany or the Netherlands might have higher Gross Domestic Products per capita than the developed areas of Italy or the UK yet be in competition with these latter countries' assisted areas for ERDF monies.

Figure 4.9 'Less Favoured Areas'
under The Common Agricultural Policy

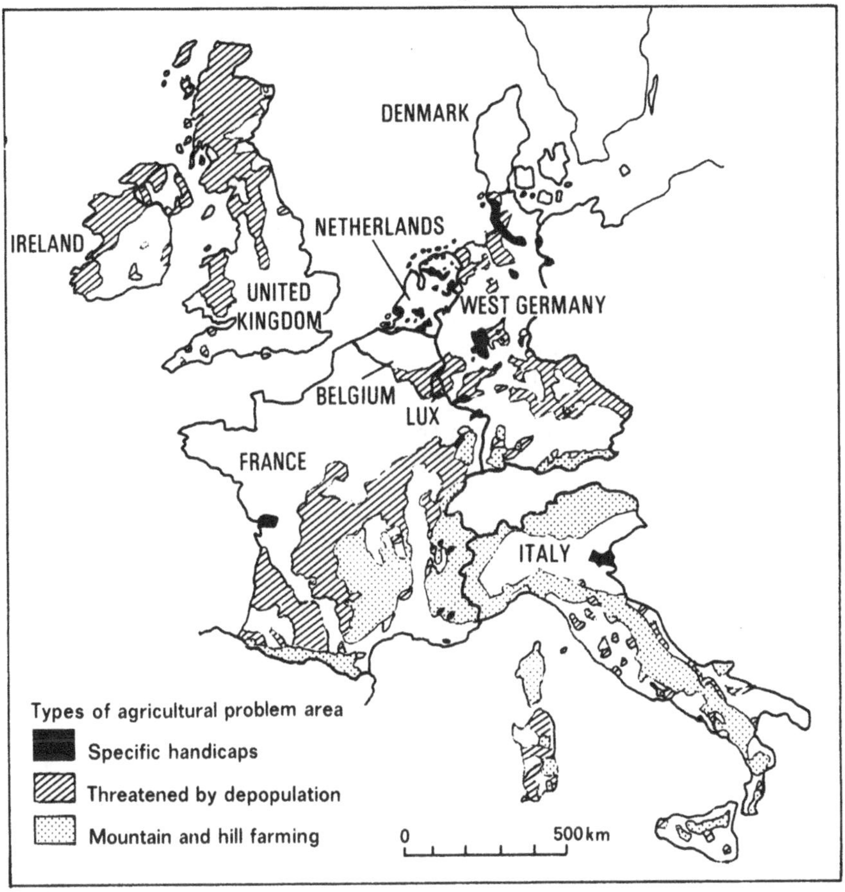

Source : Bowler (1985)

In the 1980s, the ERDF has moved towards a more autonomous role; providing funds for specific projects outside the quota allocations of aid to member states (though such non-quota funds remain a small proportion of the total budget), helping coordinate member state regional policies, and providing a more effective research base (particularly for the growing number of integrated rural development projects with which it is involved). The fundamental limitations of the Community's regional policy nonetheless persist. Ideologically, it is at odds with long-term Community goals, yet for member governments who need to address the problems of declining regions, it remains high on the political agenda.

This is not only so for each nation within the Community but also for the Community as a whole. There remains the need to redress the vast differences in incomes and living standards that exist between regions if the 'Community' is to have validity as a supra-national unit. There is, at the moment, little sense of 'one-ness' between the hill farmer of the Auvergne or the peasant of southern Greece and the grain barons of the Beauce or East Anglia. European regional policy currently has a limited ability to address the forces that exacerbate regional disparity. The dominant influence of the CAP, with funds vastly in excess of those available to the ERDF, effectively makes the latter a minor apologist for the former. Yet allowing the ERDF to grow in relation to the CAP will provoke significant conflict. Both politicians and Community bureaucrats alike are aware that market forces are moving towards larger economic operations. Electoral dictates determine that politicians cannot be openly seen to support this process nor acquiesce to the effects that process is having upon peripheral agricultural regions and small-scale producers. They must at least pay lip service to stopping this trend and ameliorating its past effects. However, the real costs of stopping this trend (given existing private sector market conditions) would be too high.* The resulting price increases (say for food) would undoubtedly have electoral repercussions. In this way, the structural framework laid down by the CAP is itself revealed to be subject to the broader structural limits of a capitalist world

---

* The same could be said of European environmental policy which, when held in comparison with agricultural or regional policy, has been developed with consummate ease. It is in the nature of concerns for the environment as a whole that people agree, for the broad goals of environmentalism command widespread support. It is, however, the detail of prohibitive, controlling and regulative policies that provokes disagreement. In its pursuit of a supra-national environmental policy, the Community has sought to avoid such disagreements, but as a consequence has been reluctant to pursue environmental goals which do not carry widespread public support. Underlying Community policy there has been concern that the spread of pollution controls will have detrimental economic ramifications, that growing popular concern will manifest itself in active protest, and that domestic member state environmental policies will hinder the smooth operations of free-trade. Thus the Community's environmental programme has addressed itself principally to the issue of pollution and has sought to promote the wide-scale introduction of environmental impact assessments. The impact of these assessments has yet to be fully judged (Lowe and Goyder 1983).

economy (Taylor 1980). That national governments can nonetheless adjust world-wide processes is clear from the dissimilar farm structures and national agricultural policies of, say, Britain and France. Structures are not rigidly determining. Behavioural variability is a significant element of causal processes.

THE BEHAVIOURAL CONTEXT

In reality the imposition of a division between structural and behavioural contexts is a somewhat false one. Structures do not exist in a vacuum. They have to be produced and reproduced over time. Hence, structural contexts are also behavioural ones. Where a dividing line can be drawn between the two is in time frame and generality of conditions. The settling of western lands in North America produced a land ownership pattern that has had consequences to this day and continues to condition any developments that occur in the region. Similarly, both the organisation of corporations on a transnational basis and the emergence of intergovernmental cooperation (like the CAP) impose restraints on governmental behaviour and helps direct broad trends in development. The determinants of these structural frameworks are in no sense separate from specific governmental acts. Such acts, as investigated here as behavioural manifestations, are nonetheless distinctive since they are oriented towards an immediate (or short-term) behavioural response. They do not seek to impose a structure on future actions in the long-term. As for the preceding section on structural conditions, our examination of behavioural manifestations is in no sense comprehensive. We are seeking to draw out the rural consequences of international relations, while also pinpointing some aspects of the causal mechanisms which underlie the behaviour described. In procedural terms, three aspects of international behaviour have been focused on. In turn, these are on international trade between East and West, international aid and immigrant labour policies. The first concentrates on the general imperatives which underlie international trade, while the second and third consider specific aspects of those imperatives; the former being linked to the altruistic intention of aiding less advanced nations, the latter intended to bring benefit to the economy of the advanced country.

Trade with the Soviet Bloc

Although advanced capitalist agriculture has long benefited from trade with communist countries, a number of researchers have concluded that processors and traders rather than farmers have reaped the most rewards (Frundt 1975; Biggs 1976). To appreciate the case made by these advocates, it is pertinent to understand how shifts in trade links between eastern and western blocs have occurred. In tracing these trade paths, we shall illustrate that while the subservience of farmers to agricultural industries exists, this is secondary to the lowly position of agriculture relative both to other economic sectors and political ideology.

Even before the 1917 Revolution, western agricultural industries provided a significant component of Russian imports (Biggs 1976). After the Revolution, western business leaders were not surprisingly opposed to the Soviet regime (which had nationalised many of their assets). Yet by the early 1920s, with unemployment increasing in capitalist nations, trade considerations had come to outweigh ideological disagreements. There was certainly much window-dressing to explain the rapid switch in many nations from active sponsorship of the overthrow of the Soviet regime to bickering over trade opportunities (White 1979). With a myopia that characterises self-interest, business leaders, the press and the governments of the West were quick to charge that the Revolution had failed; the Soviet Union, apparently, was on its way back to capitalism. Thus, Sir Robert Howe, the 1921 President of the British Board of Trade stated that: '... nothing will so upset the communist system there [the Soviet Union] as to resume trade'. Apparently, the '... only way in which you will succeed in killing Bolshevism will be to bring Russia and the Russian people under the civilising influence of the rest of the world, and you cannot do that in a better way than by beginning to enter trade and commerce with them' (quoted in White 1979:25-6). In this quest for ideological purification, agricultural industries were to play a key role. By 1927, for instance, 85 per cent of all tractors in the USSR carried the Ford emblem (this company also helping establish the Soviet car industry; Biggs 1976). Another measure of the importance of (agriculture-related) trade between East and West was the 1931 importation by the Soviets of two-thirds of all US exports in agricultural equipment and power-driven metal working equipment.

For US farmers the effects of Soviet trade relations at

this time were indirect and fluctuating. During the First World War, for instance, Turkey blocked Russian wheat shipments, inducing West European nations to turn to the Great Plains for their supplies (the result being an increase in wheat production of 250 per cent on the Plains between 1914 and 1918; Worster 1979). With the termination of hostilities, prices fell sharply for North American farm products (for wheat the fall was by 50 per cent by the early 1920s; see Wood 1924; Burbank 1971). With European nations responding to the war challenge with tariff walls to promote indigenous farm production, the post-war period saw a major decline in North American food exports (the percentage of US wheat production exported fell from 37 in 1920 to 17 in 1923 and remained near that figure until 1929; Shover 1965). Farm incomes plummeted, against a background of earlier rising prices which had encouraged operators to increase their farm size and invest in more machinery (both of which led to higher mortgage commitments). Here was the basis for increased bankruptcies, rural outmigration and, in some areas, marked social unrest (Wood 1924; Shover 1965; Burbank 1971; Hadwiger 1976).

Compared with its previous importance as an export earner, the farm sector now entered a trough. Only with the start of the Second World War and, more particularly, the signing of the 1942 US - USSR lend-lease agreement, did agricultural exports return to their original position as a contributor to the US balance of trade (Figure 4.10). Shortly following the second world conflict, a major change occurred in US - Soviet trade relations which influenced agricultural and other exports. Typified by George Kennan's 1947 'Mr X' article in the journal Foreign Affairs, and in Winston Churchill's 'Iron Curtain' speech, western politicians quickly came to question Soviet foreign policy aims. The formation of the North Atlantic Treaty Organisation in 1949 institutionalised lines of conflict that persist to this day. Of course, the existence of opposed ideological camps does not inevitably have trade consequences (as shown by the 1917 to 1947 era), yet the United States, unlike most advanced capitalist states, chose to extend its ideological dispute into trade. On the surface, the cause of US distinctiveness is summed up in one word - McCarthyism. Built on a shameful exploitation of a population already roused by anti-radical purges in the 1920s and 1930s, McCarthyism was '... notable for its crude, below the belt, eye-gouging, bare-knuckled partisan exploitation of anti-communism, usually on the basis of half-truths, warmed over revelations and plain lies' (Caute 1978:47). That this was built on the personal aggrandisement of a group of ambitious politicians is true, just as it drew its strength from the Republican Party's ability to weaken the support for the Democrats by claiming they were 'soft on communism' (Caute 1978). McCarthyism heralded some notable

Figure 4.10 US agricultural exports as a percentage of agricultural imports 1901 — 83

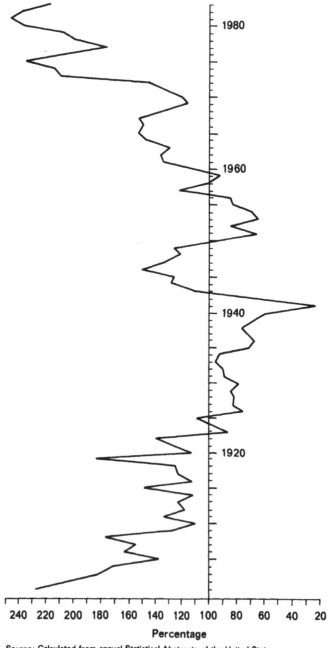

Source: Calculated from annual Statistical Abstracts of the United States

examples of anti-trade legislation (like the Battle Act of 1951 which threatened to terminate military and financial aid to countries that continued trading with the USSR). Although superficially suggesting the supremacy of ideology over the profit imperative, the real trade consequences of such legislation were less impressive. This was an era, after all, of rapid expansion for US corporations at home and abroad (Fry 1970; Vernon 1979). In toto, there were innumerable profit-making opportunities without eastern bloc trade (in addition to which the apparent ideological fervour of the Soviet leadership was an active disincentive for investment in, or trade with, Eastern Europe). It would still require severe contortions of reasoning to hold that capitalists inspired a loss of markets for themselves. When evaluating the comparative importance of governmental and business actors in the emergence of McCarthyism (and its consequent trade manifestations) the primary role clearly falls on the side of the public sector. This, however, cannot be taken to mean that capitalists were not important in the emergence of this movement. We must bear in mind that the strategic interests of capital were not challenged by McCarthyism (far from it, for it promoted anti-leftist sentiment within the United States). Consequently, there was little need for business pressure to redirect state policy.

When the 'need' for such redirection emerged, capitalists were not slow in providing it. This was seen in the emergence of a Soviet - United States détente in the latter years of the Vietnam War. As a number of commentators have pointed out, there were mutual benefits in a thawing of East - West relations (Lauter and Dickie 1975; Wilczynski 1976; Levinson 1978). For the Soviet Union, there was the promise of access to western technology. For the United States and other countries' capitalists, there was the appeal of expanding sales beyond increasingly product-saturated North American and European markets (and, for some transnational companies, by drawing on a disciplined, skilled and cheap work force, there was an added bonus in setting up production facilities behind the Iron Curtain; see Levinson 1978). In political terms, there were no particular reasons why détente should have been so strenuously advocated over this period. Neither US nor Soviet leaders were noted for an ideological commitment to improving East - West relations. The international arena, despite the Middle East and Vietnam, was not so unstable that painstaking efforts were needed to reduce tension in order to avert war (both the main conflict areas were 'shatterbelts' that had a tradition of unstable, but localised, political climates; see Cohen 1973). In addition, there were no significant military build-ups to strain national economies and the international mood (on the contrary, as Cusak and Ward 1980 have shown, the image of arms expansions to counter military increases by the 'other side' is more myth than reality; defence spending is more notably determined by internal conditions, like the supply of funds in eastern bloc nations or the desire to prime-pump the economy prior to an election in the United States, than it is to events abroad). Where pressures for change did arise was in trade. Over the 1960s and early 1970s East - West trade grew by 18 per cent a year. This outpaced both intra-COMECON and world trade levels. Transnational company pressure for improved market access led to the embargo list for US exports to the Soviet bloc falling from 1,700 to 70 items between 1970 and 1973 (Wilczynski 1976; Levinson 1978). Not only manufacturing product exports were involved here. Around 20 per cent of Poland's industrial production is now based on western licences (Wilczynski 1976). Banks were also quick to extend credit to eastern bloc nations due to their impeccable repayment records (by the time of the Polish Solidarity uprising in 1980, that country alone owed western banks $24 thousand million; Sampson 1981). Agriculture also played a key role.

The agricultural component of East - West trade growth was most spectacularly seen in the 1972 US - Soviet wheat deal. Through this, the Soviet Union imported almost 25 per cent of the total US wheat crop. The effect on income levels in nonmetropolitan areas was immediate (Table 4.4). This is not surprising, for: 'It is a well known characteristic of world agricultural markets to manifest considerable price swings in response to comparatively small percentage changes in total supplies' (Nagle 1976:52). With increased exports to China also occurring at this time, the resulting world price for wheat rose by 50 per cent between the first quarter of 1972 and the same period in 1973. For the next decade both grain prices and rural incomes were to fluctuate significantly as a consequence of governmental decisions (though these were obviously not the only factors involved). From the outset, the marked improvement in US farm incomes can be traced to a government decision. This was made in the Soviet Union. Its purpose was to maintain living standards and to prevent set-backs in meat production by importing grain rather than 'tightening belts'. The US commitment to the wheat deal was beneficial not only as an aid to detente but also in reducing federal payments to farmers (in the form of price guarantees). Continued high grain prices were, of course, contingent upon climatic factors as well. Improved Soviet yields in 1973, 1975 and 1976, and restricted North American output in 1974, all reduced US - Soviet wheat trade. Nevertheless, as Biggs (1976:544) reported, estimates have suggested that in both the 1960s and early 1970s, as much as 93 per cent of variation in US grain exports was accounted for by fluctuations in eastern bloc imports (with the USSR accounting for 82 per cent of that

Table 4.4 Farm Incomes ($000s) Per Farm By US Agricultural Region 1972–82

Agricultural Region	Year										
	1972	1973	1974	1975	1976	1977	1978	1979	1980	1981	1982
Livestock	4.96	8.77	5.94	4.61	3.92	3.86	7.19	9.45	6.50	7.01	5.71
Field Crops	5.91	14.78	9.77	8.22	9.31	8.15	11.01	13.60	1.74	13.24	9.53
Cash Grains	11.10	24.44	20.46	19.33	10.80	13.19	17.23	19.51	8.35	20.69	12.29
Livestock/Field Crops	5.92	12.32	10.36	5.30	6.22	3.44	8.25	9.58	5.99	12.18	5.49
Livestock/Cash Grains	9.27	17.77	10.46	11.39	7.13	8.14	13.18	15.87	8.90	14.40	11.08
Livestock/'Other'	6.55	10.48	6.52	9.80	8.33	8.97	13.13	13.47	10.26	14.60	12.48
Fruit or Dairy	8.09	11.52	11.16	9.15	11.57	9.52	15.70	19.34	17.60	16.71	14.08
Field Crop/'Other'	9.65	19.62	25.01	17.79	13.05	13.05	21.53	21.30	15.14	15.70	6.43
Mixed Farming	5.20	8.61	7.95	6.97	6.84	7.24	9.12	10.28	7.40	10.43	7.91

Source: Authors' research. The data base for this table comprises a one in five systematic sample of all US nonmetropolitan counties. The agricultural classification was based on the ideas of Mighels (1955) using 1978 production figures. Counties were allocated to categories based on their having at least half of all farms involved in a particular product (for single product counties) or at least a quarter of all farms (for multi-product counties). Income data was from the US Bureau of Economic Analysis's Local Area Personal Statistics (annual).

variation on its own). The connection between this export link and government policy signified not simply that wheat trade was heavily conditioned by Washington, but that farm incomes were similarly placed.

This point was to be re-emphasised in 1980. The run-up to this started on 27 December 1979 when Soviet troops entered Afghanistan to bolster a regime with pro-Soviet inclinations. This invasion (or 'invited entry' as the Soviets maintained) was answered by the United States with a 4 January 1980 suspension of all grain shipments to the USSR (beyond the eight million tons covered under a five year bilateral agreement; this blocked delivery of 17 million tons). The immediate effect on the USSR was muted. Soviet leaders responded by drawing on alternative grain sources (Table 4.5), slaughtering cattle and instituting special measures to increase animal fodder (Tarrant 1981). In the United States, there was a significant slump in rural incomes (Table 4.4). Estimates of the real cost of the embargo vary. Ayubi et al (1982:30) suggest figures of $3.0 thousand million for the lost trade revenues, with a further $2.5 thousand million in extra federal spending. The Congressional Quarterly Inc (1984) cited a figure of $11.4 thousand million. Whichever is closest, the embargo was extremely expensive for the US.

Fortunately for US farmers, the embargo had a short duration. Seeking electoral support in the Midwest, then presidential candidate Ronald Reagan promised to lift the boycott if elected. This he did on 24 April 1981. An immediate increase in US - Soviet wheat trade resulted (Table 4.5). In effect, the shortness of the embargo merely emphasised the futility of President Carter's gesture. That the agricultural sector took the brunt of the pain from this gesture is nonetheless significant. Agriculture apart, the other major exports to be restricted were in the high technology field (which was not so important as many were already banned for defence reasons). Mainstream manufacturing was largely left untouched. Banks continued to advance credit behind the Iron Curtain. Under the umbrella of anti-communist rhetoric, symbolic gestures were made (like the Olympic Games boycott), and only comparatively weak economic sectors were disrupted. Not that this was an unusual experience for agriculture. In 1973, for instance, President Nixon had suspended exports of soya beans, timber and scrap metal to Japan in order to ease inflationary pressures within the United States. Such steps have made the United States an unreliable trading partner and have encouraged nations to search for alternative supplies. As can be seen from the declining role of the US in world soya bean trade (e.g. Figure 4.11), these have been relatively easy to find. Farmers, in particular, and rural areas more generally, have lost out in

Table 4.5 Exports of Major Wheat Traders to Eastern Europe 1978–83.

| | Metric Tons Exported (000s) | | | | | |
	1978	1979	1980	1981	1982	1985
Exporter	Exports to USSR					
USA	2779	5096	1681	3878	4081	4595
Canada	2453	1376	4457	3876	6164	6390
Australia	324	937	3046	1530	2099	NA
Argentina	950	233	2292	2958	2732	NA
	Exports to Eastern Europe (non-USSR)					
USA	710	1347	1093	284	86	36
Canada	926	623	1272	1282	1607	432
Australia	0	102	0	0	0	NA
Argentina	0	0	0	0	0	NA

Source: United Nations Statistical Papers Series D
Commodity Trade Statistics (annual) New York.

the process. Agriculture has insufficient clout to avoid being used as a sacrificial pawn in the quest for other foreign policy aims. An activity in which this is clearly revealed is the granting of aid to foreign countries.

Foreign Aid

For agriculture, foreign aid became an important component of US foreign (and agricultural) policy with the passing of Public Law 480 (PL480) in 1954. This law separates eras of US aid giving. Earlier incursions into the foreign aid field had largely been oriented toward helping re-establish European economies after the Second World War. Under the Marshall Plan, the United States sought to reinvigorate advanced capitalist markets both to increase US exports and to enhance living standards so as to ward off communist insurgency

Figure 4.11 World soya bean cake exports 1962—83

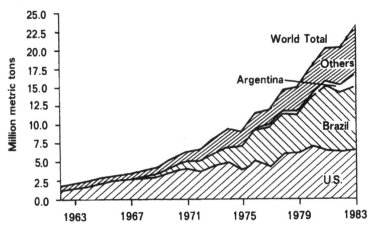

Source : World Food Institute (1985)

(Preston 1982). PL480 deviated from this pattern in two ways. First, it principally sought to aid poorer nations rather than advanced ones. Second, it was grounded in food exports rather than in promoting manufacturing and infrastructural improvements. In its origins, PL480 was not based primarily on altruism. It was not the plight of Third World nations that provided the stimulus, but the condition of US agriculture. Following a bumper wheat yield in 1953, US surpluses rose to 41 per cent of the total crop (Congressional Quarterly Inc 1984). While relative to industrial and financial interests farmers stood in a relatively weak position, their voting power, their propensity to adopt radical political stances in times of economic hardship (e.g. Shover 1965; Burbank 1971) and growing unease over their plight during the Depression (Worster 1979), had led to federal purchases of surplus farm commodities in the 1930s. In a sense, then, the commodities were already owned by the federal government. Public Law 480 was a surplus disposal policy (with a by-product of improving the international image of the United States). This constituted the basic aim of the programme until the mid-1960s (Cathie 1982).

At this point in time, food aid took on a more integral role in promoting US interests in the international arena. Renaming the programme 'Food for Peace', PL480 donations now came under the remit of the national security lobby (as evinced in its controlling agency changing from the Department of Agriculture to the State Department). As Secretary of Defence Robert McNamara declared in 1964: 'In my considered judgement ... the foreign aid programme has now become the most critical element of our overall national security effort' (quoted in Mann and Dickinson 1980). Most researchers do not see food aid occupying a key role in US foreign policy-making beyond the early 1970s (largely because sales to China and the Soviet Union reduced food surpluses), but the underlying rationale of utilising aid to promote US policy has become entrenched (in 1976, for example, Secretary of Agriculture Earl Butz was still espousing the view that food aid was a principal negotiating tool for the United States; Maddock 1978). In effect, food aid objectives became unashamedly political (Griffith-Jones 1984). Aid fostered policies sympathetic to 'the American way' (this did not mean that only 'friends' of the US received aid, for aid might help other nations 'turn around'; Wittkopf 1974; Abate 1976). Thus, aid was often conditional upon the adoption of population control policies (Frundt 1975) or on signing International Monetary Fund agreements (which, as in Jamaica, usually reduced a nation's control over its own economy and promoted US businesses; Griffith-Jones 1984). Some nations were forced to accept PL480 products they already had in surplus (like cotton to Pakistan) in order to obtain other goods they desperately needed (Mann and Dickinson 1980). In many instances, the coded message behind aid giving was muted. In others, it was stark. In the declining years of US involvement in Vietnam, for example, the Nixon Administration circumvented a Congressional veto on extra defence aid by increasing food aid to the South Vietnamese. This was then sold on the open market, with the funds obtained going into the defence budget (Cathie 1982). More 'negatively', US aid to Chile was cut when its people offended Washington by electing a socialist leader, but was restarted as soon as a right-wing coup established the present military dictatorship (Tarrant 1981). That the 'bullying' of Third World nations is still an integral part of US foreign aid is clear (The Guardian 21 December 1983, for instance, reported that US aid to Zimbabwe had been halved in retaliation for that country's criticism of the US invasion of Grenada). From our perspective, however, neither the foreign policy nor the ethical implications of this are important. What is is that agricultural incomes in the United States are substantially enhanced by such aid giving (Congressional Quarterly Inc 1984 reported that by 1980 the US had donated $11.4 thousand million in food aid and made loans of a further $19 thousand million under PL480). That this programme has been less important as a source of aid either to Third World nations or to US farmers and more of an aid to US diplomatic policy is nonetheless instructive (see Rossiter 1985). It helps emphasise that rural fortunes are extensively affected by decisions directed at non-rural outcomes.

This can be seen not simply in the way rural resources help promote national foreign policy, but also in the utilisation of food aid to provide more profitable openings for US companies. A specific case which boldly illustrates this has been described by Frundt (1975:228). This concerns India's 1964 attempt to build a fertilizer plant to service its 'green revolution' requirements. At the time, India had offers from five US corporations to build such a plant. As these carried the unacceptable condition of a guaranteed annual 20 per cent return on capital (and that laws controlling fertilizer prices and distribution be repealed), they were not accepted. The World Bank, however, refused India a loan to build the plant because it had these five offers. In 1966 PL480 entered the plot when India's harvest failed, food riots broke out and Mrs Gandhi appealed to President Johnson for food aid. Not only was this refused, but other US aid to India was suspended. Only when Mrs Gandhi acceded to one US company's offer to build the fertilizer plant was aid reinstated (and a World Bank loan granted). With events of this 'calibre', it is perhaps not surprising that researchers have often seen greater benefits in food aid for government and corporate interests than for farmers themselves. Thus, Mann and Dickinson (1980) have noted that food aid has often been given for just long enough to establish western products in the recipient nation's dietary habits, then it is withdrawn; leaving a vacuum that US corporations have filled with profitable sales (cf. the sale of high protein milk to Guatemala).

For the farming sector, the aid programme now offers less scope for income enhancement than was once thought likely. As the programme increasingly came under national security and corporate control, its use for non-humanitarian causes brought it into disrepute. When a so-called charity has a major nation requesting that it no longer be a recipient, because the costs of receiving aid were too harsh (as India did in 1974), the end of a programme - as then constituted - would seem to be at hand. In fact, by the time of this Indian request, the character of PL480 aid had already begun to change. Spurred by intense criticism of US foreign policy (most especially emanating from the Vietnam War) and by the doubts of liberals and radicals over the true aims of foreign aid, United States policy-makers sought a lower profile abroad to quell anti-US and anti-imperialist sentiments. Pressures were brought to bear on other advanced nations to increase their aid commitments. The result was that the US share of world food aid fell from 96 per cent in 1963 to 55 per cent in 1973 (Cathie 1982). Programmes were also directed away from bilateral agreements toward multilateral schemes which emphasised 'need' in aid allocation processes (Mosley 1981; Tarrant 1982). With the option of large-scale food sales to Soviet bloc countries emerging in the early 1970s, the aim of PL480 donations changed once again; this time it occupied the position of a surplus disposal mechanism (a means of disposing what could not be sold). In some respects this brought the programme back to its original purpose, except that federal priorities now emphasised trying to find foreign buyers first (and 'giving' away what was then left). For farmers, of course, no matter what the precise objective of the programme, the fact that the federal government was able to reduce production surpluses eased downward pressures on food prices, helped maintain rural incomes and subsidised the continued existence of many farm operations.

Immigrant Labour Policies

The use of imported labour to keep farm worker wages low is another aspect of the government's subsidisation of agriculture. Of course, it is not simply farmers who benefit from this policy. Indirectly all consumers are favoured by cheaper food prices and, with these easing pressure for higher wages in other economic sectors, so are industrial producers. However, although cheaper agricultural labour costs do benefit those in other economic sectors, they have negative consequences for many workers. In local labour markets where agricultural employment is significant, low farm worker wages provide a downward pull on incomes in other sectors. This is especially seen in areas which use much (temporary) immigrant labour. For one thing, downward pressures can be intensified here by the threat of replacing 'troublesome' workers, who ask for higher wages, with 'foreign' labourers (who, because they are 'foreign' and are not covered by as wide a range of safeguards as home nationals, can be more easily 'disposed' of if they prove 'troublesome' or are no longer required). For another, unless labour shortages exist (which has not been common in rural areas employing high levels of migrant labour), the introduction of foreign workers increases labour supply relative to demand, thereby enhancing competition for jobs and weakening the bargaining power of labour. The persistence of immigrant workers in labour markets thereby tells us much about the relative standing of labour in class conflict. This especially applies for short-stay immigrant workers. These are certainly important in a number of industrial sectors (see, for example, Castles and Kosack's 1974 report on their key role in West German industries), but these tend to be sectors experiencing labour shortages. This has rarely` been the case in agriculture. Immigrant farm labour is more reflective of the weak bargaining power of agricultural workers.

It should perhaps be noted that the utilisation of immigrant labour represents just one strategy which farmers employ to keep their employees' wages low. We do not intend to investigate other dimensions of this process here, since this is a task for later chapters, but it is pertinent to point out that the 'immigrant labourer effect' can be achieved without immigrants themselves. In essence, the basis of this effect is the ability to treat a specific worker group (usually immigrants) as 'inferior beings'. Inferior, that is, in that they are not covered by the 'normal' rules that restrict employer abuses of labourers. For this not to incur the wrath of labour as a movement usually requires that the targeted group is somehow 'isolated'. This clearly applies to blacks in the US South. Institutionalised racism, which creates divisions amongst the working class and maintains a cheap, subservient and easily discarded black workforce, has long brought significant benefits for southern employers (e.g. Greenberg 1981). John Steinbeck's Grapes of Wrath description of the exploitation of 'Dust Bowl' migrants in California captures the intensity of the process involved (Friedland and Nelkin 1971 provide a more analytical evaluation of migrant worker conditions). It also signifies that racial or national distinctiveness is not an inevitable component of immigrant labour processes. It is nonetheless significant that these are commonly intrinsic features of these processes.

In part, this is because the employment of immigrant agricultural labour is a particular feature of the North American farm economy. Most other advanced capitalist nations changed their employment bases from agricultural to non-agricultural sectors in a process that was marked by labour shedded from farms. In North America, by contrast, the nineteenth century (in particular) saw the contradictory pressures of industrialisation and agricultural expansion (the settling of the West) at the same time. This restricted the ability of major landowners to find workers for their farms. In most newly settled areas this helped induce a farming system dominated by family farms. However, on account of its proximity to West Coast sea routes and the coincidence of railway construction requiring large supplies of cheap labour, the landowners of California were well placed to develop a 'foreign' hired worker scheme. Fluctuating in form and intensity, this scheme persists to this day. It sets California somewhat apart in that the character of its capital-labour (agricultural) relations is somewhat different. Yet this is a difference of degree rather than kind. California growers have been able to push their interests further because of structural weaknesses in the bargaining power of their employees. These weaknesses arise not simply with reference to owner-worker links but also from the weak links agricultural workers have with other segments of the working class.

The experiences of Australia, Canada and the United States are similar in the manner in which working class people responded to the first major wave of immigrant labourers - the Chinese. In North America, Chinese labourers were initially introduced largely for railway construction. Opposition to this immigrant flow stemmed largely from small businesses, who were not benefiting from this labour pool, and white trade union labour leaders, who recognised the damaging effect such 'cheap' workers would have on wage rates. These combined forces proved sufficient to promote legislation to exclude or restricted Chinese employment (like California's 1882 Chinese Exclusion Act or Australia's 'prescribed' language test). Prior to this, violence was often resorted to, as white workers sought to exclude Chinese labourers. The behaviour patterns of white labourers, employers and governments in this regard informs us about the exercise and locus of power. In terms of national standings, for instance, the response of some 'respectable' Chinese citizens to Australia's 1878 anti-Chinese goldfield riots helps pinpoint why particular nationals have been utilised as cheap, immigrant labour:

> If such a thing had happened in China - if a
> number of English miners had been subjected to
> such a cruel and wanton outrage - every newspaper
> in Great Britain would have been aflame with
> indignation; your envoy in Pekin would have
> demanded prompt reparation and adequate
> compensation and if this had not been acceded to,
> some men-of-war would have been ordered to the
> mouth of the Pei-Ho. Our Emperor and mandarins
> would have been reminded of the solemn
> obligations they were under to be faithful to the
> treaty engagements, and they would probably have
> been lectured on the barbarous and scandalous
> conduct of those who had insulted and despoiled
> and maltreated peaceful and industrious
> foreigners. (Willard 1923:36)

One rule for the powerful and another for the weak? Certainly
this summarises the history of immigrant labour politics.

This does not mean that power is static (or that the
powerful are never-changing). White labour 'victories' were
seen in the US 1882 Chinese Exclusion Act. Similarly, the
continuous lobbying of the American Federation of Labourers
was undoubtedly a key factor in the passing of the 1924
Johnson-Reed Act, which placed limits (and then national
quotas) on immigration into the United States. Governments
are undoubtedly influenced by the threat of civil disturbance.
The fear of voting blocs like farmers and trade unionists
being mobilised can also influence government policy. Yet the
'victories' which arise from such threats are restricted in
scope by a predisposition in capitalist economies for
governments to cultivate private sector profit-making in order
to maintain and enhance national economic conditions (Offe and
Ronge 1982). This commonly results in capitalists attaining
the same end, merely by other means. Thus, when their supply
of Chinese workers was cut off, California land holders turned
to the Japanese (whose California numbers increased from 86 in
1882 to 72,000 by 1910; Baird and McCaughan 1979).
Restrictions on their importation was forthcoming but hardly
proved problematical, since the colonial status of Filipinos
meant that they provided a fluid labour force, while the
proximity and poverty of Mexicans lent a further easily
obtainable labour supply. The severe economic depression of
the 1930s certainly reduced the political attractiveness of
these labour sources, but it also provided a massive inflow of
hungry ex-farmers from the southern Great Plains (Worster
1979). These came to fill whatever low-paid, agricultural
labourer positions were available (McWilliams 1939). The
utility of immigrant labourers was starkly revealed at this
time. No longer needed by employers, Mexican workers were

deported on a massive scale (75,000 going from Los Angeles County alone in 1931; Baird and McCaughan 1979). The usual grab-bag of excuses was used (like claiming that these people were 'communists') and their lameness was readily seen when the economy picked up again. Now, with white workers finding non-agricultural jobs, agrarians again needed the help of Mexican labourers. Cries of communism were forgotten as the Second World War placed pressures on labour supplies. With the Korean War suggesting that migrant workers were to be an important component of Californian agriculture for the forseeable future, Congress passed Public Law 78 (PL78) in 1951, thereby establishing a formal programme of post-war Mexican labour imports (Baird and McCaughan 1979; Price 1983). As the programme expanded, however, so did illegal immigration. At its peak PL78 provided around 450,000 annual jobs, whereas estimates of the number of illegal immigrants sit around the one million mark (over 60 per cent in the southwestern USA are in agricultural labour; Baird and McCaughan 1979; Congressional Quarterly Inc 1984). In addition, labour was becoming more organised.

The formation of Cesar Chavez's National Farm Workers Association in 1963 was met with a negative reaction from growers. However, the mood of the nation was not in the growers' favour. Church leaders were offended by the terrible work conditions of agricultural workers. Their support made the growers' cries of 'communists!' less able to stick (Baker 1976). General public support was also aroused. The New Frontier administration of John Kennedy, followed by Lyndon Johnson's Great Society programme made the importation and exploitation of cheap agricultural workers appear out of place (Craig 1971). White trade union support was forthcoming and ultimately led to the termination of PL78 in 1964.* In addition, the public showed great hostility (including product boycotts) towards the exploitative practices of growers (as early as 1947 Mexico allowed no labourers to go to Texas on account of the racial discrimination practices prevalent there; House 1982), and favoured César Chávez with their support. The political flavour of the times boded well for agricultural labourers. Twenty years later, however, their lot has not advanced substantially (by 1975, for example, only three states offered unemployment insurance for agricultural labourers; Price 1983). Farmers were quick to respond to the gains in bargaining power of farm workers by mechanising production practices. In doing so, they signified that

---

* In 1960 the US admitted 447,207 foreign temporary workers; by 1970 the number had fallen to 47,483, continuing to decline to 16,548 in 1980 (Congressional Quarterly Inc 1984).

mechanisation was being introduced more as a mechanism for social control than as an economic or production advance (Baker 1976; Price 1983). Whether or not immigrant labour was used, farmers were able to circumvent pressures from farm workers. For present purposes, however, the point to be conscious of is the manner in which landowners were able to direct international labour in order to enhance their economic standing and weaken the bargaining power of agricultural workers.

PERSPECTIVE

In this chapter we have not sought to develop or support any general theory that ties events in rural parts of advanced economies to international structures and processes. It would certainly have been feasible to have examined the propositions and weaknesses of theories like the world-system model, pointing out what this means for advanced economies' rural areas compared with their counterparts in peripheral nations. This however would have directed our attention too far away from the specific focus of our concern; which is rural areas in advanced nations alone. As the next chapter will indicate, the socio-economic processes which bind peripheral rural areas and national cores together (in advanced societies) are similar to the ties of advanced and peripheral nations. While these are points that need developing and clarifying theoretically, they have only been in the background in this chapter. The particular messages we have sought to project in this chapter might appear to be specific to the particular contexts examined, but we do not believe this is the case. In the drive to expand the boundaries of the United States, we find the same quest for capital accumulation and the preservation (or enhancement) of national (capitalist) interests as was evident in the formation of the European Community (to maintain European manufacturers' economic position) and in the ebb and flow of East-West trade relations. As the formation of the European Community confirmed, capital accumulation is not the only determinant of national foreign policies. In this case strategic considerations were also important (as they were in aid to Third World countries). The rise of transnational companies clearly threatens the ability of national governments to adopt policies in 'the national interest' (as does the European Community); most especially when this threatens capital accumulation. Of course, it is essential to place the phrase 'the national interest' in inverted commas in the last sentence, for both the foreign aid and immigration policy sections clearly showed that policies which are justified 'in the national interest' inevitably benefit some social groups more than others.

What has been identified in this chapter by dint of specific example is: first, the strength of the capital accumulation impulse in determining foreign policy; second, the likelihood that capital accumulation per se will clash with other valued goals which national political leaders are promoting; and third, that nations cannot be regarded as unitary bodies – there are conflicts of interest within nations as well as between them. As for rural areas per se, each of our six sections indicates that there is no rural lobby which is consistently able to direct foreign affairs to its own benefit. Farmers have recorded some periods of gain (as in high European Community agricultural commodity prices and the wage cost reductions resulting from US–Mexico labour agreements), but these are temporary in duration and are set against the backcloth of industrial capital's indifference (or support). When the interests of industrial and agrarian capitalists have clashed (e.g. over East-West trade), agriculturalists have lost out. As we shall see in the next chapter, this outcome emanates directly from the structure of power within capitalist economies. Rural areas lack a powerful vested interest whose production and sales needs have spin-off benefits; the power house of capitalist production is entrenched in the city environment. Consequently, rural locales are predominantly recipients of international development impulses; they are characterised by responding to change pressures emanating beyond their reach and, while they adapt to these pressures in non-uniform ways, these pressures are essentially out of their control. Major upswings and downturns in rural fortunes occur in short time periods. The international scene imposes substantial uncertainty over the direction and duration of rural development urges.

Chapter Five

NATIONS, REGIONS AND RURAL DEVELOPMENT

In the previous chapter, specific examples were used to show how the developmental experiences of rural residents are constrained, channelled and/or promoted by international structures and processes. Below these international impulses, and more apparent to rural residents themselves, are the impositions and opportunities afforded by actions and structures at a national level. In reality, such national inter-relationships are intricately interwoven with, and cannot be disentangled from, their international counterparts. In exactly the same way, no sharp divide can be drawn between local and national spheres. Nevertheless, for analytical clarity, a distinction needs to be drawn between these levels in order to acknowledge both their partial autonomy from one another and their association with 'unique' social forms and processes. However, insofar as our ecological definition of 'rural' is locality-specific, it behoves us to explain why national-level forces should be examined as a separate section and not as a component of a section on rural localities. As for international processes, this largely arises because the national sphere provides structure and behavioural input into development processes that transcend individual localities. Spatial regularities exist in developmental imperatives which require understanding of the national context of their generating processes.

There is nothing startling in pointing out that broad similarity in development experiences provides a 'regional' character to rural development. The problems and potentials of the Highlands of Scotland (or Appalachia) are clearly not of the same kind as those of South East England (or New England), yet there is similarity across localities within these 'regions'. However, as a concept, 'region' has very uncertain theoretical properties (Urry 1981). Geographers have frequently been guilty of endowing regions (or some of them at least) with almost mystical attributes and at times seem too willing to accept the explanatory significance of 'region' per se. To clarify our stance, the concept region is herein used as a descriptive device to pinpoint the existence of similar development pictures across spatially contiguous localities. Used in this way, no starting assumption of causal significance is implied. Nevertheless, the underlying belief that a regional patterning exists in rural development characteristics is in keeping with the general tone of literature on rural areas (as evinced in the broad similarity of alternative classifications of rural locales - Figure 2.1). Most striking in these categorisations is their division between 'pressured' areas (in our terminology, metropolitan regions) and 'non-pressured' or 'peripheral' areas (i.e. non-metropolitan regions). As a starting point in accounting for dissimilar rural development experiences, we do not find this division very persuasive. It provides a description of actuality, but that is all. However, this descriptive framework does highlight divisions in region-based theoretical models of rural development. Where these models differ is in their explanation · of how regional dissimilarities in rural development emerge. To open this chapter, the explanatory accounts of these models will be evaluated.

REGIONAL MODELS OF RURAL DEVELOPMENT

In reviewing alternative perspectives on rural-urban (developmental) linkages, Buttel and Flinn (1977:256) made the obvious but necesary point that '... a theory depends as much on normative evaluations of underlying assumptions as on detached analysis of explanatory efficacy'. This merits emphasis because the widely differing characters and assumptions of region-based development models coexist alongside evidence which partially supports all perspectives. Judicious selection of evidence is not required to orient researchers towards a single theory. Disagreements abound over 'acceptable' mechanisms of evidence gathering and large gaps in our knowledge inevitably induce heroic (or jaundiced?) jumps in evaluative logic. Perhaps it is on account of the intensity of these disagreements that there are so many alternative region-based development models (e.g. Cooke 1983). Within each of these there are sub-divisions which expose nuances in rendition. For our purposes, however, three major types of model exist. These, we will argue, can really be reduced to two fundamentally different categories. Each rests on a different set of evaluative and normative foundations, and each is flawed in its own particular ways. Yet, even when viewed in the broadest outline, these models provide a foundation for appreciating the very dissimilar (and changing) fortunes of the rural areas in different regions.

The Diffusion Model

This model rests on an assumption of 'cultural' dissimilarity. Essentially derived from a functionalist, integration theory perspective on society, the diffusion model suggests that development occurs where people are most willing to grasp opportunities and take risks. With regard to metropolitan — nonmetropolitan differences, it is posited that cities provide a social environment conducive to innovation, whereas remoter rural areas are dominated by traditional, slow-to-change beliefs and social practices.* This model is not only seen to be applicable to socio-economic development (Berry 1973) but also to the emergence of new patterns of political behaviour (Rokkan 1970). Within the economic sphere, size also brings advantages of stability. For one thing, size is usually believed to be tied to employment diversity, which means that economic fortunes are not dependent upon high prices being sustained in a single or few economic sectors. This feature clearly distinguishes metropolitan economies from remoter nonmetropolitan ones (Simmons 1976). Further, the size of cities makes it more difficult for one firm to dominate employment or sales markets, thereby inducing more competitive environments and increased efficiency (so in any general downturn in the economy city outlets are most likely to survive). With enhanced economic security and innovatory potential, cities are the power-houses of modern economies. Assuming innovations find acceptance there, rural areas benefit as these new practices trickle-down the urban hierarchy (Berry 1970). Through this urban-centred diffusion process, rural areas in close proximity to metropolitan centres receive their positive inputs more rapidly than places

---

* As Fischer (1975) explained the situation, social change (including entrepreneurial innovations) emerges from the unconventional behaviour of a few and then spreads. Independent causal effects for settlement size and population density arise because a 'critical mass' is required to nurture unconventional subcultures (put another way, a certain size ensures that there are sufficient people to sustain new or 'unusual' behaviour patterns, yet these people are a small enough sub-group - at least initially - not to threaten prevailing local social norms). On this account, cities should always be in an advantageous position. Even as remoter rural areas make general a normative order or innovation, new value changes have already emerged in cities which again place remoter areas in a 'retarded' position.

in peripheral nonmetropolitan zones. Indeed, through regular commuting, other visitations and the immediate impact of metropolitan newspaper, radio and television outputs, such proximate rural areas are effectively integral parts of metropolitan-dominated regions (Green 1971). The most important geographical division is therefore not between rural and urban, but between nonmetropolitan and metropolitan.

From the outset, it should be noted that there are many studies which have reported weaknesses of innovative, competitive behaviour in nonmetropolitan regions. In the outer Hebrides, for example, Caird and Moisley (1961) recorded that changes in the crofting system, which were supported by government and favoured by the majority of crofters, were not implemented because social norms stressed following local elders' views and they were opposed. For many non-metropolitan entrepreneurs, the goals of commercial existence are as much social as economic. Thus, Dunkle et al (1983) found small-town Indiana merchants placed residence within their existing locality as a more important goal than business expansion or continuing in one line of operations. A heavy regard exists for personal reputation, so entrepreneurs are reluctant to take business risks in case this leads to loss of face, if unsuccessful, or jealousy, if successful (e.g. Chadwick et al 1972). At times open resistance to 'outside' interventions bears witness to an adherence to 'traditional' values. In the Central Basin hollows of Tennessee, for example, 'strangers' who entered local land auctions were 'talked to' - informed of a secret bidder, who would outbid them, or that the land might be withdrawn - or might even find themselves embroiled in a brawl (Matthews 1965). Frequent conflict between 'indigenous' inhabitants of villages and recent urban in-migrants - over 'inadequate' local facilities (Hennigh 1978) or leadership positions (e.g. Forsythe 1980) - further illustrates the cultural divide which can exist.

Emphasis in this last sentence must be given to the words can exist. The social baggage urban in-migrants bring into nonmetropolitan settlements are not uniform. This has been ably portrayed in Forsythe's (1983) six categories of urban-rural migrant. The message of Forsythe's work is that the ease with which in-migrants are integrated into existing rural social structures depends upon their existing social, economic and ideological ties with their receiving localities. The further one progresses down the hierarchy of questions in Figure 5.1 without a positive response, the more likely the migrant will destabilise existing social patterns. This picture, as Forsythe (1983) made clear, is not a complete one. Although implicit throughout, no explicit consideration of social class is found in this schema. Yet class represents a significant element in conflict between in-migrants and

Figure 5.1 Schema for urban — rural migration

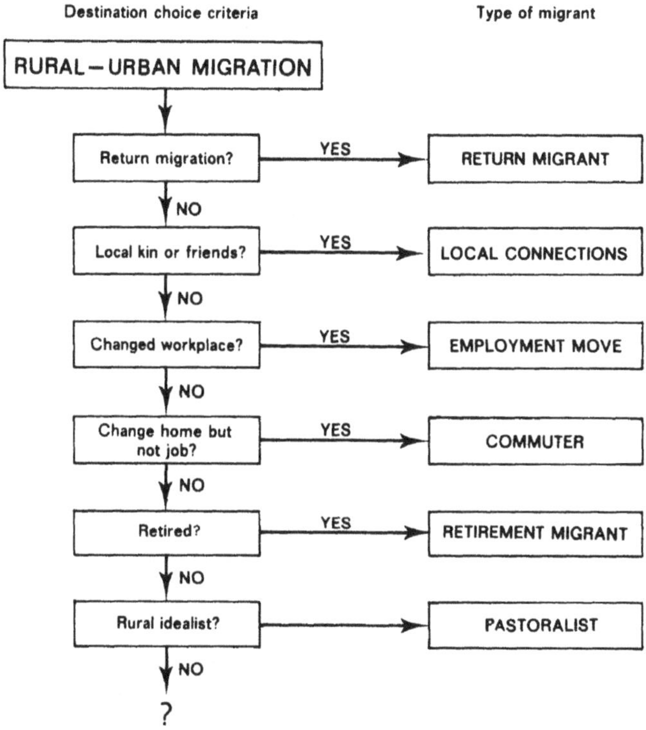

Destination choice criteria   Type of migrant

```
┌─────────────────────────────────┐
│   RURAL — URBAN MIGRATION       │
└─────────────────────────────────┘
              │
              ▼
   ┌────────────────────┐   YES    ┌──────────────────────┐
   │  Return migration? │ ───────▶ │   RETURN MIGRANT     │
   └────────────────────┘          └──────────────────────┘
              │ NO
              ▼
   ┌────────────────────┐   YES    ┌──────────────────────┐
   │ Local kin or friends? │ ─────▶ │  LOCAL CONNECTIONS   │
   └────────────────────┘          └──────────────────────┘
              │ NO
              ▼
   ┌────────────────────┐   YES    ┌──────────────────────┐
   │ Changed workplace? │ ───────▶ │   EMPLOYMENT MOVE    │
   └────────────────────┘          └──────────────────────┘
              │ NO
              ▼
   ┌────────────────────┐   YES    ┌──────────────────────┐
   │ Change home but    │ ───────▶ │      COMMUTER        │
   │ not job?           │          └──────────────────────┘
   └────────────────────┘
              │ NO
              ▼
   ┌────────────────────┐   YES    ┌──────────────────────┐
   │     Retired?       │ ───────▶ │  RETIREMENT MIGRANT  │
   └────────────────────┘          └──────────────────────┘
              │ NO
              ▼
   ┌────────────────────┐          ┌──────────────────────┐
   │  Rural idealist?   │ ───────▶ │     PASTORALIST      │
   └────────────────────┘          └──────────────────────┘
              │ NO
              ▼
              ?
```

Source: Adapted from Forsythe (1983)

'indigenous' nonmetropolitan residents. What is more, these conflicts are not endemic to remoter areas but are equally, if not more, characteristic of rural areas in metropolitan regions (e.g. Dobriner 1963; Sinclair and Westhues 1974; Green 1982) and indeed of cities themselves (e.g. Collinson 1963; Robson 1982). Such conflicts are then in no sense peculiar to remoter rural habitats. This is to be expected, for research confirms that, once age and socio-economic factors are taken into account, value differences between rural and urban residents tend to disappear (Glenn and Hill 1977). Of course, the fact that nonmetropolitan regions have a disproportionate share of lower-standing social groups is understandable from a diffusionist perspective. However, the assumption that this arises from inadequately 'advanced' values or an unwillingness to see social change, is suspect. After all, prior to the civil rights campaigns of the 1950s, a notable feature of North American radical social movements was their rural orientation (Lipset 1950; Burbank 1971). It was town dwellers and industrial workers who were more likely to be stalwarts behind preserving the status quo. Although the situation was more complex than this last sentence implies, the broad implication, that nonmetropolitan residents per se are not opposed to change, does not alter as fuller complexity is introduced. This is indicated by rural areas having high rates of new firm formation (Fothergill and Gudgin 1982; Gould and Keeble 1984) and rural-based enterprises proving to be as innovative as their urban counterparts (e.g. Markusen 1985; Pellenbarg and Kok 1985).

The diffusionist model is seriously flawed in its assumption of inevitable nonmetropolitan backwardness (as suggested by nonmetropolitan areas commonly having the strongest growth rates in population and manufacturing employment over the last decade - e.g. Johansen and Fuguitt 1984; Keeble 1984). This flaw has serious implications, for it carries a damaging message for policy-makers; one that is based on fallacious reasoning. In essence, low levels of development in nonmetropolitan areas are projected as area residents' 'own fault'. It takes no giant leap for this to be translated into policy recommendations which intensify nonmetropolitan problems. Consider, for example, the following value-laden suggestion from Brian Berry (1973:158):

> Efforts to develop economically or culturally poor areas may not produce long-term advantages for either the nation or for more than a small proportion of the inhabitants of the region. On the other hand, regions in decline but possessing useful social overhead facilities and good location may be candidates for significant public action.

Such statements are certainly brought into question by the so-called rural turnaround which has characterised most advanced capitalist societies since the 1970s. Quite simply, diffusionism lacks a dynamic component which allows for variability in the locus of primary economic growth. This is principally because this model neglects the power relationships which underlie all social structures and processes. It does so because its explanatory framework is ahistorical. One example is sufficient to illustrate this.

Following the US Civil War, the black population was technically free and (under law) equal with contemporary whites. Of course, with prior restrictions on formal education and with work experience largely restricted to the lowest paid jobs (as agricultural labourers and domestics), the black population was at a disadvantage in the labour market. Given equality of opportunity, however, these imbalances should have been overcome in a generation or two. In actuality, the Civil War was followed by exclusionary work laws (the Black Codes), which placed restrictions on black labourers' geographical and occupational mobility (Silk and Silk 1985). Although declared unconstitutional in their most overt forms, even struck-down passages emerged in other guises within new legislation. With an active policy of disenfranchisement for blacks (96 per cent of Louisiana's blacks lost the right to vote within two years of the state's new 1898 constitution – Nelson 1979), the black population could not rely on political pressure to help mitigate economic woes. Even without such legislated restrictions, blacks would have found it difficult to advance without elite support. As Bruce Catton (1963:351) has explained, blacks entered the labour market at the lowest level, competing with whites who were (economically) uncomfortable already and did not want a new worker influx. When dock workers in Cincinnati rioted in 1862, they drove out blacks who worked for lower wages, with the police merely standing by. In effect, official sanction was given to white working class violence. Black workers were driven into a corner. To survive, they were forced into a compliant, non-changing, low-wage and predominantly agricultural lifestyle. White pressure of this ilk was later honed and institutionalised in the Ku Klux Klan. The reign of terror this organisation perpetuated against blacks was, on the surface, a response to lower class white fears. However, the periodic intensification in Klan activities (e.g. in the 1920s and 1950s), and the broadening of its bigotry to attacks on Jews and 'communists' (Trelease 1981), revealed the hand of the southern land owning elite (Greenberg 1981). Not simply could the Klan be used to maintain the inferior position of blacks, but it also provided a base for attacks on any group which appeared to challenge that elite's hegemony. The southern elite sought to perpetuate the myth – also implicit in diffusionism - that economic deprivation was due to innate inferiority. Thus, a deliberate (tax reducing and profit-enhancing) strategy of promoting a low wage economy, wherein public sector improvements and socio-political reform were restricted, was rationalised under the guise of 'market forces' and lack of initiative (Nicholls 1969). The long-term effects of such a deliberate policy are clear. As Caird and Moisley (1961) reported in their outer Hebrides study, the lack of initiative of community elders must be understood in the light of a long regional history of proprietary subjuga-tion of workers. If the experience of history reveals that initiative and calls for change are frequently met with repressive responses, over time the likelihood of such calls declines markedly (e.g. Gaventa 1980). 'Status quo' attitudes do not arise in a vacuum, but are commonly the result of social conditioning which favours existing elites (e.g. Newby 1977).

A further major flaw in the diffusionist case emerges from its static character. The model lacks a sense of dynamism (the sloth of trickle-down is the nearest it approaches). So the revival of nonmetropolitan economic fortunes since the 1970s (e.g. Keeble 1984), the associated decline of central cities in traditional manufacturing belts and population turnaround (and its associated implications) are not comfortably explained by this model. Diffusionism provides an account of observable differences in economic activities in the 1950s and 1960s. It lacks appreciation both of underlying power relationships and of the interactions of space and social structure (which are cause and effect of each other's form). Diffusionism presents an historically flawed, spatially deterministic picture. Its empiricist account falls short because it does not start from an appreciation of the power relations which are integral to social structures and practices. Neither internal colonial nor uneven development models fall into this trap, even though they offer dissimilar accounts of social reality.

Internal Colonialism

As put forward by theorists like Casanova (1965), certain (peripheral) regions can be regarded as (internal) colonies of more dominant (metropolitan) national cores (much as neo-colonialism has been described as the condition of Third World nations vis à vis advanced capitalist countries). In short, a condition of monopoly is believed to exist, whereby core area institutions control peripheral economic affairs. The (peripheral) 'colony' is said to accommodate itself to the economy of the metropolis (often depending heavily on one product or market), with military, political and administrative dominance by core institutions perpetuating (or even enforcing) the 'colonial' relationship. The publication of Michael Hechter's (1975) <u>Internal Colonialism</u> drew much attention to this model. What Hechter offered, in a more forceful manner than in earlier conceptualisations, was a justification for viewing particular regions as internal 'colonies'. The distinctive feature of these regional areas was their dominance by ethnic minorities. The regional separatist movements these minorities often spawned in a sense represented the advanced economies' equivalent of Third World independence movements (Williams 1980). Although readers should be wary, since separatist movements have emerged in relatively wealthy regions (e.g. Payne 1971; McRoberts 1979), the juxtapositioning of peripherality, economic backwardness and separatist sentiments in areas like Brittany, Corsica, Scotland and Wales enhanced the appeal of the internal colonialist perspective (e.g. McRoberts 1979; Reece 1979; Mughan and McAllister 1981). Except as they bear witness to a sense of injustice and exploitation felt in peripheral areas (e.g. Overton 1979; Claval 1983), studies of regional protest movements per se will not specifically concern us here. However, the criticisms of internal colonialist interpretations which have emerged from these investigations do provide pertinent insights.

An obvious problem with this model revolves around what actually constitutes an internal colony. Regional separatist investigations have focused on areas of ethnic distinctiveness, which in some cases (e.g. Quebec, Scotland and Wales) are formally distinguished by a pre-existing element of political separation. From this perspective, internal 'colonies' emerge because one ethnic group attained an initial superiority (by whatever means) and systematically used state power to enhance its superior standing (Hechter 1975). As a general conceptualisation, this falls somewhat flat. There are many areas which are not distinguished by ethnic distinctiveness, yet area residents nonetheless feel a similar sense of core exploitation, as well as standing in similar socioeconomic positions (if not worse) as areas which have nurtured protest movements (e.g. Overton 1979; Claval 1983). As has been argued for Appalachia, it is not unrealistic to contend that:

> The parallels between Third World, non-socialist,
> underdeveloped countries and advanced capitalist
> nations on the one hand, and Appalachia and
> metropolitan America on the other, are striking.
> The continued exploitation of our region is
> probably best understood in terms of those
> economic relations which typify imperialist

> exploitation of the Third World: the terms of
> trade, absentee-ownership, capital outflows, and
> the widening income gap. (Dix 1973:25)

Yet if areas which are not ethnically distinctive are to be
conceptualised as internal colonies, we need a clear-cut
criterion (or criteria) which distinguish 'colonies' from
other areas. The mere existence of economic backwardness is
most definitely insufficient in this regard, even if it makes
good rhetoric for aspiring local political leaders. In
particular, even for ethnically distinct regions, the utilisation of an 'internal colonial' label must be treated with
caution, for it implies that unity exists in both core and
peripheral areas. This is a false premiss.

Within 'colonial' territories, class divisions are
fundamental elements in development processes. A key feature
in understanding Third World neo-colonialism, for example, is
the alliances which are forged between elites in the 'colony'
and transnational corporations (Evans 1979). These alliances
enhance the wealth and power of Third World elites, often at
the cost of intensifying transnational exploitation of the
masses (e.g. in higher prices, lower wages or legislation
which restricts worker organisation). Although the exploits
of the former Philippino leader might tempt us to refer to
such alliances as 'The Marcos factor', we must be aware that
this process is not limited to Third World economies. As
Grant (1965:9) recorded for Canada:

> Our ruling class is composed of the same groups
> as that of the United States, with the signal
> difference that the Canadian ruling class looks
> across the border for its final authority in both
> politics and culture. (and indeed in the economy
> - see Clement 1977)

The importance of 'internal' social divisions is further shown
in another Canadian setting - that of Quebec. Any justification that exists for tarring this province with an internal
colonial tag owes more to conditions within the province than
it does to its relations with other parts of Canada (Lord
1979; McRoberts 1979). As a province, Quebec has
traditionally comprised part of Canada's core and its
relations with other provinces (such as Newfoundland - Crabb
1973) have cast it more as exploiter than exploited. This
might seem strange given the separatist stance of the
province's Parti Québecois, but it is readily understandable
given that party's belief that Anglo-Canadian dominance in
Quebec depends upon the support of the national (Anglo-
Canadian) power elite. Break those ties and French Canadians
could gain control of their 'own' province.

Significantly, a major factor in the rise of Quebec separatism has been the emergence of a new Francophone elite. Anglo-Canadian dominance in the province's economy was consequent upon 'accommodations' reached in the late 1700s which divided spheres of influence into an agricultural sector, which was 'granted' to the Church and large-scale landowners, and an urban (industrial and commercial) sector, which fell into Anglophone hands (McRoberts 1979). Changes in the character of advanced capitalist economies have everywhere seen a substantial growth in government employment and activity over the twentieth century. In Quebec this provided a new channel for Francophone social mobility and the power base from which an alternative Francophone leadership emerged (as late as the 1950s, the practice of branding critics of the Roman Catholic Church as revolutionaries and communists was still common - Lord 1979). The 'colonial' status of French Canadians came not from the actions of Anglophones in other provinces, but from a Francophone elite's willingness to subjugate its own people for the sake of preserving (and enhancing) its own power.

The coincidence of industrial-commerial and landowner elite preferences in Quebec provides just one example of a class alliance which results in urban-based elites supporting their land owning colleagues in preserving low wage, exploitative rural social structures. Of course, land owning elites must offer something in return. In the United States, for instance, the Civil War might have ruptured the compromise between northern-industrial and southern-land owning elites, but the alliance soon reappeared following that conflict. Quite simply, industrialists depended too heavily on their land owning associates to allow the 'agreement' to lapse. Thus, with populist attacks on monopoly power threatening the 'special' position of major manufacturers, northern elites were extremely grateful for southern landowners' efforts in thwarting populist urges (Schwartz 1976). In this class-based action, which pitched elites against masses on a broad front, the utility of a subservient (low wage, initiative-deprived) southern population was clearly evident to northern industrialists. As the following editorial extract from the Mobile Daily Register (30 October 1892) makes clear, the peculiar power structure of the South enabled its elite to cajole the population:

> We warn the coloured voters that when they unite
> with the white Third Party they invite a catas-
> trophe from which their race cannot recover in
> 100 years ... we say now, clearly, pointedly and
> with full deliberation, and knowledge of the
> weight of our words, that so certain as Alabama
> goes for Weaver [the populist candidate] on the

> eighth day of November, we do not intend to wait
> for a force law after next March to tie us hand
> and foot and to deliver us over to such black
> leaders as Wickersham, Booth and company, but the
> Alabama legislative meeting in November, will
> before a new year sets in, take negro suffrage by
> the throat and strangle the life out of it. The
> coloured voters can now take their choice.
> (quoted in Schwartz 1976:283)

Similar landowner-industrialist compromises can be found in
the founding of modern Italy. Here, in return for an
enlarged market for manufacturing goods, southern landowners
received the support they required (including military
suppression) for maintaining their subjugation of a poverty-
stricken (and increasingly disgruntled) peasantry (Mingione
1974; Graziano 1978). Students of Scottish history will
recognise a comparable situation following the 1745 defeat of
Bonnie Prince Charlie at Culloden. Here, the change from a
Scottish to an anglophile elite led to altered regional social
structures. Former relations of mutual obligation which bound
landlord and peasant together were replaced by landowners
seeking maximum profits in order to maintain their costly
style of living south of the border (Carter 1974; Mewett
1979). The removal of 'rebel' Scottish lairds in favour of
English lords perhaps meant that 'a man of the people' was
replaced by a representative of an 'alien' power, yet the
post-rebellion behaviour of those Scottish lairds who sided
with the victorious House of Hanover was no different from
their newly acquired English counterparts. National origins
were not fundamental. Social class was.

The implications of this for the internal colonial model
are significant. As Ragin (1977) has pointed out, ethnic
distinctiveness does enhance the locational salience of
political issues, as well as the ease of political
mobilisation. But it does not constitute the fundamental
condition that distinguishes social relations across areas.
Thus, core-periphery upper class alliances have been forged
with elites in peripheral areas of ethnic distinctiveness
(e.g. Quebec and Scotland) and in peripheries which are not so
distinguished (southern Italy and the US South). Similarly,
challenges to existing regional elites have at times been seen
through the emergence of regional separatist movements (as in
Quebec), but these also arise amongst people who cannot use an
ethnic reference point for a clarion call (as in the anti-
farmer, environmentalist challenges that new middle class
residents pose for the traditional farmer elite in lowland
England - Newby 1980a).

Calling deprived regions dominated by ethnic minorities

'internal colonies' might be a catchy description, but it disguises unity in underlying causal processes which extends to deprived areas largely populated by the ethnic majority. As a general explanatory framework the internal colonial model is thereby flawed. The end-products it portrays are subsumed under more general uneven development models. These also provide more convincing accounts by not assuming that ethnicity per se has causal developmental powers and by providing a clear framework which makes the position of dependent elites intelligible.

Uneven Development Model

Theory from the uneven development mould starts with the premiss that class structure has primacy as an explanatory mechanism (Buttel and Flinn 1977). Within capitalist society, the underlying assumption is that competition does exist amongst capitalists, but that this is subsidiary to capital-labour struggles. Uneven development models thereby stand firmly within the orbit of conflict-based social science theories. Most usually they are linked to neo-marxist theorising, although associated core-periphery models (like that of Myrdal 1957) are not so oriented. Further, the broad position outlined by these models could perhaps be reached from general, non-marxist theorising on competition between elites and masses (e.g. Pareto 1901).

The underlying logic of (neo-marxist) uneven development models is that spatial differences in socio-economic fortunes emerge principally from the drive for capital accumulation (e.g. Massey 1984; Markusen 1985). This is held to orient producers toward points of maximum profitability. These are acknowledged to change over time, so that temporal variability in the geographical locus of economic growth should occur. On this point, these models stand in clear contrast to the static conceptions of core and periphery in diffusionist and internal colonial models (and indeed in many core-periphery models – Cooke 1983). Contrary to the mechanistic assumptions of neo-classical economic models, uneven development accounts recognise that adaptability is required if capitalists are to maintain their profitability. In a sense, capitalist production contains the seeds of its own destruction within its accumulation process. To circumvent this, constant adaptation is required. One component, or perhaps more correctly strategy, in this process is geographical relocation.

To illustrate this feature of capitalist societies, the experiences of cities as investment sites can be drawn on. In

the nineteenth century, cities provided an ideal environment
for industrial and commercial profit-making. On the one hand,
they provided large markets, so the chances of products being
sold locally or, due to better transport facilities, being
delivered to purchasers more quickly, was enhanced. This led
to city entrepreneurs receiving payment for their products
(i.e. returns on their initial capital investment) more
quickly. Because the circulation of capital in cities was
faster, the same initial outlay would in the long-term produce
higher rates of profit (and so put city enterprises in an
advantageous position with respect to their rural
competitors). On the other hand, city entrepreneurs were
advantaged by the availability of a large workforce. With
many under-employed or unemployed people at hand, owners
gained from worker competition for those jobs which were
available. Without trade union organisation, which
capitalists resisted for obvious reasons, workers had to
bargain for wages and work conditions from an extremely weak
position (where they fought against one another as well as
against their employers). The larger (and with rural-urban
and international migration influxes, more rapidly growing)
city labour forces thereby provided a further attraction for
investors. Threats to social disorder which emerged from such
concentrations of 'exploited' workers, changes in technology
requiring more skilled workforces, social reforms inspired by
vote-catching political opportunism and genuine upper-class
concerns about working-class social problems, all conspired to
improve workers' living conditions. However, these same
factors also weakened the pull of city locations. Higher
wages and stronger trade unions in cities, new technologies
favouring land-extensive production facilities and improved
social welfare programmes (which have significantly raised
urban tax rates above those in the countryside), have induced
this effect. In response to these changing circumstances,
manufacturers have sought new production sites. One
consequence is inner city blight. An associated product is
rural 'regeneration' (Urry 1984).

Of course, the recent re-emergence of rural locales as
loci for manufacturing output growth represents just one
(temporally specific) outcome of capital-labour-state
relationships. Suburbanisation is an earlier adaptive
strategy employed by capitalists. Moving production
facilities to Third World nations - the so-called internation-
alisation of capital - is another recent strategy (Evans 1979;
Fröbell et al 1980). More in line with the focus of this
chapter, so is the regional restructuring of capital. By
'restructuring' we refer to locational adaptations (very
possibly with technological or production procedure changes)
which companies engage in to enhance their profitability.
Geographical shifts in economic activity do occur without

restructuring. Manchester's loss of its dominant economic standing in the late nineteenth century, for instance, was not on account of manufacturing decline or capital restructuring, but arose from London's re-emergence as a major commercial centre (Garside 1984). In recent years, however, restructuring has become a marked feature of capitalism.

The precise form such restructuring takes varies. The drive for lower business costs is evident; as is an associated desire to 'discipline' workers. What particularly strikes a European in the United States at this time is the openness and lack of subtlety that surrounds this process. Analysts have long observed the tendency for large corporate enterprises to threaten to relocate their facilities if local governments impose anti-pollution legislation (see Crenson 1971; Phelan and Pozen 1973). This, however, was a 'behind the scenes' event, that was not brought to the immediate attention of the general public. In the present intensely pro-business atmosphere in the United States, the need for discretion appears to have gone. Manufacturers openly announce that their employees have to take wage cuts, work in much poorer conditions and have more insecure contracts or the company will move elsewhere.* That this pattern represents no idle threat is evinced in the 1985 decision of Mack Trucks to relocate its plant out of Allentown, Pennsylvania, into the South (because the wage cuts workers accepted were not large enough). This process is not restricted to city locations. Enforced wage reductions were accepted by all of the Hormel Meat Packing Company plants except that at Austin, Minnesota. The resulting labour-management dispute aroused surprisingly little violence, but the company still asked the National Guard to guarantee that non-union strike-breakers could cross picket lines. Where worker organisation is weak, wage reductions can be enforced and relocation is unnecessary. Where trade union power is strong, as in Allentown and Austin, a general message can be sent to workers – if they do not accept the owners' demands, they could lose their jobs. How they lose them will vary. In Hormel's case, the Austin facility was critical, since it was their 'show' plant. In the Mack Trucks case, the Allentown factory was relatively old and was discarded. Of course the blatant nature of such 'attacks' on workers' living standards is more likely where labour is weak (though there is some temporal variability in this condition).**

In truth, however, instances of 'covert' restructuring are by far the most common. This is clearly seen in the

---

* The process is not limited to manufacturing, as the 1986 TWA flight attendants' strike vividly portrayed.

emergence of the southern states as economic growth foci. Krumme (1981), for instance, has noted that only 1.5 percent of the North's employment losses between 1969 and 1972 were due to the outmigration of firms (and these represented only 1.2 percent of the South's employment gains – see also Breckenfeld 1977). In times of expansion, plants are opened in new locations (the South) or existing plants which are expanded tend to lie outside the old manufacturing core (the North). In economic recession, where excess capacity is identified, or as technological improvements reduce machinery and labour requirements, plants with less recent investments (i.e. northern ones) are most likely to close (as they are likely to be the least efficient). Although this process has intensified over the twentieth century – as corporate mergers and takeovers, and the increased incidence of new production technologies, has resulted in a glut of spare capacities – it does occur on a continuous basis, even within the same city. What we need to explain in accounting for the growth of the US Sunbelt is why restructuring took place on a regional basis (i.e. North to South). First, we should recognise the strong anti-union orientation of southern work environments. This has not arisen by chance, for it is a deliberate outcome of elite-dominated southern social structures (Greenberg 1981; Cobb 1982). Second, there is the effect of federal government policies. The precise governmental role here has been the subject of some debate. Indicating where he put the primary 'blame', Kirkpatrick Sale (1975:65) concluded that: 'Washington rather than Wall Street is at the core of cowboy

---

**In Britain and the United States, the corporate restructuring process has been accomplished after much locational bargaining. The apparently 'restrictive' burdens of British land-use planning controls have regularly been shown to be paper tigers. Corporate pushes for an increasingly hard-won profitability, accompanied by local policy-makers' concerns over area unemployment levels, have led to land-use planning restrictions (Elson 1986) and environmental controls (Blowers 1983) being discarded in favour of new investment. In this process, central government measures have also been instrumental. As Barrett and Healey (1985:348) have shown, there has been an increasing centralisation in government policy-making around restructuring objectives (viz. Department of Environment Circulars 9/80 and 22/80), as well as a relaxation of locational restraints on industrialists and residential developers. Even so, the spatial surfaces of unemployment and desired capital investment sites are uneven. Since they do not mesh together exactly, some areas still fiercely resist new industrial plant proposals, while for other places the wish for new investments is not satisfied.

growth'. There is some evidence to back this claim. Bolton (1966), for instance, concluded that without defence spending 1962 personal income levels would be 17 per cent lower in Utah and 11 per cent lower in California (see also Nash 1985). Yet the gains which some states would make if there was no defence spending, while being evident in the North (like Iowa at 10 per cent, Vermont at 12 per cent and West Virginia at 15 per cent), were equally evident in the Confederate states (with Tennessee at 11 per cent and Mississippi at 13 per cent). Certainly, as Mollenkopf (1983) has observed, War Production Board grants provided the means for new industries like electronics and aircraft to establish themselves in the South; as well as providing much of the infrastructure (like water projects) which made expansion feasible (especially in the South West). Yet, with the clear exception of some specific areas (like California), it is now evident that there has been a more equitable distribution of federal funds than analysts like Sale (1975) suggested (e.g. Archer 1983); if only because (northern) Democratic Party efforts to channel funds into the major cities conflict with Republican (and southern Democrat) demands for a 'rolling back' of the state to 'free' the energies of private enterprise (Mollenkopf 1983). As a third element in the formula, we must include state governments. In recent years, southern states have offered major incentives to industrialists (some states having 'hit lists' of firms to be encouraged southwards - Breckenfeld 1977). Tax reductions, grants, free training of staff and cheap loans make up programmes that have seen half the US states now maintaining permanent overseas offices for enticing investment by trans-national companies (Fry 1980). Undoubtedly, the attractions of southern locations vary across employment sectors. Clear skies and warm temperatures are significant for aeronautical and space industries, as well as for tourism and those serving retirement migrants. Proximity to (Oklahoma and Texas) oil resources has helped promote plastics and electronics enter-prises. Anti-unionism and low wages have acted like magnets for many labour intensive manufacturers. For some, then, the attractions of the southern social environment have been almost coincidental (insofar as locational requirements are attached to particular aspects of the physical environment which are over-represented in southern locations). For others, the attractions have been closely tied to the social relations of regional class structures.

The social mechanisms through which capital-labour relations are conditioned has a significance which as yet has not been convincingly translated into an uneven development framework. Perhaps the most widely acclaimed representation of alternative social mechanisms is that of O'Connor (1973). As he portrayed the situation, advanced economies are charac-terised by three major divisions:

The Monopoly Sector. Here production units are typically large, with markets of a national or international scale, where wages are high and barriers to entry for new firms are extreme (due to overheads, government regulations, high capital costs and brand loyalty). Although it would be more accurate to refer to this as the oligopoly sector (Bloomquist and Summers 1982 refer to it as the 'concentrated sector'), on the grounds that a few companies (rather than a single firm) dominate an economic sector, 'monopoly' does emphasise that market power is critical. As O'Connor stressed, neither prices nor wages are determined by market forces in this sector. With relatively few producers in a market, prices can be 'fixed'. Wages are higher, since costs are readily passed on to consumers and stability in the workplace is preferred to disruption from worker unrest (hence unionisation is often encouraged).

The Competitive Sector. Here production is more likely to be small in scale, relative to the total size of the market (i.e. there are a large number of outlets). Frequently larger concerns are slow to develop, for product substitution is easier than in the monopoly sector, so the advantages of sectoral dominance are smaller. Markets are more likely to be local or regional in scope. They are characterised by (relatively) low ratios of capital to labour, low wages and a poorly developed labour movement. In a rural context, agriculture represents the most notable example of a competitive economic sector (retailing is another). Although large-scale corporations are increasingly dominating farming, their total impact is still small compared with those of, say, General Motors or Ford in the US and British car markets. Units in this sector are price-takers rather than price-makers. Entrepreneurs are much more inclined to squeeze their labour forces to extract as high a work rate as possible for as little pay as possible. Since prices cannot be controlled, much more attention is devoted to cost minimisation.

The State Sector. This shares many common strands with the monopoly sector, insofar as it is usually characterised by (relative) stability of employment, large 'production' units (which foster the growth of labour unions) and an

ability to 'fix' prices (i.e. revenues).
However, it also shares some features with the
competitive sector (both are labour-intensive,
for instance). In addition, despite its seeming
monopoly position, its controlling direction is
(potentially at least) influenced by competitive
urges (i.e. competition for elected positions),
with significant public pressures existing for
reductions (or limited increases) in the size of
this sector.

On its own, this classification is more of an empirical
observation than a theoretical accounting framework for
advanced economies. However, it does provide a useful
heuristic device which helps us understand the counter-
currents and main threads in regional development processes.

Empirically, there is much evidence which backs the basic
divides in O'Connor's schema. Quantitative analyses of the
major 'production' features of United States industrial
sectors identify clear groupings which separate monopoly and
competitive sectors; and in these analyses the state sector
does reveal its somewhat anomalous status (e.g. Oster 1979;
Tolbert et al 1980).* While the empirical existence of such
distinct sectors alone does not attest to their theoretical
utility, this scheme can help us understand processes and
patterns of rural development. To appreciate this we must
first take a further dimension of economic structure into
account; namely the corporate organisation of economic
sectors. In the last chapter, Hymer's model of corporate
organisation was presented (Figure 4.5). The basic divisions
of this model between corporate head offices, regional or
functional headquarters and production or line units, are
equally applicable for transnational and intra-national
organisations. Rather than higher-paid, innovative and
decision-making functions being in core nations, within
states they tend to be located in major cities and their

---

* As presented above it might appear that O'Connor's schema is
static and unchanging. This is not in fact the case. The
'monopoly' position of the major car manufacturers in both
Britain and the United States has been affected by growing
imports of foreign (and particularly Japanese) vehicles.
This has not fundamentally challenged the monopolistic
character of this economic sector, but it has weakened the
market dominance of single firms. A draw-back in the schema
is that its focus is heavily on national economies, at a
time when national markets are becoming more international.
This will certainly disrupt previously dominant national
market structures.

suburbs (for proximity to specialised services and face-to-face contacts amongst senior executives). London alone, for example, houses over 70 per cent of the head offices of Britain's largest 1,000 industrial companies (Goddard and Smith 1978), while over 90 per cent of the largest corporate head offices in the United States have central city addresses (Burns 1977). Despite some suburbanisation in both nations, the locus of corporate control has altered little over the twentieth century (Goddard and Smith 1978; Cohen 1979). At both corporate and regional/functional office levels, rural areas stand in a position of marked inferiority. Of course, this provides no guarantee for the future. But we have been hearing about the major decentralising effects which tele-communications will herald for a long time (Memmott 1963). Decentralisation there has been, but rural and small-town locations have seen only the movement of production and line outlets from this. As evidence on inter-locking directorships and membership of social clubs signifies, top corporate executives still place a high premium on face-to-face contacts (Clement 1975; Dye 1976; Useem and McCormack 1981). This undoubtedly favours cities as sites for top corporate offices. Perhaps, for rural localities, hope exists in two features of job creation and product innovation processes: first, there is evidence that small firms tend to be at the forefront in generating new jobs (e.g. Birch 1979); and, second, the dominance of smaller firms in rural areas does not seem to be a handicap in product innovation (Markusen 1985; Pellenbarg and Kok 1985). Realistically, the potential for enhancing income levels and decision-making power which this offers for rural locales is slight. Evidence clearly reveals that growth in smaller firms characteristically leads to their being 'absorbed' into large corporations. A usual consequence is the loss of both high income and senior executive positions to (urban) corporate headquarters (Leigh and North 1978; Smith 1978). Once again the divide between rural areas in metropolitan and nonmetropolitan regions is pertinent. Places within commuting distance of cities are able to derive direct headquarters' benefits. More peripheral locations lose out.

In the main, peripheral locales are dominated by line or production units. Where the size of companies is small, however, higher income jobs (i.e. those of owners and managers) are present. At this point, the relevance of O'Connor's (1973) schema becomes clear. It is not true that size of a corporation is directly equatable with its monopoly or competitive sector position (many large corporations are, after all, involved in a wide variety of economic sectors - like Unilever's involvement in food processing, transportation and detergents or Boeing's role as aircraft manufacturer and farmer). It is nevertheless true that more peripheral locations have economies in which the existence of senior positions is most closely tied to small enterprises in competitive sectors. Income generating potential is thereby reduced, since local economies are dominated by either the lower-income end of the monopoly sector or by price-taking enterprises (for which maintaining low wages contributes a primary entrepreneurial aim).

This has important implications for the relevance of uneven development models in explaining patterns of rural advancement. First, it should be clear that the uneven development implication that capitalists shift points of production to areas of 'maximum' profitability largely holds for larger corporations. Quite simply, small-scale entrepreneurs do not have the resources to engage in wholesale geographical relocations. Second, it follows that the necessity to relocate is itself conditioned by the sectoral characteristics of production. Industries in the monopoly sector are less concerned with minor cost differences. Hence, except where new establishments are set up to exploit geographically specific resources (as in mining), the flow of monopoly sector industries into more peripheral areas often carries social implications (in that the locational choice can be influenced by the 'attractiveness' of a settlement – Spooner 1972). In the competitive sector, however, cost criteria are to the fore. This has undoubtedly been one of the main attractions of the southern United States (Cobb 1982). As a local counterpart to competition between nations for transnational corporate investment, local (and state) governments in peripheral areas frequently offer financial aid packages to entice investment in. This is most likely to prove successful for competitive sector outlets, given that they are so cost-conscious. But the low wage orientation of these enterprises brings benefits far below those local residents often expect (e.g. Summers et al 1976). Indeed, many researchers have contended that government financial incentives are frequently greater than the benefits eventually received (e.g. Summers et al 1976; Dillman 1982). This results in the fiscal position of localities worsening. However, poorer peripheral locations are often desperate for new employment. To attract and keep enterprises, governments appear to seemingly forego 'reasonable' checks on the use of their money or on the viability of the business itself. As Mathias (1971) has shown in a series of Canadian case studies, this can lead to a disastrously costly and ultimately doomed use of public funds. Moreover, since competitive sector enterprises are responsive to cost fluctuations, their very life in one location can be of short duration (heavy government subsidies reduce the time firms need to reap a profitable return on their investment). Numerous studies indicate that the life-span of branch plants in rural locations is shorter than that of locally-owned firms; albeit these investigations have not distinguished monopoly and competitive sector enterprises (Clark 1976; Anderson and Barkley 1984). How long plants need to stay in one location for a municipality to 'profit' from its investment will depend on the size of incentives given (Shaffer 1974, for one, concluded that a positive return was achieved after only a few years in a sample of Oklahoma counties), but the desperation of poorer peripheral local governments to enhance both their tax bases and their local employment often induces extremely generous offerings (in the case of Newfoundland's Stephenville linerboard plant, for instance, Canadian Javelin had to put up only $C11.5 million out of the total plant cost of $C143.6 million - Stewart 1969).

The implications of this are clear. Whether in the monopoly or the competitive sectors, outside control of local enterprises induces instability in economic fortunes and dependency. As the uneven development model suggests, there is a dynamism to processes of income generation which involves significant shifts in geographical patterns. The focal points for these shifts still appear to be cities, though the arena on which income planes oscillate is universal. The absence of a rural base for major power centres means that the 'cream' will not be found there. How much 'milk' is obtained is variable in time and space. Yet cream could emerge from locally controlled enterprises. However, with agriculture, retailing and government occupying disproportionate shares of the rural labour force, their effects will be determined by the dictates of the competitive and state sectors. On account of the (relative) geographical immobility of locally-owned enterprises and their neglect of political party or electoral decision criteria, uneven development models do not provide us with a wholly satisfactory framework for analysing these sectors. Uneven development ideas nonetheless help explain monopoly sector contributions to rural development. In the next sections we shall examine how somewhat different processes in the competitive sector and the state sector further contribute to rural development.

THE COMPETITIVE SECTOR

Capitalists are divided by their national, regional and local allegiances, by their economic sectors and sub-sectors and by the size of their enterprises. Protectionist measures which flood world markets with cheap European Community butter (or iron and steel) hurt North American producers in much the same way as higher feed prices harm the profitability of hog farmers. Competition induces conflict, but also cooperation. Thus, producers in one locality (or nation) often combine to promote their mutual interests. The cross-currents of conflict and cooperation are multitudinous. But out of the complexity, in rural areas the dominant conflict within the bourgeoisie casts entrepreneurs in the competitive sector against those in the monopoly sector. In part, this arises because of the relatively small size of rural competitive sector enterprises. This, combined with inter-sectoral rivalry (especially agricultural against industrial), casts elements of the competitive-monopoly division in the same mould as labour-capital struggles.* To understand the character of the rurally-based competitive sector, an appreciation of the competitive-monopoly divide is required. This can be most vividly obtained from an examination of agrarian protest movements.

Agrarian Protest

It is the winter of 1985. The place is Hills, Iowa. A farmer enters his bank, shoots and kills the bank manager, returns home to take the lives of his family and then kills himself. The news media react uncharacteristically. They forego declaiming this as an instance of personal depravity. Instead they portray the tragedy as symbolic. The farm 'community' is described as being in crisis (e.g. one-quarter of all Farm Home Administration loans had payments overdue by 1982, up from 12 per cent in 1979; Congressional Quarterly Inc 1984). The farmer's actions are seen as an expression of despair. That the (initial) target of that despair was a banker is also symbolic. Bankers have been hate characters in agrarian circles ever since family farmers shifted their orientation from a subsistence to a market economy. Not that bankers have stood alone in this regard. In North America, family farmers have traditionally stood in opposition to monopoly capital. Hills, Iowa, provided one expression of that resistance. Over time, more socially significant and broader social resistances have emerged in the periodic rise of agrarian protest movements.

To comprehend these movements one needs to understand the agrarian-monopoly capital divide. As with other forms of competitive-monopoly sector divisions, competitive producers react to the relative insecurity of (or inability to control) their work environment and, especially in times of economic hardship, to their dependence on monopoly ventures (like

---

* As Nelson (1957:22) has reported: '... for the American farmer to choose whether to call himself [sic] worker or businessman is not so simple'.

banks).*   Key factors that distinguish agriculture from other competitive sectors are: first, that its producers comprise a larger, more coherent group around which much public sentimentality and support exists (compare the image of family farmers with that of painters and decorators for instance); and, secondly, that the farmer - non-farmer divide is seen (by farmers) to have much greater pertinence in conflict generation than, say, the painter - non-painter separation.

It is important nonetheless to bear in mind that agrarian - non-agrarian conflicts have developed into agrarian protest movements only in temporally and spatially selective instances. It is not without significance that late nineteenth and twentieth century protest movements have been more characteristic of the open plains of North America than of either the British countryside or North American metropolitan regions. On the Prairies and Great Plains, the representatives of monopoly capital (banks, grain elevators and railroads) have a more visibly distinctive presence (in an otherwise small business environment). Isolated from the plight of urban dwellers who concurrently shared their misfortunes (e.g. in the Great Depression), farmers were inclined to view their position as unique (significantly, as Cox and Demko 1969 found for the 1905-7 Russian peasant disturbances, and as applies to agricultural labourer radicalism in Newby 1977, protest is more likely the farther the area is from alternative 'distracting' opportunities in urban centres). The Jeffersonian idea that democracy is dependent upon the existence of family farming helped persuade farmers that theirs was a righteous cause. An undercurrent of disquiet over high railroad charges and the questionable grading of grain yields by the elevator companies has been a permanent feature of plains agriculture. What raised this sea of discontent into a storm was the manner in which banker encouragement to take out farm loans for capital investment rapidly turned into demands for immediate repayments

---

* Some readers might object to the suggestion that banking sits in a monopoly position. This objection would be difficult to sustain in nations like Britain and Canada, where a very small number of institutions dominate, but it is a point with relevance in the United States. Despite various insurance schemes to protect investors, it is still the case that there is an alarming rate of bank failures in the United States. Yet even within the US, monopoly sector sits comfortably as a descriptor. Most especially in nonmetropolitan regions, banks frequently have a local monopoly as a lending source. Further, they have legal recourse to recall loans, with borrowers having little flexibility of action when this occurs.

when grain prices began to fall (Goss et al 1980 provide a figure of 25 per cent for the proportion of farms whose mortgages were foreclosed between 1930 and 1940; see also Shover 1965; Worster 1979). Perhaps not surprisingly, therefore, when agricultural depressions occurred (in the 1880s and 1890s and again the 1920s and 1930s), farmers were at the forefront in demanding restraints on monopoly capital (like the 1933 Farmers' Holiday Association demand that control of the national monetary system be taken away from bankers) and in the promotion of egalitarian measures (like graduated income taxes, the direct election of senators and the use of referenda; McConnell 1959; Shover 1965).

That early agrarian protests were not successful is evident if we compare the US situation with that of France. Largely on account of the power of the French farmers' lobby, French governments have granted the Crédit Agricole privileges which are not extended to other French banks (like exemption from corporate taxes and subsidised credit). Now taking around one-quarter of all French savings, the position this bank occupies as the champion of the small farmer stands in marked contrast to the place of banks in most US farmers' minds (Sampson 1981). These cross-national differences were not tied to major dissimilarities in the importance of agriculture in their respective national economies. The Crédit Agricole was founded in 1892, at which point US agrarian protest was at its peak and over 60 per cent of the US population lived in rural areas. In significant ways, however, this 1890s protest movement bore the seeds of its own ineffectiveness within it. Indeed, this protest era provided an early insight into a feature that now dominates agricultural political activity; namely, the importance of class divisions within agriculture. As Schwartz (1976) has shown for the South, the drive against monopoly power included an attack on the southern tenancy system which brought plantation owners into conflict with the movement's basic aims. Yet these large landowners were not excluded from the movement. They used their influence (inside and outside its orbit) to thwart, disorient and restrict its effectiveness.

Large-scale industrial concerns were equally opposed to the anti-monopoly sentiments of (agrarian) populism. But the concerns of industrialists were broader than this. They were particularly worried about the 'inefficiency' of agriculture. At the dawning of the twentieth century, industrialists were perturbed by the relative decline in US food exports (Figure 4.10), which appeared to offer a threat to the nation's foreign currency earnings. In addition, a 'rapid' rise in food prices appeared to threaten to increase pressure for higher industrial wages (indicative of this situation, between 1900 and 1910 farm production rose by only 9.3 per cent while the US population grew by 21.0 per cent). Around this issue an urban-based coalition of industrialists and senior government bureaucrats came together to form the Country Life Movement. Its aim was to 'industrialise' agriculture (Danbom 1979). Farmers in general were reluctant to adopt either the spirit or the specific suggestions of the Country Lifers. Until the 1914 War this Movement was largely ineffective in implementing its goals. It was more successful in establishing an ideological orientation in government circles which favoured industrialised farming. With the opening of hostilities in Europe, the farmers' ambivalent social class position again revealed its importance. The 'working class demands' of the 1890s were now forgotten as farmers rushed to expand their operations to capitalise on the (war-induced) inflated prices they could obtain for their products. The resulting push into land and machinery purchases pleased the Country Life Movement as this moved farmers more towards 'efficient', industrialised production practices. What it also did was to put farmers into debt and bind them more intricately into an economy dominated by industrialists. At the same time, mechanisation increased farm output, providing a strong downward pull on prices. Then, through Country Life inputs, and pressures from agents supporting large-scale farmers (like the American Farm Bureau Federation and its county agents), the mythology began to develop that the only way to improve farm incomes was to produce more (and then more and then more again - see McConnell 1959). When the inevitable and dramatic drop in farm commodity prices came, aggrieved farmers were not in a strong position to mount a protest campaign (albeit throughout the 1920s and 1930s various attempts to do precisely this were made - Shover 1965; Burbank 1971). In particular, the influxes of southern and eastern European immigrants into the country in the period since 1890 had brought new urban-oriented political alliances which weakened the farmers' position. The ideological tenor was less amenable to the claim that monopoly capital was 'attacking' the family farm and was undemocratic. Farm failure was now more likely to be seen - as the Country Life Movement intended (Danbom 1979) - as resulting from farmer 'inefficiency'. In addition, in this post-war era, farmers were to be significantly divided by the commodity they produced. Corn and cotton prices were low in the early 1920s, but the price of cotton recovered and even regained its war time position (before plummeting again in 1926). With corn and wheat prices showing less variability, the chances of uniting farmers across region, commodity and class were severely reduced. Yet this was the last real opportunity for farmers to challenge the industrialists' hegemony. By the 1980s, when penny auctions* again appeared on the Great Plains (Smith 1983), farmers constituted a small segment of the economy. Even the seeming 'victories' of populist appeals -

like US anti-trust legislation - were having little effect and some of the cherished aims of early populists (like graduated income taxes) were introduced at other times for other reasons. Although agrarian protest movements attempted to enhance rural development, they were largely ineffective. Their existence nevertheless bears witness to a conflict of interest that separates competitive sector farm enterprises from monopoly capital.

Industrialised Farming

Statistics that signify the 'industrialisation' of agriculture are startling. As an example, in 1930 it required 127 farm labour hours to produce 100 bushels of corn. Today it takes six hours (Price 1983). Figures of this kind can be repeated across all sectors of farm production. Underlying these changes has been a consistent movement towards the control of agricultural fortunes by industrial corporations and national governments. The combination of relatively low returns on capital, long periods for product maturation (slowing the circulation of capital), uneven labour requirements and unpredictability of yield (due to weather and pests), has retarded the direct involvement of industrial capital in farming. As the Country Life Movement has symbolised, however, there was always an interest in farming in government and industrial circles. The alliance they formed is preserved to this day. The goal of cheapening food costs, though laudable in the abstract, has had markedly detrimental effects for most farmers. For one thing, there has been a major decline in farm numbers (Figure 5.2). For another, largely through the medium of government inspired research, farmers are constantly presented with a variety of new machines, chemicals and hybrids which increase yields. If these are not adopted, they could provide other farmers with sufficient productive edge to render a non-adopting producer 'inefficient'. Increasingly, then, farmers are pushed into a state of subservience to the money lenders to whom they must appeal for assistance (whereas in 1950 US farmers could pay off their

---

* 'Penny auctions' were characteristic of 1930s agrarian protests. Forced farm sales (due to non-payment of mortgages) were met by a large farmer presence at the sale, ensuring by threat of violence or other means, that the property was purchased by its original owner (for a low price). Not surprisingly, banks commonly retained ownership of the land rather than see this happen, and in some instances even preferred to issue a moratorium on farm foreclosures.

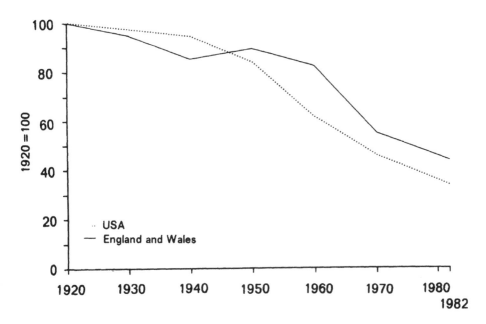

Figure 5.2 The decline in farm numbers

debts with 75 per cent of one year's average income, it would now take them over three years' income; Price 1983; also Figure 5.3). Dependency on banks leads to lack of autonomy in production decision-making (Perelman 1977, for instance, reports on mid-western banks forcing farmers to continue using chemicals rather than adopting an organic approach to farming, and Green 1984 has found bankers taking a more active role in the day-to-day and strategic decisions of farmers - like whether to buy a new machine or use conservation practices). There is nonetheless much variability in farmers' dependence on 'regular' money lending institutions. For some, an alternative outlet has been the companies that process their commodities. By accepting 'contract farming' arrangements (i.e. guaranteeing sales to one processor), farmers receive benefits of financial support, technical aid and fewer worries over marketing destinations or times of sales. What the corporate world gives with one hand, however, it takes back with the other. Contract farmers must produce in the manner designated by contracting industrialists, at controlled prices, with heavy penalties for failure to produce yields which meet the company's quantitative and qualitative targets. As Vogeler (1981:138) observed: 'Contracts give agribusiness the advantage of treating farmers as employees without the responsibility of paying them as employees'. The risks are still the farmers. Of course, there are other options. Perhaps the most notable of these is multiple job holding. In 1929 just 11.5 per cent of US farmers reported working off their farm for more than 100 days. By 1982 the figure was 43.0 per cent (in the European Community 64.0 per cent of farmers worked off-farm in 1980; Hill 1984). Opportunities for multiple job holding do nonetheless vary signficantly; as they do for contract farming.* The picture so far painted is therefore exaggerated. But as we fill in the detail to clarify the basis of exceptions, the essential story line does not alter. The evolution of agricultural change owes its roots to industrial capitalism, not to agriculture per se.

Pressure for low food prices comes not simply from industrialists and governments, but also from farmers themselves. This does not arise because farmers deliberately seek to reduce their own income, but from a desire to increase production levels. For single farmers, production and sales increases obviously lead to higher returns. Once multiplied

---

* Thus, Hightower (1973) reported that 75 per cent of processed vegetables, 70 per cent of citrus products, 55 per cent of turkeys, 40 per cent of potatoes and 33 per cent of vegetables in the United States were produced under contract. Hart (1978) reported that 40 per cent of farm production in Britain was undertaken under contract.

Figure 5.3 Components of US farm production 'costs'

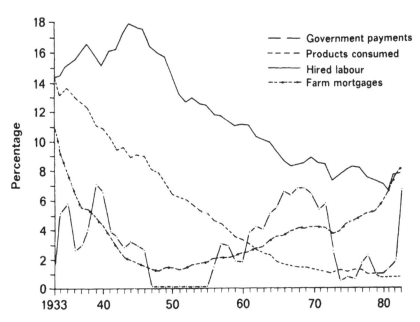

Source   Statistical Abstract of the United States (Annual)

over the whole farm sector, however, increased yields flood the market and force down commodity prices. Hence, income advantages largely accrue to the first innovators. Catching a yield-enhancing innovation late means that capital costs are incurred at a time when prices have already begun to fall (Price 1983). Since innovations are (increasingly) costly, the resources available to large-scale farmers puts them in an advantageous position in this innovation-yield prices cycle.

In the main, however, even large-scale farmers are dependent on 'outsiders' for innovations. Research and development is expensive and over time has become more and more costly in real terms. We need to go back to the nineteenth century to find a time when farmers per se were significant innovators. Then, the extreme wealth of large landowners in nations like England, and the comparative lack of technical sophistication in innovations like the drainage of Midlands clays, helped cultivate a significant farmer input. With increased mechanisation, more intense chemical applications and 'improved' breeding and crop yield innovations, research costs have gone beyond the reach of farmers. The profits which can be garnered from sales to farmers have nonetheless long provided a major incentive for industrialists to develop the 'necessary' new products (Banaji 1976). That these innovations offer the promise of lower costs and/or higher yields is sufficient to encourage their adoption (even without industrial company sales efforts). It is no surprise that the number of tractors in the US increased from 10,000 in 1910 to 1,600,000 by 1939. Costs of crop production were more than halved by this machine and, by reducing the need for horses, 50 million acres were released into production, so offering the further promise of higher farm incomes (Frundt 1975).

It should not be assumed that technological innovation is a 'natural' process occurring at equal paces in all farm sectors. As a number of analysts have noted, mechanisation has traditionally been used by large-scale landowners in the western United States as a means of 'disciplining' their work forces (Friedland 1980; Price 1983). As agricultural labour power increases, landowners have turned to technological innovation to reduce their workforce needs. If cheap labour could have been secured for longer, innovation would have been unnecessary. Of course, the California situation is conditioned by a social environment dominated by very large farms and by particular agricultural commodities. In the southern USA, where large farms were equally evident, the wage reductions required to match the savings tractors brought to cotton production would have been unfeasible to enforce. Instead, landowners resorted to ejecting their workers. Thus, 100,000 southern share-cropper families were displaced in the

1930s (Price 1983). For the single family farm with no employees, the choice of weighing labour against capital costs was not available. Here innovation adoption (machinery or whatever) was necessary to stay in business. Yet the demands of innovations were not simply those of capital. To make full use of machines - and attain the economies of scale they increasingly required - necessitated larger holdings (spreading capital costs over larger annual production levels). The spiral of capital investment pressed ever upwards. So, whereas in 1940 the (US) farm input distribution was 54 per cent labour, 17 per cent land and 29 per cent capital, by the mid-1970s it had changed to 16 per cent labour, 22 percent land and 62 per cent capital (Meier and Browne 1983).

The interaction between agricultural-input manufacturers' and farmers' pressures for higher yields was advanced by government actions. Over time, we have seen considerable government intervention in agricultural production for the purposes of promoting 'public welfare'. A characteristic step was the decision to disallow the use of cans for storing milk, in favour of bulk storage cooling facilities. While beneficial in health terms, this step forced farmers who wished to stay in dairy production to invest heavily in new machinery. To warrant so big a capital outlay, the farmer needed to have a large herd (Dorner 1983). Implicit within this change then was a bias towards large-scale operations. Not that governments neglect smaller farm enterprises. Government programmes like Britain's Payments to Outgoers Scheme (1967 Agriculture Act), Canada's ARDA (Bunce 1973) or France's SAFER scheme (Clout 1968), all of which aid small-scale farmers to change occupations, reveal a social welfare concern for people in an 'impossible' position. Yet it would be too charitable to believe that this is the principal impetus behind government endeavours. This is readily seen in Hightower's (1973:27) evidence on the distribution of government-inspired research on rural areas. Drawing on the inputs of state agricultural experimental stations in the United States, Hightower contrasted the 1129 standard worker years which were spent in 1969 on improving the biological efficiency of crops with the 18 days spent on improving rural income. Compared with 842 standard worker days devoted to controlling insects, diseases and weeds in crops, 95 were allocated for insuring that food products are free from toxic residues from agricultural sources. The 200 standard worker days given over to ornaments, turf and trees for natural beauty, make a stark contrast beside the meagre seven days afforded to rural housing (and 60 per cent of research on rural houses was concerned with their physical fabric rather than with those who lived in them; Hightower 1973:53). In this we see that the long-term government goal is to reduce farm prices, rather than increase rural welfare.* Opinion surveys have clearly shown that the public does not appreciate the food product changes that have occurred as a result of agriculture's industrialisation. As Handy and Pfeff (1975) have reported, consumers complain about food being 'tasteless', of 'rubber tomatoes' and 'oranges that do not taste like oranges'. This is no sentimental cry for 'the good old days', for there is much truth to these claims. Government funded and inspired research has been behind the move toward more 'resistant' (i.e. thicker skinned) products like tomatoes, in order to better allow for harvesting mechanisation. We should be very clear in our recognition of the bias in government policies which favours both industrial/commercial capital over agriculture and large-scale producers over small-scale operations.

If governments were concerned about food prices per se they could readily attack this problem. A variety of estimates exist, but there is much agreement over the inefficiency and waste in food processing and retailing. Parker and Connor (1979), for instance, put their 1975 estimate for the overcharging of consumers due to seller concentration at $10-15 thousand million for the US market. Cross (1976) suggested 25 per cent of total sales prices as the correct figure, but reported that other estimates have gone as high as 40 per cent. She also found that:

> It is noticeable that the highest rates of overcharge, in terms of percentage of industry sales, occur in soft drinks, beer and liquor. This supports the FTC's [Federal Trade Commission's] conclusion that industries dominated by a few firms make more, waste more, advertise more, and are then able to hide some excess profits through bookkeeping exercises. (Cross 1976:177)

Why should such over-charging occur in what appears to be a competitive sector (retailing, that is). The answer lies in market power.

Over time, agricultural processing and food retailing have moved inexorably into a monopoly sector position. In terms of specific processing sectors, for instance, Hightower

---

* What is also significant is the slight attention agribusiness puts into welfare concerns. Frundt (1975), for instance, reported that under 10 per cent of food industry research and development costs are devoted to the nutritional content of food. Over 90 per cent goes on consumer appeal and quick preparation.

(1975) reported that four companies - Coca Cola, Pepsico, Royal Crown and Seven-Up - control around 90 per cent of the North American soft drinks market. Three companies (Borden, Carnation and National Dairy) account for 60-70 per cent of dairy product sales, with 80-90 per cent of breakfast cereals in the hands of General Foods, General Mills, Kellogg's and Quaker. More generally, Vogeler (1981) reported on the Federal Trade Commission finding that 24 manufacturers control 60 per cent of all US processed food sales. As Hightower (1975:14) observed when presenting his figures: '... such statistics are difficult to come by. The concentrated food industry, practically immune to public inquiry, keeps the lid on as tight as possible'. This is hardly surprising given the estimates of waste previously noted.* The picture for retailers is less extreme, but the pattern of movement is toward further concentration. Thus, in 1972 the top 20 retailing companies in the US accounted for around 70 per cent of food retailing profits (up from 53 per cent in 1963; Frundt 1975), while more than 50 per cent of food sales in Britain go through the major supermarket chains (Body 1983). Such concentration at the sales end is matched by domination by a few firms at the agricultural input stage. In the previous chapter we noted that this has significant effects on the cost of farm inputs (as for tractors - Canada Royal Commission Farm Machinery 1969). At this stage we merely wish to re-emphasise this point.

The cost-price squeeze in agriculture is imposed by two arms of monopoly capital. The filling for this sandwich is competitive sector agriculture. Pressures from these monopoly components make the continued existence of competitive agriculture open to serious doubt. Numerous fears have been expressed about the movement of large corporations into farming and, in particular, about the danger this presents for the family farm. Such fears seem to us to derive too strongly from sentimentality over the family farm as an institution. If we examine the actual involvement of large

---

* The authors have not come across comparable figures for Britain or other countries. But even casual checking of shop shelves for the few names listed above is sufficient to reveal the international character of this pattern. The precise names do of course vary. Bachelors, Bird's Eye, and Walls are more dominant in Britain - though all are subsidiaries of Unilever - and government agencies have a stronger role in the UK (thus the Milk Marketing Board owns 75 per cent of manufacturing capacity for butter and 50 per cent for hard pressed cheese - though this only translates into 22.5 per cent and 25 per cent of the consumer market; Gilg 1980).

Figure 5.4 The Del Monte Corporation

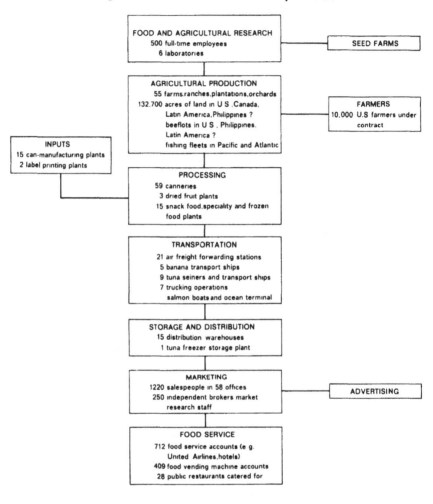

Source : Hightower (1975)

corporations in farming, a few points become obvious. First, corporations often use farming as a means for achieving other goals. Thus, the fears of the early 1970s over insurance companies rushing into agricultural land ownership in Britain were much over-blown (Munton 1977). Simply put, with land values rising very rapidly, the ownership of land was merely a

logical business response to a short-term opportunity to make
bumper profits through purchase and later resale. Substantial
proportions of farm land have traditionally been owned by non-
farmers (Healy 1980 puts the figure at 44 per cent for the US,
with Harrison et al 1977 giving an 11 per cent figure for
public institutions alone in Britain). The frequent mass
media references to corporations like Tenneco owning huge land
masses should not confuse us into believing that these are
major farmers. Most of Tenneco's land is rented out and was
only purchased to safeguard reserves of oil, gas and coal
(Frundt 1975; Cortz 1978). In addition to which, evidence on
a more general front indicates that even the largest farms
still tend to be family controlled (e.g. Schulman 1981).
After all, why should processors become involved in farming
(even more so those not in the farm sector). Corporations not
used to this form of production have previously revealed that
failure to appreciated the nuances of agricultural production
can lead to some spectacular failures (Cortz 1978). Besides,
vertical integration through land ownership ties up capital
which could be used more profitably elsewhere (though not when
land values are appreciating rapidly). The advantages of
controlling all stages of production and sales processes is
appealing to some companies, for there are advantages in
vertical integration. However, the owned farm production
component of such vertically integrated corporations tends to
be small (Figures 5.4 and 5.5). Quite simply, ownership is
not needed in order to dominate farm production. Farmers have
been so squeezed and have become so reliant on fresh infusions
of capital, for productive 'improvements', that they
constitute a ripe market for contracting arrangements. For
corporations, contracting is much 'cleaner' and more flexible
than farm ownership. If alternative, cheaper markets can be
found, contracts need not be renewed. The farmer, like the
industrial worker, can be discarded when not required.

Farmers are not totally defenceless in the face of
monopoly capital pressures. A key tactic in circumventing the
full force of the farm cost-price squeeze is multiple job
holding. Indeed, the multiple job holder has now become a
key, if not even the dominant, farmer-type, across a wide
variety of nations (Buttel 1982a). For the USA, for instance,
Deseran et al (1984) reported that only five per cent of farm
families now received only self-employed farm earnings (by
1982, 36.4 per cent of farm operators worked off-farm for more
than 200 days in each year). Opportunities to engage in
multiple job holding are nonethless affected by job skills,
farm commitments and opportunity. Hence, higher incidences of
part-time farming have been recorded amongst those with more
formal education (e.g. Gasson 1966; Perkins 1972), who produce
less time-demanding commodities (Layton 1978; Buttel and
Gillespie 1984) and live closer to urban centres (Fuguitt

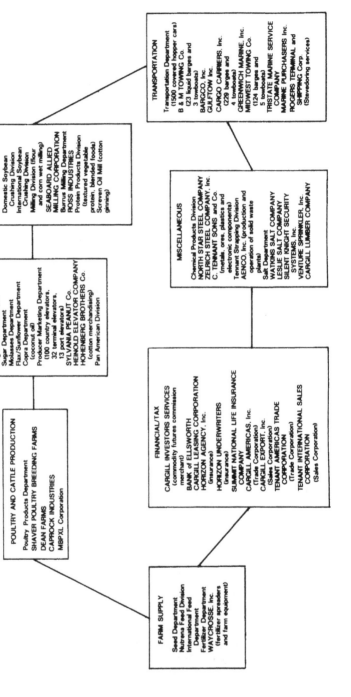

Figure 5.5 The Cargill Corporation's products and services

Source : Adapted from Gilmore (1982)

1959; Gasson 1966; Mage 1974). This sets up a pattern of considerable spatial variability in the incidence of farm-related multiple job holding (Mage 1974). This is further complicated by differences in the character of multiple job holders (Buttel 1982a). Mage (1976), for instance, identified six basic classes: (1) the small-scale hobby farmer, with a low attachment to farming, but a desire for particular aspects of farm living (like horse raising), for a 'rural lifestyle' or the status of farm ownership (Flinn 1982); (2) the aspiring full-time farmer, who is a part-time farmer for economic reasons, but seeks to reach full-time status (increasingly an unattainable goal); (3) the persistent part-timer, who has held two jobs for a long period (at least 8 years), and needs the second job to stay in agriculture (although, over time, the high costs of farming relative to its returns often leads to farming assuming a secondary importance; Mage 1976); (4) the sporadic part-timer, who tends to be older and engages in off-farm work as an income supplement, on chance opportunities and in small quantities; (5) the prosperous, large-scale hobby farmer, who often uses farming for tax loss purposes (as Coughenour and Swanson 1983 reported, one-quarter of all residents on farms in 1975 could report farm income losses which they could set against non-farm earnings);* and (6) a 'unique' group which is often comprised of specialists (say in poultry or eggs). Even from the brief descriptions provided here, it is clear that multiple job holders exhibit quite different characteristics. At one end, there are those who see farming as their primary occupation and use multiple job holding as a means of clinging to it. At the other end, there are those for whom farming is a 'play thing'. This last group of people is important in numerical terms. Further, these 'agriculturalists' partly account for a new feature in farming; namely, a growth in small farm numbers (e.g. Harper et al 1980). More so than at any time in the past, farming has bifurcated into a large-scale, more commonly full-time group and a small-scale, usually part-time category (Buttel 1982). Numerically, the latter is most important. In terms of production and political power, however, the former is dominant (in 1982, for example, 3.9 per cent of US farms produced 31.8 per cent of farm output by value).

A factor to acknowledge if the aim is to understand how government policy contributes to perpetuating the existing structural weaknesses of agriculture is the divisions which exist within the farming sector. Larger farms are

---

* A similar case for entry into farming comes from considerations of inheritance, since agricultural land is commonly assessed at very favourable rates for death duties (e.g. Lyson 1984).

increasingly components of an intricate web of legalised interconnections between themselves and large industrial corporations. For the very largest farms, agriculturalists are often processors in their own right (Smith 1980). More commonly, disguised by owner-occupation of their farms, producers have succumbed to the position of labourer on behalf of industrial capital. Whatever the precise arrangement, this group stands in a different corner from the small-scale farmer. Although a few operators of small concerns aspired to their larger bretheren's productive patterns, more commonly a sharp difference exists between these two groups in their orientations toward farming as an occupation (e.g. Gasson 1969; Newby et al 1978). The so-called hobby farmer is clearly not intent on being drawn into a status of subservience to the processing industry. Elderly farmers often adhere to 'traditional' husbandry practices which predispose them in the same way (Gasson 1969; Buttel and Gillespie 1984). An increase in part-time farming thereby lowers the overall commitment to (or perhaps more accurately dependence upon) farming as an occupation. This in turn helps enhance the dominance of large-scale, corporate-connected farms in determining the future of agricultural production.

The Agricultural Lobby

In a competitive market situation, where profitability is squeezed between two virtually immovable, and certainly uncharitable, arms of monopoly capital, farmers have increasingly turned to governments to ease their plight. In truth, as many analysts have recognised (e.g. Meier and Browne 1983), farm operations do not arise from immutable economic laws but emerge from the economic principles that governments allow to operate. This is true of all economic sectors, but has particularly overt and immediate effects in agriculture. In part this arises because of the strategic importance of food. With Europe experiencing two widespread and devastating wars in this century, it is no surprise that government intervention has an eye to strategic (including balance of payments) considerations. However, there is widespread agreement that the farm sector has an influence in national policy-making which is disproportionate to its influence in the economy (Wilson 1977; Newby et al 1978). This does not come from the size of the farm population. Although, in nations like France (Wright 1955), the political clout of the farm population does partly emerge from its share of the total population, this is far from the case in most nations (in the USA, for instance, even when farmers comprised the population majority, they adopted the political tactics of a minority group; Hadwiger 1976). Simple numbers provides an

Figure 5.6 Proportion of the US population on farms in 1980

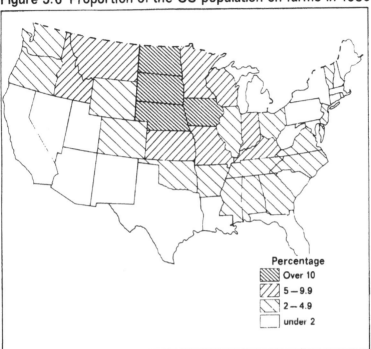

Percentage

- ▨ Over 10
- ▩ 5 – 9.9
- ◰ 2 – 4.9
- ☐ under 2

insufficient explanation.

In the majority of advanced capitalist countries, farmers constitute a small component of the voting population. This means that even if they voted as one, they constitute a small and declining electoral block (e.g. Howarth 1969). To some extent of course the electoral virility of farmers depends upon the arrangements under which elections are held. The federated government structure of the United States, for instance, enhances the potential impact of farm voters. With two senators per state, irrespective of population size, and a spatial concentration of farm voters (Figure 5.6), agrarian interests have the potential to 'unduly' influence the selection of both senators and state governors. At times this electoral condition has seemingly had a significant impact on national policy-making. In the 1950s, for example, many analysts reported that farmers were using their voting strength to penalise the Republican Party, as a protest over Secretary of Agriculture Ezra Benson's efforts to reduce commodity price supports (Vidich and Bensman 1968; Hadwiger 1976). More recent investigations have found little support for the continuance of this picture (see Knoke and Long 1975; Sigelman 1983). This cautions against the assumption that residence per se is a sufficient cause for a political lobby. Thus, from his investigation of state government politics in the 'rural states' of Mississippi, Montana and Vermont, Bryan (1981:254) recorded that:

> From an empirical search of 8000 contested roll calls in three states over 44 years one may, quite literally, count on the fingers of two hands the times rural delegations 'ganged up' on their urban colleagues and won.

Further, legally enforced reapportionment in the 1960s and 1970s has weakened the previous biases in state legislatures which gave an over-representation of rural districts (Rose-Ackerman and Evenson 1985). Yet rural-related political influence can still be strong. In behavioural terms, for instance, Tickamyer (1983) found that rural legislators were amongst the most powerful in the North Carolina General Assembly. As with the strength of southern representatives in Congress, this partly comes from years of tenure in office (Johnston 1980), which in itself is partly dependent upon an absence of political controversy or conflict within a district (that this is more common in remoter and agrarian districts is evinced in the frequent absence of political debate which analysts report for rural communities - e.g. Vidich and Bensman 1968; Wild 1974 - and the high rates of unopposed elections recorded in the same areas; Bealey et al 1965; Madgwick et al 1973). Such situations can nonetheless change quite radically as population composition alters through in-migration (Hennigh 1978). More significant, because it is a structurally imposed advantage for agrarian (though not rural) interests, is the existence of independent national government departments for agriculture. That such agricultural departments exist (to promote and protect this sector of the economy) when similar departments do not exist for other major production sectors (like car manufacturing or coal mining), provides a vivid symbol of the privileged status of agricultural endeavours.

For most of the twentieth century, a key underlying base for the political strength of farm interests has been the sentimentality which surrounds farming as an occupation. In truth, this is not a feature that is restricted to agriculture, but extends to rural areas in general. Although in recent decades inroads have been made into their positive public image (Lowe et al 1986), for most of this century farmers have been seen as the embodiment and the protectors of superior living styles. This sentimentality is not recent in origin. Emerging from early constitutional discussions in the USA, Thomas Jefferson's ideas about the special moral values of farm people have long been influential.* Added to this, the writings of Wordsworth and American Romantic writers, like William Cullen Bryant of the New York Post, all gave leads in the development of a public taste for wildlands, nature and rurality in general (Nash 1967; Keith 1974). These are now integral to national self-images. As Lewis (1972) expressed it, small towns are as central to the American experience and self-conception as Paris is to the French or the green countryside is for the English. For the English in particular, rural residency is also aligned with social status. Jilly Cooper (1980:233), when commenting on her childhood residential history, vividly portrays this aspect of the British class system:

> Where you live is just as important as what you live in. I myself keep very quiet about having been born in Hornchurch, always justifying it by saying we were only there temporarily ... With relief, after two years we moved back to Ilkley in Yorkshire, where my father was born. Even here the Middleton side of the valley was much smarter than the town side, because it was greener and got the sun all day; but at least Ilkley was smarter than Otley because the inhabitants were richer and because it was further away from the industrial towns like Bradford and Leeds; and Harrogate was much smarter than Bradford and Leeds because it was more rural and nearer the East Riding, which was much smarter than the industrial West Riding, but not as smart as the [even more rural] North Riding.

---

* The essence of Jefferson's view is encapsulated in the statements that farmers are 'the chosen people of God, if ever He [sic] had chosen a people', whereas cities were 'essentially evil' and 'ulcers on the body politic' (quoted in Key 1964:21).

The status and romantic appeal of rural imagery is commonly employed by television advertisers, to persuade people that giant corporations producing butter, breakfast cereal or canned soup are no different from homely family farms (Bouquet 1981; Goldman and Dickens 1983). Residential developers employ the same imagery, in road or housing estate names like Blackfriars, The Meadowlands, Copse Close or Hill Chase; the aim is to surround their mass-produced suburban subdivisions with the ambience of the countryside. Along similar lines, it is significant that the rock concert Live Aid has now been followed in the United States by two Farm Aid concerts, though not by a single Public Transit Aid or Textile Manufacturers' Aid. Such blatant use of the sentimentality surrounding rural and farm living is based on sound principles. Evidence from opinion polls consistently reveals a majority preference for rural or small-town residence (e.g. Carpenter 1980). This translates into an overt preference for policies which nurture the visual appeal of country areas and preserve their 'traditional' institutions. Opinion polls in the United States, for instance, repeatedly find strong support for the preservation of family farming and for restrictions on corporate control of agriculture (Flinn 1982). As we shall see, this disposition can raise problems for farm leaders – the bulk of whom come from the large-scale farm sector – but this underlying sentimentality can provide a powerful political tool if it is used 'properly'.

In practice, in many nations, the need to draw on this stock of public support is diminished by a favourable predisposition toward agriculture and landowners amongst political leaders. Norton-Taylor's (1982) inventory of the agricultural links of Mrs Thatcher's early Cabinet makes this abundantly clear:

Lord Carrington:	1200 acres in Buckinghamshire and Lincolnshire.
Willie Whitelaw:	estate in Dunbarton and partner of a farming company in Penrith.
Francis Pym:	500 acres in Bedfordshire.
Michael Heseltine:	400 acre farm near Banbury.
James Prior:	300 acre farm in Cornwall.
John Nott:	130 acre farm in Cornwall.
Ian Gilmour:	land in Middlesex.

Michael Jopling: farm in Westmorland.

Peter Walker: 200 acres in Shropshire.

George Young: share of 1,000 acre farm in Shropshire.

Lord Thornycroft: 44,000 acres in Stirlingshire.

This strong farmer/landowner presence is perhaps to be expected in a Conservative Cabinet. But the Conservatives have formed the British Government in 25 of the 41 years since 1945. Further, even in the supposed proletarian ranks of the Labour Party, the likes of James Callaghan and Richard Crossman provide examples of powerful figures who owned farms. Of course, even the opulence of Conservative Members of Parliament pales into insignificance compared with the riches of the United States's Congressional representatives. As Ralph Nader's Citizen Action Group found for 1975, the US Senate's 100 members included 21 millionaires but only five members who shared the material condition of 99 per cent of US citizens; that is, an annual income of under $50,000 (Vogeler 1981). While wealth per se is not equatable with agricultural interests, it is commonly associated with land holding and has been clearly linked to Congressional voting records which favour land owning interests (Vogeler 1981). Moreover, amongst state governments, at least, there is evidence that the agrarian links of legislators are associated with higher public expenditures on farming (Rose-Ackerman and Evenson 1985).

Yet no matter what the underlying and internal support for agricultural endeavours, this has not proved enough in the long-term. Compared with the relative unity on monopoly-capital, farmers are too divided by region, commodity and size to force legislation in their favour. Although the level of state support for agriculture is generally high, this acts more as a subsidy for industrialists who provide the inputs and process the outputs of farms. Incomes and prices favourable to farmers are generally low and are highly unstable (e.g. Figure 5.7). Economic powerlessness (plus internal divisions) within the farming sector reduces the direct political effectiveness of agrarian interests. It is here that public support and sympathy comes into its own. As Lipsky (1968) cogently argued for relatively powerless groups in general, if the support of powerful third party lobbies can be captured, even seemingly weak groups can have a substantial policy impact (also Dearlove 1974). For long-term effectiveness, the ability to maintain such political efficacy depends on the maintenance of a tightly-organised, disciplined and skilled lobbying machinery. How far national farm lobbies

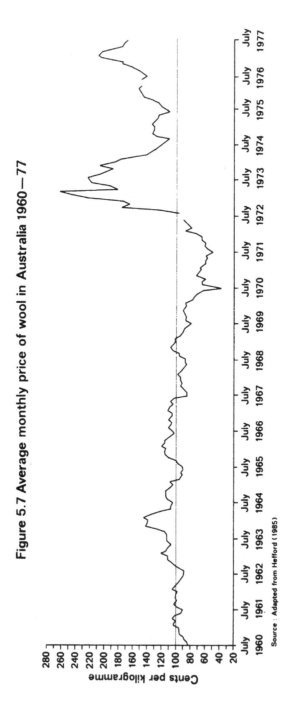

Figure 5.7 Average monthly price of wool in Australia 1960—77

Source : Adapted from Hefford (1985)

meet these conditions varies across countries and over time.

The epitome of a well organised and (relatively) effective agricultural lobby is Britain's National Farmers' Union (NFU). A commentary on its lobbying commitment is embodied in the observation of Robert Cryer, minister for small businesses in the Callaghan Government: 'Anything happens - next morning there's a paper from them on your desk' (quoted in Elliott et al 1982:84). Indicative of this commitment are Newby et al's (1978) figures for NFU activities in 1975. In this year, the NFU gave 12 major press conferences, issued 162 press notices, provided press release and editorial material for over 1,000 local newspapers, was involved in 200 special newspaper articles and had representatives participate in 100 local radio programmes. At the national level, its significance is seen in the Ministry of Agriculture, Fisheries and Food bestowing a virtual monopoly on the organisation for negotiating agricultural policy (though entry into the Common Agricultural Policy has weakened the potency of this relationship). Undoubtedly, the narrow brief of the national ministry helps here, for its jurisdiction is restricted to agriculture and not to broader rural development issues. Hence the Ministry is not divided over its support in, say, the industrialised agriculture versus conservation debate. Partly, but only partly, on account of this, there is also a helpful unity of disposition and purpose between the NFU and Ministry senior officials. Helping garner both bureaucrat and politician support in this regard is the NFU's preference for private, restrained negotiation (Wilson 1977). This enhances that organisation's image of 'responsibility', encourages trust and helps the organisation to be drawn into early stages of policy formulation.* This was clearly seen in the legislative formulation of the 1981 Wildlife and Countryside Act. This measure formed one component of the Conservative Government's attempt to instil a stronger pro-business and pro-growth impetus into British society. In essence, the Act reduced restraints on farm activities in that if farmers wished to destroy environmentally sensitive landscapes (for farming

---

* The operations of the NFU and the reaction of Ministry officials are in this regard consistent with the broad outlines of organisation theory (see Thompson 1967). Especially for large bureaucratic concerns, organisation theory stresses that institutions seek stable operating environments rather than conflictual or uncertain conditions. This leads institutions to 'incorporate' units they regard as 'responsible'. Such units must then 'behave' in order to maintain their favoured status, but have the compensation of influencing organisational policy.

purposes), they could only be stopped if conservation agencies (in particular the Nature Conservancy Council) paid them for not doing so (Lowe et al 1986). The favoured status of both the National Farmers' Union and the Country Landowners' Association was clearly seen in the policy generation process here. Both were instrumental in early stages of drafting this legislation, whereas conservation groups had to await publication of legislative proposals for their input. As Cox and Lowe (1983:49) explained:

> Failure to be closely involved with policy formulation at its formative, pre-public stage often means that environmental groups are faced with an uphill campaign, at later stages, against a course of action to which officials and major interests have been committed.

The critical role of this pre-presentation access is revealed in Griffith's (1974) evidence on the effectiveness of Parliament in adjusting legislative proposals once presented. For the three years he examined between 1967 and 1971, only 11 bills were markedly affected in their passage through the House of Commons. Out of 1503 amendments to legislation proposed by opposition members at the committee stage, only 131 were agreed to. On only 26 occasions could the Government be considered to have been defeated at the committee stage. As these defeats could be reversed or amended at later points in the legislative process, only nine found their way into legislation. Early access to the policy-making process thereby opens the door for significant agrarian input (see also Saunders 1984 on farmer - water authority interlinkages).

Farm lobbyists must nevertheless be conscious of the somewhat fragile position they are in. A monopoly negotiating position carries expectations of responsibility toward the general public. In fact, public support for British farmers has seemingly declined over the last decade (Lowe et al 1986). Butter (and various other) 'mountains', higher food prices, conservationist attacks on 'prairie landscapes' (i.e. the removal of hedges and trees), the destruction of wildlife and delicate biological habitats, and the increasing use of steroids and chemicals to increase yields, have all brought the farm sector's concern for general public welfare into question. That farm leaders are aware that they must show concern for the general public is also clear; as evinced in the NFU's support for Farming and Wildlife Advisory Groups, set up in the early 1980s. In Cheshire, Walters (1975) found that local NFU leaders, when faced with the option of paying either domestic or industrial electricity rates for their farms, selected the more expensive household charge structure for fear of negative public reaction to their having low cost energy. Especially important in preserving public support for farm lobbying organisations has been the maintenance of a public facade of unity amongst farmers. Analysts on both sides of the Atlantic have commented upon the remarkable success of farm organisations in this regard (e.g. McConnell 1959; Wilson 1977). Cross-national differences do nonetheless exist in this regard. The United States farm lobby has traditionally been more divided than that in Britain. Further, it has been weakened by the party political affiliations of lobbying institutions. Whereas Britain's NFU projects itself as a non-partisan farm organisation, the American NFU is aligned to the Democratic Party and often conflicts with the Republican-dominated American Farm Bureau Federation (Wilson 1977; Tarrant 1980; Stokes 1985). With the Grange and the National Farmers' Organisation as further lobbyists, the US has a less united farm sector representation than is found in many European countries.

This division in agriculture-related lobbying is multi-faceted. The existence of a variety of organisations which, if only to justify their separate existence, must have some different goals, is important. So are the dissimilar lobbying aims of farm and agribusiness sectors.(Stokes 1985). Further, with changes in tenure practices, divisions are often engendered within the farm sector per se. Thus, increases in corn prices have now become an anathema to livestock farmers, since this crop is used as feed. Such commodity divisions have long proved significant, but as the demands of corporate contracts and the economic realities of machinery costs intensify, farmers are increasingly pushed towards mono-culture. This intensifies pressure for stronger commodity organisations and weakens farmer attachment to general farm institutions (Meier and Browne 1983; Stokes 1985). Intertwined with this organisational complexity is the effect of farm size. Particularly evident in the United States, where American Farm Bureau Federation attacks on agencies and programmes favouring small farm operations are notorious (e.g. McConnell 1959), the division between large and small farms has taken on an enhanced importance over time. This is also found in other nations. In the Netherlands, for example, a farmers' political party has emerged to support smaller holdings and favour policies aiding security of farm tenure (Nooij 1969). In West Germany, the weakening of the Deutscher Bauernverband (the German Farmers' Union) emerged from that organisation's stronger support for full-time, large-scale segments of the farm sector (Andrlik 1981). In Britain, the emergence of the Farmers' Union of Wales was a consequence of disgruntlement over the NFU's bias toward lowland, grain-producing and mainly large-scale English producers (Madgwick et al 1973). On the surface, small-scale farmers have strong grounds for concern over the policies promoted by dominant farm organisations. For one thing, as Coughenour and Christenson (1983) found amongst 1,640 Kentucky farm families, there is opposition amongst large-scale, full-time farmers to policies which explicity support small farm operations (see also Commins 1982 for Ireland). Further, on account of a variety of programmes, including the tying of subsidy payments to production levels, and the offer of grant assistance for items large-scale enterprises can best make use of (like machinery and new building investments), there is a consistent pattern of government aid unduly assisting bigger farm operations (e.g. Josling and Hamway 1976; Mann and Dickenson 1980; Vogeler 1981; Bowers and Cheshire 1983). Some caution is nonetheless needed before interpreting these points in size terms alone. In the USA, for example, the bulk of federal farm aid has traditionally gone on just three products - cotton, feed grains and wheat (Bonnen 1972). Similar patterns of product bias are equally evident in other nations (Bowers and Cheshire 1983). There are undoubtedly size-related differences between favoured commodities and others, but farm sectors exist (like US cattle ranching) which are dominated by large-scale enterprises, and yet receive comparatively little assistance. Further, as Bloomgarden (1983) found amongst cotton growers, it is feasible that within commodity groups, there is little conflict between small- and large-scale producers. Yet to assume that commodity-type and not size is critical would be a misconception. Both are important, even though size is most generally reported as the dominant factor.

Anthony Sampson (1982:321) has pointed out that amidst the bountiful homage paid to the free enterprise spirit of Adam Smith's writings, rarely have advocates drawn attention to Smith's warnings about the potential dangers of an economy driven by the profit urge. In particular, reference is sparse to Smith's warning that:

> people of the same trade seldom meet together, even for merriment and diversion, but the conversation ends in a conspiracy against the public, or in some contrivance to raise prices.

The agricultural sector abounds in examples of this. Of course, a small group of farmers is little able to affect prices per se, but then they no longer operate within a private enterprise environment. Farm product prices are increasingly determined by political decision rather than market forces (albeit the two are connected).* As John (1981:256-7) stated it:

> ... the hidden factors of subsidies, grants, tax allowances, loans, guaranteed prices, etc., seldom appear in the calculations of profit-

ability, and milking 100 taxpayers makes larger profits than milking 100 cows.

Why this is so important is on account of the biases which exist in farm organisations. In general, with the USA as something of an exception, farmers' organisations report very high levels of farmer membership. In West Germany, for instance, 90 per cent of farmers belong to the Deutscher Bauernverband (Andrlik 1981), with 86 per cent of British commercial farmers joining the NFU (Wilson 1977). Traditionally, however, farmers have seen these organisations as instrumental; the more they believe attendance and participation is essential for benefits, the more they attend (Warner and Heffernan 1967; Andrlik 1981). As V.O. Key (1964:30) saw it: 'When the pain left, the farmer quit squawking and went back to the plough'. All that is except the large-scale producer. They have shown a consistent tendency to be active participants in farm organisations (McConnell 1959; Bloomgarden 1983). The significance of this situation emerges from a clear tendency for farm organisations to be controlled by a small, active elite (e.g. Walters 1975; Wilson 1978). At this juncture, Adam Smith's stricture on producer efforts to 'fix' markets in their favour becomes more evident. In a production sector where political participation is dominated by self-interest, where an elite controls the main organs of lobbying, where this elite is biased in favour of large-scale operations, and where there are dissimilar policy aims between large- and small-scale producers, it is no surprise that farm organisations have traditionally favoured big producers. It is also no surprise that, with increased specialisation in production, and a growing involvement of 'urban' middle class professionals in (hobby or part-time)

---

* The evidence which supports this is so overwhelming that one is at a loss over where to begin illustrating that it is so. A few examples that demonstrate the point are the marked changes that usually occur in farm production practices following a change in government regulations. Thus, Lee and Helmberger (1982) showed how midwestern production levels for corn, feed grain and soya beans were intricately linked to government acreage allotment, acreage diversion and set-aside programmes. While Dilamarter (1971) clearly demonstrated that the 1958 decision of the Ontario Wheat Producers' Marketing Board to pay for transport-to-market costs led to a sharp increase in marketed wheat (and a decline in its use as feed) and a significant geographical shift in production towards areas distant from elevator points. On the income front, Richard Body (1983) has estimated that 166 per cent of British farmers' net income comes in the form of government subsidies.

farming, pressure would arise for alternative representation modes.* Yet, whereas commodity groups are likely to obtain support amongst legislators - spatial coincidence of commodity production zones and constituency areas aiding this process - associations representing small-scale farm operations should find less joy. Basically, the big guns are already entrenched and have shown a clear willingness to resist programmes which aid small farm units (McConnell 1959). Existing biases in favour of larger producers should in fact intensify over time, since major producers of farm inputs and processors of farm outputs are adopting more vocal lobbying on behalf of the large-farm sector (Body 1983). In this, of course, these industrialists are pushing competitive agricultural producers further towards the status of piece-work labourers for monopoly capital.

The Non-Agrarian Competitive Sector

A considerable section has been devoted to the farm sector, with scant attention to other employment areas. On the surface this bias is unjustified, given that agriculture accounts for a relatively small segment of rural labour forces. Indeed, it might appear that we have fallen into the trap of associating 'rural' with 'agricultural' and have neglected other employment sectors. This is not so. Manufacturing now employs more rural residents than farming in most nations of the advanced capitalist world. Yet manufacturing, as with mining, is largely dominated by monopoly capital, not competitive enterprises (Tolbert et al 1980). Indicative of this is the dominance of large corporations in (rural) industrial employment. From a survey of four US states, Erickson and Leinbach (1979) reported that 84 per cent of employment growth in nonmetropolitan manufacturing came from branch plants. The question then is whether farming merits the attention it has received within the competitive sector alone. The reasons we feel that it does are straightforward: first, agriculture is the dominant employment sector in the rural (and especially the non-metropolitan rural) areas of many nations (in France for example 32 per cent of the rural population was farm-related in 1975; Bontron and Mathieu 1982); second, there is a basic dependence of many other activities on agricultural endeavours

---

* The scale and diversity of farm operations in the USA has made feasible the long-term existence of 'alternative' farm groups - like the National Farmers' Union - though the farm lobby is still dominated by large-scale producer organisations (in particular the American Farm Bureau Federation).

(illustrative of this, Bible 1978 reported on an investigation which found one non-farm business closing for every six families leaving agriculture between 1958 and 1963); third, on account of the interdependency that exists between many activities and farming (recreation is a good example, with farm-based holidays, land sales for second homes and second jobs in vacation facilities standing out); and, fourth, other major competitive sector enterprises reveal very similar developmental consequences as farming, but in a less 'sophisticated' manner.

This last point requires some clarification. What we are essentially contending is that there is much similarity across rural competitive sectors. Take retailing as one example. Here, like farming, enterprises are predominantly small and family-owned. Their numbers have been declining rapidly (Johansen and Fuguitt 1984 reported that three-quarters of US retail functions were present in fewer nonmetropolitan places in 1970 than in 1950; a similar picture of decline is reported in detail for a selection of British counties in Standing Conference of Rural Community Councils 1978). What is more, in parallel with the rise of hobby farming, the character of many retail enterprises is changing. There is a trend toward serving a more urban, middle class clientèle (thus, the number of rural craft and antique stores continues to grow, in part contributing to the demise of outlets serving local populations; Dawson 1976). Furthermore, we find clear-cut differences in value dispositions between larger and smaller units. Larger units are frequently components of major supermarket chains in which profits are a primary goal (Safeway for instance had sales equivalent to the 15th largest industrial corporation in the USA in 1984). Smaller, independently-owned outlets on the other hand commonly stay in operation for social reasons, as well as economic (Molyneux 1975; Dunkle et al 1983). The desire for companionship, the wish to provide a service, and the goal of staying within the same community, often provide justification for continuing an operation which on strict economic grounds might not be viable. In this manner, many rural retail outlets are similar to small-scale farm operations (especially if they are run by elderly owners), and part of the reason for a decline in retail (and farm) numbers is an unwillingness of younger people to take on low yield enterprises once operatives retire or die.

Where significant differences do arise between farming and both recreation and retailing is in the degree of political organisation of their lobbyists. Compared with the efficient and uncompromising machinery which backs farmers, retailers represent a naive and politically impotent sector. In Britain, for example, it was only in 1976 that the Independent Small Businessman and Retail Trades Association was formed to protect the interests of small retailers. Similarly, although represented through a variety of organisations (like the British Tourist Authority and national parks authorities), rural recreation stands in a much weaker position than farming in political influence. For one thing, as Gilg (1984, quoting Brotherton) has reported for Britain, political party considerations have a role to play (as in the Thatcher Government 'stacking' national park authorities with 'political appointments' drawn largely from farmers). Openness to such overt political influence can nevertheless be a two-edged weapon. In their conflicts with the US Forest Service over national park designations, for example, conservationist and recreational interests have undoubtedly benefited from the empathy of senior federal officials (Twight 1983). In the state sector, decision criteria are likely to be quite different from those in the monopoly and competitive sectors.

THE STATE SECTOR

Such decision criteria differences are nonetheless more short-term than long-term in character. This arises on account of the structure of social relationships which condition state behaviour. Precisely how this structure affects government behaviour is not universally agreed. Two main schools of thought exist (with innumerable sub-schools). These correspond with the major divide in social science research between integration and conflict theories. As described in Chapter Three, we find the underlying premises of, and empirical justifications for, integration theory far-fetched and primarily normative. This is not the place to delve into a detailed critique of alternative theories of the state, but readers can refer to a wide variety of reviews which present the basic features of this debate (e.g. Greenberg 1974; Alford 1975; Taylor 1983). The understanding of the state which informs this discussion is most closely aligned with the theorising of Claus Offe and Frederick Block (e.g. Block 1980; Offe and Ronge 1982). These theorists recognise the long-term importance of capitalist social structures on state behaviour, but insist on the decision-making autonomy of state institutions from capitalist control. Put simply, it is held that capitalists direct the long-run path of state actions, but not short-run (or specific) state decisions. As Offe and Ronge (1982) put the situation, the main elements of capitalist states are: (1) property and, significantly, wealth generation are predominantly in private hands; (2) government dependence for service provision upon the ability to raise taxes (i.e. to draw on private sector wealth); and (3) an inevitable reliance of governmental programmes on the health

of capitalist profit-making.   Since, to maintain public
support, programmes must be sustained (and perhaps expanded),
for self-preservation (both for politicians and bureaucrats),
the state must promote capital accumulation.   Of course this
structural relationship does not exist in a vacuum.   Neither
does it arise on account of some immutable natural law.
Social relationships and structures must be reproduced over
time.   Changes in the basic structural condition can occur.
State agencies could, for example, take a more active role in
wealth generation.   However, as our discussion on
transnational corporations stressed, capital is very mobile.
If a state did adopt such a role (except, say, in the
'emergency' conditions of war), it would soon feel the
opposition of capitalists (especially as it would be either
raising taxes 'unduly' to pay for its expansion or else would
be impinging on capitalists' profits by excising profitable
sectors from the private realm).   Capital would move to sites
offering higher returns, leaving the state with a declining
resource base; thereby reducing the ability to promote and
fund state policies. This process would likely be a slow one.
Without a revolution to restructure basic social relations (à
la 1917 Russia), state 'encroachments' into the private realm
will consequently usually be small-scale and incremental in
character.   Similarly, the response of capital will not be
dramatic.   Business leaders - as in the Confederation of
British Industry - are regularly threatening the collapse of
national economies consequent upon new government legislation.
Trade unions, income taxes and corporate taxes, votes for
women, unemployment insurance and old age pensions have all
been portrayed as destroyers of society and the economy.   Yet
once introduced, capital accumulation continues apace.
However, if too many such reforms are enacted, over too short
a time period, without due regard for the criticisms of major
business leaders, then capital flight can result (one only has
to observe the reaction of British capital to the election of
a Labour Government to find confirmation of the sensitivity of
capitalists even to the expectation of 'too much' social
reform).

     Every individual act of a government contributes to the
image it projects to private investors.   In deciding on
specific courses of action, however, there is considerable
state autonomy. This does not mean that entrepreneurs are not
influential in deciding specific acts.   Rather, it recognises
that   there is considerable conflict amongst capitalists
themselves over how the state should act in many instances
(for example, even if there is agreement that the government
should subsidise capital investment, there will be disagreement over which sectors of the economy should obtain the bulk
of the benefits and the procedures through which these
benefits should be distributed). State agents can utilise

this conflict - by playing off one opponent against another - to help promote the goals favoured by an agency's leaders. These goals incorporate much more than support for capital accumulation. For politicians, voters must be 'cultivated' if re-election is desired. For bureaucrats, if higher salaries, more promotion, greater work flexibility and job satisfaction, along with enhanced social status, are to be achieved, then distinctive institutional objectives must be pushed for (these favour organisational growth and at times seek new work relationships). Even without pressures from the general citizenry, these three sets of social agents can produce substantial conflict over policy goals. This should not confuse us into conceptualising the state in pluralist terms. Overriding any competition over specific policies is a long-term drive for capital accumulation. Further, within the structure of individual behaviour modes is an advantaged position for politicians, bureaucrats and capitalists in the policy-making process. There is elitism intrinsically inter-twined into the behaviour of state institutions. This severely restricts their permeability to the general public's influence (e.g. Hoggart 1984).

Nevertheless, the alternative aims of these decision agents do lead to complexity in policy enactment. It is certainly not possible to talk about a rural policy, even within one nation. Dissimilar objectives within and amongst national (provincial or local) agencies mean that a single, coordinated policy will not emerge. At the international level, this was seen in the dissimilar goals of development advocates and foreign policy diplomats in formulating US foreign aid policy (e.g. Rossiter 1985). Similar conflicts arise over internal policies. Conservationist support within (parts of) the Department of the Environment commonly clashes with the 'pro-efficient' agriculture stance of Britain's Ministry of Agriculture, Fisheries and Food, while the overtly grass-roots orientation of the Tennessee Valley Authority (required to maintain local elite support for the institution) has led to conflict with the more publicly insulated federal agricultural department (Selznick 1949). Policy currents and counter-currents ebb and flow with variable intensity and direction, gouging new paths, collapsing old landmarks, and creating new structures, with a complexity, dispersal of goals and inconsistency of objectives that defies easy analysis. In part, the lack of clarity in government policy direction comes from the indirect impact of many state agency acts. As Clark (1982) has observed, for instance, by providing services like electricity for a uniform unit cost, agencies effectively provide hidden subsidies for rural areas (as costs of installation and maintenance are higher in the countryside). Such subsidies are counter-balanced by cost structures unfavourable to rural areas. Representative of this is the cost of petrol (and especially the European practice of heavily taxing gasoline). The statement of David Penhaligan, Member of Parliament for Truro, that '... if you really want to harm the economy of rural areas, there is nothing more effective than putting up the price of petrol' (quoted in Gilg 1982:23), might contain an undercurrent of voter appeasement, but the sentiment has considerable substance. What is more, the fact that increased (or decreased) petroleum costs are usually called for for seemingly aspatial reasons (like strategic concerns, conservation and revenue needs) is symbolic. Too frequently, government policies with spatially discriminatory effects are debated and introduced almost in ignorance of their geographical consequence (on these effects see Shucksmith 1980). Buttel (1980), for one, is correct in pointing out that the pro-defence, anti-social services mentality which underscores both the Reagan and Thatcher administrations has serious consequences for nonmetropolitan regions. Defence spending is heavily concentrated in large cities and their suburbs, while an over-representation of the poor and the elderly in rural locales means that welfare payment reductions lead to a redistribution of government spending away from the countryside. Added to this, the high and volatile interest rates which monetarist theory has justified for these governments has detrimental consequences. Amongst price-takers like farmers, high interest rates cause considerable financial woe (Boehlje 1985); most especially given the strong pressure on farmers to take out large loans to expand their operations (i.e. buy more land) and enhance their productive efficiency (i.e. buy more machinery). It is not that governments are unaware of the geographical discriminations that such policies impose. Rather it is that they do not hold them in high regard.

Government policy-making inevitably introduces gains and losses in dissimilar measure for different groups in society. If the group penalised is sufficiently powerful that it can harm political leaders (i.e. their re-election), the likelihood of being listened to is enhanced. In a nutshell, this causes a bias in state actions in favour of urban populations (and capitalists). As an illustration, the US urban riots of the 1960s led to a wide range of programmes intended to help the urban poor. These programmes largely ignored more intensive and (extensive) rural poverty (Huddleston and Palley 1981). The challenge the riots presented to both the image and the actuality of US democracy spurred national political leaders (with business support) into providing new 'appeasement' programmes. That this was an appeasement measure is revealed by programmes ignoring the more scattered (and so less prone to riot) rural poor. In truth, rural areas did benefit in exactly this way when their residents were a more powerful lobby (relatively at least). The regulation of railroads following populist unrest in the 1890s provides one example of this. The introduction of marketing agreements, price controls and special agricultural assistance in the 1930s is a more recent response to farmer insurgency (Shover 1965). By awarding partial, limited concessions to militant opponents, major power interests have cooled sentiments favouring radical change.

This pattern of response to crisis can be of major importance in changing the basic legislative framework in which rural policy is enacted. We should be clear, however, that such basic changes do not simply emerge from crises (whether directed against current capitalist social relations or the ruling government itself). Significant changes can emerge from a genuine desire to help specific social groups. Institutions like the Highlands and Islands Development Board and the Tennessee Valley Authority certainly contain elements of this. Yet, one of the main themes that runs throughout this book is that such governmental responses to 'need' tend to be short-duration, poorly financed, cautious and ultimately largely ineffective instruments. Overwhelmingly, dominant trends in government favour major producer groups or 'buy' the support of antagonistic factions (the state must not only aid accumulation but also promote public acceptance of prevailing structural arrangements; O'Connor 1973). Wholesale support for farmers, for example, has not continued because governments feel that farmers are a particularly meritorious group of people who deserve government patronage. Rather, it has emerged from strategic considerations and from the pressure of industrial capital for low food and manufacturing input prices (Frundt 1975; Body 1983).

This viewpoint might be objected to, as an irrelevant criticism of governmental policies, given the (relatively) healthy economies of rural areas in recent decades. On the face of it, rural areas in general have been net beneficiaries of the corporate spatial restructuring process (Fothergill and Gudgin 1982). This is not because they possess any specifically geographical advantages (Urry 1984:55, for example, maintained that international capital is largely indifferent as to where industrial plants are located within nations), but arises because rural labour pools are both cheaper and less well organised (Urry 1984:57). Nevertheless, although capital may be increasingly indifferent to location, or is at least less constrained by traditional agglomerative economies, in periods of growth, governments in most advanced capitalist nations have sought to direct private investment toward 'depressed' areas. When crises of accumulation (or legitimisation) develop, the market is usually given freer rein to dictate the location of investment and the state tones down its regulatory function. Note, for example, the decline in the potency of Industrial Development Certificate restraints on British industrial location decisions throughout the 1970s, with their eventual complete removal in 1981 (see Moore et al 1986). Likewise allocative planning aims, based upon household consumption objectives like amenity, have given way to goals aligned with the stimulation of the economy. In Hertfordshire, for one, restraint policies conceived in the 1950s as a basis for channelling and controlling growth, have, despite their proven success, been blown off course by the demands of macro-economic trends and restructuring goals (Elson 1985). Where this is important for (particularly peripheral) rural areas is that many localities have not benefited from restructuring processes. Although only marginal in effect,* government policies which direct economic activity to 'backward' areas can still have a significant local role.

In the sphere of rural development, state interventions to encourage capital investment in peripheral areas have a long history. In the United Kingdom, the Development Commission was formed in 1909, in Italy, the Cassa per il Mezzorgiorno in 1950, in the United States, the Tennessee Valley Authority in 1933 and in France, the Délégation de l'Aménagement du Térritoire et l'Action Régional (DATAR) in 1963. In simplistic terms, such policies came into being with social welfare intentions; in that they sought to ameliorate regional economic disparities and improve the living conditions of peripheral residents. The understanding was that these areas were unable to prosper without external assistance. To some extent this assistance was motivated as much by political as by social or economic goals. It is nonetheless evident that early 'regional' policies were prompted largely by a concern for the welfare of a (relatively) impoverished labour force. In the United States, for example, the depression years of the 1930s provoked the first federal rural development programmes as part of the New Deal. Included amongst these programmes was the Farm Security Administration (FSA). This agency was responsible for commissioning some of the most poignant images of rural

---

* Disputes do exist over the effectiveness of regional aid programmes. Some researchers - like Moore et al (1986) - have found local benefits accruing from subsidies for locating in 'preferred' (or 'deprived') areas. However, other analysts report that most companies would have located in (or near) these places anyway and use the grants as a windfall subsidy (e.g. Springate 1973). Differences will exist across companies, economic sectors and locations, but the general theme which emerges is of slight positive repercussions from such aid programmes.

Figure 5.8 Assisted areas in the U.K.

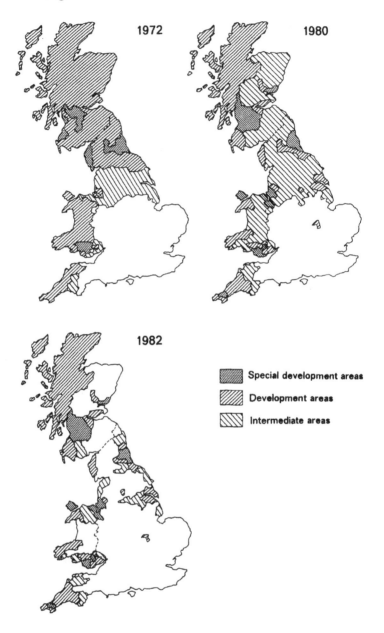

deprivation in the photographs of Walker Evans and Dorothea Lange. While seemingly a minor step, such images ultimately helped promote nation-wide awareness of the plight of many rural inhabitants. This invoked widespread support for measures aimed at easing rural poverty, which not only helped them pass previous Congressional barriers but also had a recognisable impact in sustaining the lives of peripheral settlements (Worster 1979). In addition, backed by public support, the FSA was able to increase its role in directing rural development efforts. This raised the ire of large-scale farm operators, since one raison d'être of the FSA was to help small-scale farm units and agricultural labourers. A strong campaign was mounted by these large-scale producers (through the American Farm Bureau Federation) which culminated in the dissolution of this agency in 1946 (see McConnell 1959).

Clashes in the objectives of different power interests are also visibly evident in the regional development programmes which have existed in Britain since 1970. In particular a sharp contrast must be drawn between the aims of the 1974-9 Labour Government and its post-1979 Conservative counterparts. Labour's commitment to helping poorer regions was evident in its establishing a series of development-oriented institutions during its tenure of office. These included the National Enterprise Board, the Scottish Development Agency (both 1975), the Welsh Development Agency and the Development Board for Rural Wales (1976). In the face of a worsening economic recession, however, Labour's commitment shrank after 1976-7 (thereafter concentrating upon the protection of existing jobs and services rather than on the creation of new opportunities). Starting from an already depleted base, the Conservatives have further weakened regional policy (as part of a general policy to reduce governmental intervention in the private sector). This has not simply been seen in the level of financial contributions and in the tightness of regulatory controls (Moore et al 1986), but has also been evident in the geographical areas which are now eligible for 'special' assistance. Even by 1980 there was a considerable shrinkage in the areas eligible for the highest levels of assistance (e.g. Figure 5.8). Just two years later, the regional policy programme appeared to have been decimated. Many remoter areas (like central Wales and the Pennines) were now removed from even the weakest category of special assistance status.

On its own, this need not have been a fundamental problem. Whereas regional policy in Britain has been concerned principally with promoting industrial growth, there are specific rural development agencies (e.g. the Highlands and Islands Development Board). These institutions have broader aims (Clarkson 1980) and quite extensive geographical

Figure 5.9 UK rural development agencies and programmes

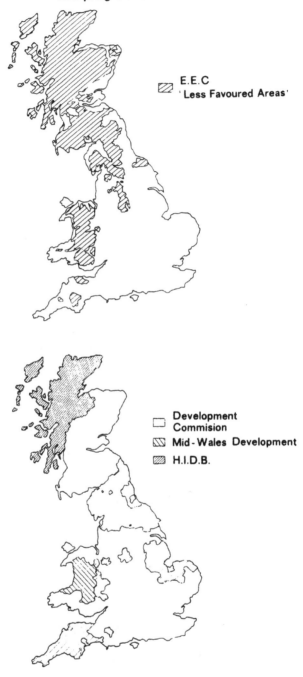

E.E.C
'Less Favoured Areas'

Development
Commision

Mid - Wales Development

H.I.D.B.

coverage (Figure 5.9). These institutions and programmes have not simply sought improvements in local economies through the stimulation and encouragement of private investment, but have also favoured diversification of employment opportunities and the provision of infrastructure and better services. In some instances these goals have even been shared with a desire to increase the political and social self-determination of peripheral rural populations (Nicholls 1976). To be fair, however, it should be noted that Williams (1985) saw an ulterior motive behind the formation of the Highlands and Islands Development Board (HIDB); namely, a central government attempt to diffuse pressures for political devolution to Scotland. That notwithstanding, the explicitly social goals of rural development agencies distinguish them from regional policy in general. Whereas the latter seeks social benefits through the 'trickle-down' of growth effects, the former address social needs on a more direct front. As such, these agencies have tended to work closely with local authorities within their target region (Williams 1985). Thus, the establishment by the Development Commission of Special Investment Areas and 'pockets of need' in the mid-1970s was partially provoked by a desire to integrate initiatives into local planning processes, so as to present a cohesive set of policies which link planning authority goals with those of the Commission (Hillier 1982).

Initially, though, with a concentration on factory unit provision, production and marketing advice, loans and management training, the Development Commission and its affiliated Council for Small Industries in Rural Areas (CoSIRA) differed from the more holistic developmental aims of region-specific rural development boards. Yet, the focus of both the Commission and CoSIRA have subtly evolved in recent years. Today, their roles include not only factory and workshop provision (which by 1984 had amounted to a total of 3,299,642 square feet of approved factory space – Minay 1985), but also support for voluntary bodies, such as Women's Institutes and rural community councils (Williams 1985:259). In its growing social policy thrust, the Commission deviates most significantly from its functions of 30 years ago and most closely resembles agencies promoting an integrated rural development policy. The coordination of CoSIRA policy with the emerging rural community councils has provided the Commission with a number of entry points into such rehabilitative policies as low cost rural housing (Williams 1984), infrastructure provision and access to health care (Development Commission 1985). Although these social policy interventions are too recent to allow a convincing assessment of the Commission's achievements, it does not bode well to have to report that the effects of the long-standing job creation programme have only been modest (Minay 1985:61).

Furthermore, clear evidence has yet to be produced to show that factory unit provision has helped halt rural service decline (Joint Unit for Research on the Urban Environment 1983; Minay 1985). Yet one needs to be wary of expecting too much from the Commission (its target since 1975 has only been to create 1,500 jobs a year in its designated recipient areas). What can be said for the Development Commission is that, in spite of the political environment not being conducive to the intervention of government agencies, it has maintained its role and has not shied away from addressing the need it saw for a stronger social policy. Nevertheless, were the Development Commission to offer more than a temporary antidote for minor maladies, one finds it hard to imagine that it would not have suffered the same fate as regional policy. Again, therefore, we return to emphasising that such agencies mainly offer conscience-absolving palliatives for the problems of peripheral rural localities, rather than serious attempts to alter their socio-economic standing.

We do not wish to make this point with any sense of ethical or moral criticism. Far from it, for this observation is not based on any sense of 'outrage' but emanates from a realistic assessment of the power structures of advanced capitalist economies. Any agencies which directly set out to 'redirect' private sector tendencies inevitably exist (in some sense) on sufferance and so need to 'purchase' supporters. Hence, it came as no surprise to us to read Williams's (1985) suggestion that the HIDB was created partly to allay devolutionist pressures. Similarly, for the Tennessee Valley Authority (TVA) to win support from local elites, the Authority essentially traded control over its large agricultural programme in return for political support for the organisation's energy goals (Selznick 1949). Once garnered, however, this support must be maintained. Thus, the HIDB, ever conscious of the opposition of large landowners to its formation, has never used its powers of compulsory purchase (Shucksmith and Lloyd 1982). Despite the fact that a realistic push for long-term Highland development will necessitate improved land tenure practices over much of its territory, the HIDB has made no convincing attempts to stop landowners deliberately keeping land undeveloped in order to provide themselves with hunting grounds (see McEwen 1977; Hunter 1979). Conceivably the Board has acted judiciously in this regard. After all, the weaker rural development boards for Wales and the northern Pennines were, in the first case, killed at birth, and, in the second, dissolved after only 16 months, largely due to the (supposed) threats those institutions posed for local landowners (to see how trivial such 'threats' were see Childs and Minay 1977). Consistent with a policy of not offending major economic power interests, the HIDB has promoted a growth centre strategy which has largely provided cheap infrastructure for some highly speculative, large-scale industrial concerns (Davies 1978). As Carter (1974:356) contended:

> It is clear that economic criteria have primacy over non-economic, and when a conflict occurs - as between the need to maintain population levels in 'remote' areas and the pull of migration implicit in a growth centre policy - it is the economic policy that wins.

The price of this expensive industrial-centralist policy - as seen in the closure of the Invergordon aluminium smelter and the Fort William papermill (Shucksmith and Lloyd 1982) - has ultimately been failure. Even had these prestige projects succeeded, the Board really has insufficient funds to make a major impact on the region's future (in 1982 for example it only offered £16.7 million in financial assistance). As in Canada (Brewis 1969), France (Ironside 1969) and the United States (US Department of Commerce Economic Development Administration 1972), these regional programmes are only band aids. Innumerable schemes exist for promoting development in nonmetropolitan areas, but their lack of success, and the continued dependence of these regions on non-governmental impulses for major socio-economic changes (like finding North Sea oil) bears witness to the general ineffectiveness of geographically-based government policies to aid poorer regions.

To help such areas in a significant way would work against existing trends in capital accumulation. This would involve either huge investments of public funds (requiring extra taxes) or alterations in national legislation affecting private investment. Whichever the case, the end result would 'offend' dominant business sectors. The question, then, is for what advantage? By their very nature, these areas contain few voters. Electorally, there is little to be gained from offending the business lobby, which could damage other dimensions of government policy.* Electoral considerations are critical and do have significant effects on government policy. For one thing, politicians prefer easily identifiable symbols of their potency to show voters that their election (or rather re-election) can bring visible benefits. As such, a dominant characteristic of government programmes is that they are not people-oriented, but are physical artifact oriented. Nebulous conceptions of development which link it

---

* In addition to which, governments have many (more viable) problems to confront. Peripherality in a geographical sense reduces the chances of problems being seen as central.

to social circumstances are ignored in favour of development being interpreted as number of new jobs (of any kind), miles of paved roads and number of electricity grid hook-ups (Copp 1972). In this regard, politicians are actively supported by bureaucratic pressures. When it comes to promotion, a bureaucrat who controls a highway construction project has more visible signs of achievement than one who helped generate a sense of self-awareness and fulfilment amongst a local community's residents. Symbolic of the effect of this joint preference pattern are governmental attempts to resettle inefficient communities, no matter what their social vitality (Matthews 1976), and the strong emphasis placed on physical infrastructure in development efforts (for example, in its first six years, the Appalachian Regional Commission spent $2.1 thousand million on highways but only $121 million on health; Claval 1983, see also Copp 1972).

Yet there are important differences in the objectives of politicians and bureaucrats. Of major importance here is the effect of politicians' desire for re-election. This certainly makes them vulnerable to 'outside' influence. A clear example of this was President Nixon's granting of substantial milk price supports following the Associated Milk Producers' 1971 pledge of two million dollars to his presidential re-election campaign. With much lower electioneering costs and more stringent controls on candidates' expenses, British elections are less prone to such overt attempts to 'buy' support. This does not mean this practice is not present. The suggestion that Britain's present Conservative Government has 'paid off' large contributors to party funds by 'awarding' knighthoods indicates that the practice is at least believed to be in operation (see 'Honours lists linked to Conservative Party cash', The Guardian, 6 December 1983:5). Yet with electoral costs as low as they are, this probably is a minor component of the British political scene. More significant is government 'purchasing' of the voters' choice.

The practice we are referring to here is generally known as 'pork-barrelling'. This is the deliberate orientation of government programmes to bring benefits to specific geographical areas; viz. to localities in which the ruling party or individual politicians need to garner votes. For Britain, Johnston (1983) lists three examples of this in action: Labour's promise to construct the long-planned Humber Bridge at the time of the 1966 Hull North parliamentary by-election; the Labour Government's 1975 decision to invest money in the ailing Chrysler Corporation to stave-off Scottish Nationalist Party electoral gains; and, less clearly, a tendency for marginal constituencies not to lose out when Labour introduced a large programme of teacher training college closures in the 1970s. There is abundant evidence from other nations of this practice in action. In Canada, for example, winter unemployment programme payments have been shown to favour federal cabinet ministers in marginal constituencies (and any Liberal in Quebec; Blake 1976), just as highway investments in British Columbia were biased towards provincial cabinet members' ridings (Munro 1975). Similarly, Lewis (1971) has reported that the locations selected for government land, water and abattoir developments in rural New South Wales appeared to owe more to the location of votes than to logical strategic considerations. In comparison with these somewhat specific examples, the lack of a distinctive ideological premiss for the main political parties, the high cost of electioneering and the expectation that politicians will 'perform' for their electoral district, all help promote the view that pork-barrelling is most highly advanced in the United States. Specific studies of federal policy-making certainly suggest that this is the case (the most notable of which is provided by Ferejohn 1974). However, in the most comprehensive analysis of this phenomenon (admittedly only at the state level), Johnston (1980) concluded that pork-barrelling effects were not substantial enough to affect the dominant geographical patterns in federal expenditure distributions. This does not mean that the geographical allocation of national government expenditures is not closely aligned with considerations of political advantage. Recent decades have seen quite distinctive expenditure preferences for the Democrat and Republican parties in the United States and these have clearly been tied to a dissimilar geographical spread in federal outlays (Mollenkopf 1983). Similarly, the election of a Conservative Government in 1979 led to an immediate and pervasive shift in fiscal impacts in favour of areas dominated by Conservative Party supporters (e.g. Johnston 1979). As well as maintaining prosperity within the capitalist class (which induces a distinctive spatial patterning to taxation and expenditure policies), governing elites pursue policies which promote their own electoral well-being. These carry within them other, possibly contradictory, tendencies in the geographical allocation of revenue collection and programme spending. Where rural localities stand in a disadvantaged position in this regard is in their small overall contribution to political party electoral success.*

PERSPECTIVE

In this chapter we have utilised O'Connor's (1973) empirically-based classification of capitalist economies into monopoly, competitive and state sectors as a heuristic device to illustrate the different characters of rural development processes. In theoretical terms, there is little available to the rural analyst which helps explain national developmental patterns. The most coherent and illuminating theoretical insights come from uneven development models. Yet these are dominated by considerations of the forces of monopoly capital. On the surface this is no particular handicap. As we have sought to show, the competitive sector is significantly squeezed by monopoly capital. For rural areas, this sector offers little hope for substantial improvement in living conditions (relative to the potential that exists in the other main sectors at least). As for the state sector, its long-term impacts are intricately intertwined with those of the capitalist class. Ultimately, this means that state sector actions are also linked to development paths favoured by monopoly capital. However, to see that the general trend in development pursues constrained targets, does not mean that the specific form that these developments take will not be open to considerable variation (both across and within nations). There is no doubt that flexibility exists in the development patterns of both competitive and state sectors. While monopoly capital exerts substantial influence, institutions in these other sectors have autonomy of action over development paths.

In terms of the broad (regionally-articulated) pattern of rural development within nations, this means that general patterns are predominantly directed by monopoly capital, with locally significant deviations arising from decisions in the other sectors. Indicative of this monopoly capital dominance, it is pertinent to note that the commonplace division of rural

---

* At times the 'favouritism' afforded to urban locations is not deliberate. In terms of a return-on-investment formula, for example, higher yields can be obtained from programmes catering for spatially concentrated populations than for scattered target groups. Hence, with limited resources available, it can be more effective to target urban populations for treatment first. In addition, size per se has an effect on the ability of areas to make use of federal programmes. As Huddleston and Palley (1981) have reported, many smaller local governments do not have the resources to hire the specialist staff required to apply for and run 'voluntary' federal grant programmes. However, it is also the case that the conditions associated with many such programmes specifically exclude rural authorities. Moreover, were rural locales more electorally significant, legislators would ensure that these programmes would be made suitable for application in rural areas (significantly, in the 1930s it was state governors in the US Midwest who supported farmers' calls for reform of farm legislation).

areas into metropolitan and peripheral regions for a long time effectively distinguished localities which were proximate to the establishments of monopoly capital organisations from those that were at some distance from the same. Today, the movement of monopoly capital ventures into peripheral locales has been substantial (albeit only at the lowest levels of the organisational hierarchy). This has helped induce a spatial redistribution of socio-economic conditions in favour of rural sectors. More importantly, it shows two things. First, it reveals that the geographical patterning of socio-economic surfaces is not fixed. It is not possible to predict such geographical surfaces far into the future, since technological advances, capital-labour relations and inter-capitalist rivalry can all induce sudden changes in locational preferences. Second, it indicates how improvements in socio-economic standing are significantly tied to changes in the monopoly sector. The closer an economic activity approaches a 'pure' competitive sector state, the less likely it can be relied upon to promote significant advances in living conditions.

This is not to say that there are not important dimensions to competitive and state sector activities with geographically-based socio-economic repercussions. Within agriculture, for example, it is significant that West European cereal farmers have been able to secure higher returns from their production than sheep raisers. Undoubtedly, there has been considerable political input into this outcome, but this is to be expected in a competitive arena in which one section carves out advantages for itself. The role of this political factor is nonetheless significant, for it introduces elements into the decision framework which are not directly tied to questions of economic competitiveness. This is directly seen in the concerns of state agents to reward their supporters or to promote electoral support through bestowing 'gifts'. However, it occurs on a much broader front. Pensions for example have notable impacts on rural living conditions. This is not so much because rural residents extend their income earning beyond their work years, but also on account of migrant pensioners retiring into rural areas. Government policies with seemingly aspatial intentions frequently result in spatially discriminatory effects (as with defence spending and taxes on gasoline). With the relatively minor political clout of rural lobbies, specifically rural impacts are often largely unintentional rather than deliberately sought. The end results are dynamic and complex. National effects on rural locations are neither unidirectional nor are they unchanging. These impacts are seen in broad regional impositions and, as the next chapter will show, are also discriminatory in their effects on localities within single regions.

Chapter Six

LOCAL COMMUNITY DEVELOPMENT

At both the international and national levels a common process in advanced capitalist societies has been centralisation. A few giant corporations increasingly control larger shares of economic activity. National governments have become more integral components of everyday lives. From Moose Jaw to Acapulco, Alice Springs to Reykjavik, Godalming to Majorca, cultural norms are coalescing as the spread of Hollywood, Coca-Cola and McDonalds pushes us towards an increasingly uniform social experience. This basic centralisation process – whose hypothesised end-point is termed 'mass society' – has affected rural localities in much the same manner as urban places. For many commentators, however, it is in the rural realm that this process has progressed least and has met its fiercest resistance. That rural localities have (in varying measure) succumbed to the forces of centralisation has been the cause of much sentimentality and reproof. Centralisation has too often been seen to result in the decay, decline and disintegration of rural social life. Wylie (1974:371) provides an apt commentary which illustrates the underlying tone of sadness which permeates much academic writing on the issue:

> Peyrane has ceased to be a tight-knit little community in which such a group [the noon apertif] plays an essential role. The Peyranais no longer feel themselves to be – in fact no longer are – a unit functioning as autonomously as possible in defence against the Outside World; they have become an integral part of the world they once staunchly resisted.

As Tilly (1974) has asserted, however, loss of autonomy in no sense means community disintegration. Social vitality can take on new forms and change might invigorate the social and economic life of a locality (e.g. Brandes 1976). Further, it should be abundantly evident that as social entities,

182

localities are not passive, inert objects that are 'acted upon' by broader societal forces. Localities adapt to extra-local impulses. Our purpose in this chapter is to assess both the character and the effectiveness of these adaptation processes. Local and national (or international) influences interact and conflict as the forces of centralisation and localism exert pressure in ways which sometimes correspond, are occasionally tangential, but more usually clash. In these interactions, a distinctive (inter-local) geographical patterning is produced in rural development trends. The subject matter of this chapter is how such inter-local differences emerge.

LOCAL MICROCOSM AND NATIONAL MACROCOSM

The growing economic, political and social integration of rural localities into mass society has occurred through direct and indirect channels. A straightforward case of direct influence is seen in the growth of cities which reach out to swell the physical fabric of nearby rural villages. Occasionally this leads them to coalesce into a recognisable urban form (e.g. Spindler 1973), but more commonly restructures and reorients village socio-economic life city-ward. In a subtle (or indirect) manner, improved telecommunications have enhanced the homogenising tendencies of popular culture, creating what Marshall McLuhan described as '... a planet reduced to village size' (1973:161). Such indirect effects are in many respects more significant than the more visually striking impact of direct urban encroachments, since their influence covers a broader geographical area. Consider the consequences of television as an illustration. The 'Plainville' resident who grumbled, 'We never go visitin', always stayin' home glued to the "idiot box"' (Gallagher 1961:27) provided a lay person's heartfelt commentary on a device analysts have blamed for 'destroying' previously dominant visiting patterns (Forsythe 1980), eroding local values (Healy 1968), and increasing dissatisfaction with village life (Christian 1972; Brody 1973). One must not become sentimental about such changes and idealise life before the 'idiot box'. As Brody (1973) has made abundantly clear, such innovations have allowed rural inhabitants more flexibility in their behaviour patterns. Without television, people were largely dependent upon neighbours for their entertainment. With television, they had choice. That they chose to weaken their communal activities in favour of television says something about their evaluation of dependence on neighbours.

Conceptually, the impact of television has induced a reorientation in social linkages from the horizontal to the vertical (Warren 1978). Intra-local social ties (horizontal linkages) have been replaced with ties to extra-local persons and institutions (vertical linkages). For many social analysts, social change in rural areas has largely comprised replacement of the horizontal with the vertical. In fact, however, it remains unclear whether an increase in vertical linkages is both a necessary and a sufficient cause for a corresponding decline in horizontal linkages (Richards 1984). For this to be so, the sum total of social life in rural settlements would have to be fixed. We would be faced with a zero-sum game, whereby gains in vertical linkages have to be matched by the same losses in horizontal ties. This is certainly not an assumption that is universally substantiated. To illustrate, in numerous studies, researchers have reported that the introduction of 'outsiders' into rural locales has invigorated social life and had very positive (horizontal linkage) benefits, in spite of the associated strengthening of vertical linkages (e.g. Morin 1970; Moore 1976; Hill 1980). Of course, such 'outsider' interventions can also be found to have serious detrimental consequences. Perhaps most representative of this is the 'Gillette syndrome', which has been found in many (energy and construction related) 'boom town' rural settlements in the western USA, and is characterised by the four d's of drunkeness, depression, delinquency and divorce (Freudenberg 1982; England and Albrecht 1984). The consequences of increasing vertical linkages can be positive or negative. No one has so far presented a theoretical framework which accounts for the circumstances under which positive or negative outcomes result, but a firmly grasped awareness of the potential for both is essential if such theoretical insights are to be forthcoming. In truth, however, many of the participants in the theatre of rural development do act as though such changes in socio-economic linkages were unidirectional in effect. For national legislators and urban dwellers, the prevailing orientation is toward a favourable interpretation of enhanced vertical linkages. For long-time rural inhabitants, a decidedly negative slant exists in the same evaluations. Not unexpectedly, the real situation is more complex than this dichotomy suggests. At times, rural inhabitants have sought to foster stronger vertical ties (as in attempts to attract employment to stabilise local populations), while urban in-migrants in the countryside are often the most vociferous opponents of a further strengthening in vertical linkages. This double dichotomy of favourable and opposed sentiments towards enhanced vertical ties, expressed through local residents with a 'local' and a 'cosmopolitan'* orientation, provides the organisational framework for this chapter. To begin, attention is devoted to the resistance of 'locals' to change. Most visibly, this is evinced in traditional rural communities in peripheral regions. Following this, the chapter is composed of sections on local support for improved vertical linkages, outsider impositions of such linkages, and, finally, 'outsider' opposition to enhanced vertical ties.

'TRADITIONALLY RURAL'

The desire to resist change, seek stability and promote self-determination is a deep-seated sentiment in humans. It finds root as much in people's psychological predispositions towards certainty (cf. Maslow's hierarchy of needs) as it does to an adherence to tried and tested social formulations. Not that all people have similar reasons for not wanting change. Amongst members of 'traditional' rural settlements, for instance, a clear divide must be drawn between the local socio-economic elite and the mainstream population. Frequently, local inhabitants see these two groups as having common interests in opposing 'outside' interventions in local affairs. In many cases, even if overt conflict between these two factions is not present, the benefits from such a united stance are uneven. The extent to which conflict exists in either purpose or long-term benefit is significantly affected by the character of the local system of social stratification.

In the hill farming areas of mid-Wales, there has been consistent opposition amongst agriculturalists to attempts by the Development Board of Rural Wales to introduce industrial enterprises (Broady 1980). A similar situation exists in East Anglia (Rose et al 1976). There are significant differences between these two areas. In East Anglia, there is a rigid social stratification system, with a clear separation of farmers and agricultural labourers (Newby et al 1978). For large-scale farm enterprises, the introduction of alternative employment opportunities threatens to lure away their poorly paid workers. Resistance to industry emerges from self-interest amongst the local socio-economic elite. This is not the case in mid-Wales. Here farmers operate with small numbers of agricultural labourers. Opposition to industrial

---

* The concept 'cosmopolitan' is used here to refer to those whose social horizons extend beyond their locality. This concept is a significant one insofar as it identifies a clear basis for division within localities (in behavioural and attitudinal terms) between those whose social existence is locally concentrated and those whose social concerns are geographically broader (see Merton 1957; Martindale and Hanson 1969).

endeavours is extensive (reports are available of headmasters and teachers in local schools projecting negative attitudes about manufacturing employment; e.g. Wenger 1980). Further, most especially on account of fears over the diminution of Welsh culture and language (consequent upon the expected inflow of English migrants accompanying industrialisation), there is a widespread desire to restrain socio-economic change. In a similar manner, in Isabella County (Michigan), the need for more jobs to alleviate economic depression was felt to be less significant than the preservation of traditional ways of life, which the new jobs might threaten (Browne 1982; see also Wallman 1977). The significance of these examples lies in the fact that area residents were quite prepared to forego material improvements in living conditions in order to preserve a particular lifestyle. That these desires to preserve an existing social arrangement have been based on the widespread support of area residents is evinced in proposals which have been defeated in referenda (see Vincent 1978 for an analysis of this phenomenon in St. Maurice, Italy).

Popular resistance to change on the grounds of preserving traditional lifestyles is seemingly idealistic, but carries (often more prominently) a pragmatic rationale. Thus, the concept of a traditional, rural way of life can mask a deep sensitivity for economic security. The survival of the European peasantry, for one, can be attributed to a tenacious and unyielding hold on the private ownership of small holdings (Newby 1978:10). In the face of governmental pressures for plot consolidation, these small-holders have frequently proved stalwart opponents of enhancing agricultural 'efficiency' (Franklin 1969). As Gasson (1969) has reported for small-scale East Anglia farm operators, the reasons for small farmers not leaving agriculture are as much connected with a preference for farming as a way of life as they are to an inability to find alternative work or a belief that financial gain would not result. As many farmers have found, economic conditions can induce changes that go against basic ideals. Guither (1963), for example, recorded that most of his sample of Illinois ex-farmers would not have left agriculture had their farm circumstances been more satisfactory. What is meant by 'satisfactory' in such a context does vary with people's socio-cultural background. Worster (1979), for one, has reported that highly cohesive neighbourhoods provide a social environment conducive to immunity from the disintegrating effects of severe economic depression (additionally noting that Mennonite farmers required much smaller farms, and lower economic returns, for a 'satisfactory' lifestyle than non-Mennonites). Along similar lines, it is instructive to consider the conclusions of van den Ban's (1969) comparison of rates of agricultural innovation across religious groups in the United States. For Calvinistic Dutch farmers, farming was seen as being God's steward of the land. This meant that changes in farming practice and risk-taking were discouraged. Even insurance was eschewed as it suggested a lack of faith. By contrast, farmers of a German Lutheran persuasion adhered to religious norms which emphasised personal progress. They followed this philosophy through in their high rates of innovation adoption. While such innovations are seemingly induced by individual decisions, they do have a locality-wide effect. Religion in general and church attendance in particular are most commonly significant and highly valued components of living in rural localities (Barker et al 1967). They help induce a locality-wide response to 'external' stimuli. A 'friends-and-neighbours' effect will also exist in innovation diffusion, thereby enhancing the locality specificity of reactions to socio-economic change.

Where local social systems are highly stratified,* the chances of implicit (and explicit) conflict intensifies. Thus, behind the guise of rural conservation, local employers can disguise the real aims of their opposition to job creation. New industrial jobs, for instance, can bring various 'unwanted' changes to local labour markets. First, they can intensify (or perhaps even originate) pressures for higher wage payments. Second, as the experience of the US South readily demonstrates (Cobb 1982), one component of opposition to new work opportunities is the spectre of unionisation. In barely unionised localities, trade unions represent the potential threat of more coherent working class organisation, increased employee militancy and threats to local capitalist dominance (e.g. Encel 1970). Thirdly, where new job openings challenge the overriding dominance of one economic sector, it is more likely they will also influence local power structures. Quite simply, if the local political elite has a weaker grip on local employment, its ability to resist challenges from below is diluted. At the same time, the willingness of non-elite residents to provoke challenges is intensified (two examples readily illustrate this: first, Salamon and van Evera's 1973 investigation of black voter registration in Mississippi showed that job dependence on

---

* Social stratification implies a hierarchical ordering of people, mainly along the dimensions class, power and status. We do not wish to suggest that significant social divisions in localities are solely of this kind. Golde (1975), for example, recorded that in addition to the Protestant-Catholic separation he set out to investigate, periodic revivals of Nazi - anti Nazi divisions also occurred in his West German village site.

white employers does reduce black electoral participation; and, second, the clear positive association that exists between external ownership of local businesses and the openness of US local governments to the influence of the general populace - Hoggart 1984). In combination these different reasons commonly prompt local elites to claim that new jobs are not needed, even if other representatives strenuously deny this charge. Thus, as the Suffolk County Structure Plan was in preparation, two starkly contrasting statements about employment were made by local organisations. From a Babergh District Councillor the view was:

> There is little real pressure on unemployment or rural problems here, so we are able to concentrate on other things. (Buchanan 1982:14)

Whereas a 1975 submission by the Association of Suffolk Trade Councils portrayed a quite different view:

> The overall fear of our movement is that it would appear that the final plan will commit us to a level of development that will continue to restrict job opportunities and therefore wage levels and standards of living for the vast majority. (quoted in Buchanan 1982:14)

It is of course not simply in the labour market that the interests of elites and non-elites diverge. Low rates of public housing construction in rural England have clearly had detrimental effects on workers' bargaining positions in labour markets, since they have increased workers' dependence upon employer provided (tied) accommodation (e.g. Rose et al 1976). That there is a need for more low-income housing is readily apparent to local political leaders. For one thing, the waiting lists for public housing places are huge compared with availability (Clark 1982a reporting on the English Lake District found waiting lists holding the equivalent of 56.2 per cent of the total publicly-owned housing stock in 1980; see also Penfold 1974; Phillips and Williams 1983). Local politicians also reveal their awareness in seeking to preserve the (admittedly meagre) stock they currently possess (thus, a total of 130 self-proclaimed rural districts applied for exemption to the stipulation of Britain's 1980 Housing Act that public housing be sold to tenants who wished to buy their home).

Both housing and manufacturing restrictions are commonly justified by local elites on the grounds of preserving a rural way of life or maintaining the sanctity of the country landscape. These avowedly conservationist goals have been interpreted quite differently by academic commentators. For

Buchanan (1982:17), they represent an attempt to legitimise 'conservative' and 'convenient' land-use planning. For Saunders et al (1978), the implications are related more to social stratification; the ideology of rural conservation masks the reinforcement of locality consciousness, the continuance of status (rather than class) discriminations and the maintenance of stability in local elite structures. Certainly, where major economic enterprises are production-oriented, the adoption of simplistic conservation aims (i.e. leaving things as they are) overwhelmingly benefits existing socio-economic elites (i.e. major landowners). This is almost inevitably so, since change promotes grounds for challenging their position of dominance.

Where local elites are dependent upon economic returns from consumer consumption, a quite different situation exists. Here, if stability exists in population size and income, the standing of local elites relative to those in other areas will decline. This follows because the returns from their economic activities will be (relatively) stable or will even be in decline, while others are increasing. The effects of this, when placed alongside rising cost structures, are readily apparent (thus, three-quarters of the retail functions found in nonmetropolitan US settlements in 1950 were found in fewer places by 1970; Johansen and Fuguitt 1984). Increased isolation from retail, recreational or other outlets does provide some cushion against competition. Yet this is an uncertain defence. As Borich et al (1985) recorded for Sioux County (Iowa), the unexpected construction of a new opportunity (in this case a shopping mall) can lead to marked changes in consumer behaviour, with severe repercussions for previously dominant enterprises. For these reasons, it is to be expected that stability of population will be less than desirable for consumption-dependent elites. Where this is true for areas of population stability, it is even more so where depopulation is the norm.

Depopulation is a familiar experience for many nonmetropolitan areas. Indeed a brief glimpse at textbooks written in the 1950s and 1960s reveals that it long occupied a prominent place in academic interpretations of rural life. In many instances, depopulation resulted from the direct 'push' effects of more dominant economic interests. Perhaps the major example of this is labourers' job-losses and outmigration consequent upon the increased mechanisation of farming (e.g. Saville 1957). Also significant, however, has been the desire of non-elites to better their own socio-economic standing. Economic causes were again prominent here – with many leaving for higher wages (or better job prospects) elsewhere (e.g. McIntosh 1969; Drudy 1978). Nonetheless, it would be too simplistic to hold that economic reasons dominated out-migration decisions. Very clear cases have been documented of migrants leaving their rural dwelling place for substantially lower incomes in urban areas (e.g. Greenwood 1976). There are a substantial number of studies which have reported that in-migrants (and return migrants) will often accept considerably lower wages in order to move into (or return to) rural settlements (e.g. Stevens 1980 found that 50 per cent of household heads took an income reduction to move into the Oregon localities he investigated; with Ploch 1978 reporting similar findings in his Maine study). More strictly social in causal effect, Douglass (1971; 1976) has reported that the low status of farm work helped push young villagers into leaving their locality. Similarly, Mitchell (1950) found that the social cohesion of a rural settlement had a significant impact on the probability of residents leaving. This observation mirrors the conclusions of many researchers that a settlement with social vitality can (psychologically) overcome many of the detrimental impacts of a poor economic standing (e.g. Matthews 1976).

Evidence is pretty clear that the economic return from rural-to-urban migration is usually financially beneficial for migrants and their families (see Tweeten and Brinkman 1976 for a review of these studies in a US context). What is more, village residents are aware of the economic benefits that can accrue from outmigration. As one example, Schwarzweller et al (1971) recorded that 72 per cent of their respondents in 'Beech Creek' (Kentucky) agreed that people who had left the area for urban centres were 'better off', especially in a financial sense. Frequently, the merits of this belief are reinforced through return migration. The economic returns from outmigration might well be high, but there are also social considerations to be taken into account. As Richling (1985) has reported for Newfoundland outports, to leave and not return is commonly seen as a sign of failure. 'Temporary' absences from rural locales for money gathering purposes are nonetheless an acceptable strategy in the quest for survival in an inhospitable local economic climate. Further, these return migrants commonly set an example for other local inhabitants. As Gmelch (1986:168-9) reported for western Ireland:

> ... the returnees probably encourage emigration more by example than by verbal persuasion. Their presence in the community - more prosperous and more 'worldly' than local people - presents an attractive role model for the young. In the eyes of many youths, the returnees are living proof that it is possible to have the best of both worlds: to live abroad for a while and have adventure while holding a better job and making

more than one could in Ireland, and then with savings in hand, returning home to the security and companionship, at least ideally, of family and old friends.

The kind of example that these return migrants set is variable. Took (1986), for example, found return migrants to the Italian province of Chieti exerting a conservative effect on their home areas. Their economic gains were invested in visible signs of their new found consumption status, but they made little use of new talents they had developed and offered no leadership (or even participation) to help raise standards in the locality as a whole. By lapsing into the old ways of living (at a higher living standard) they implicitly projected the message that changes were unnecessary. This tendency for return migrants to use few of their new found skills has been reported by analysts in a number of countries (e.g. Lijfering 1974; Douglass 1976; Gmelch 1986); this situation in some instances arises because returnees' skills are dwarfed by the overwhelming structural restraints of their area's undevelopment. In what seems to be a small minority of cases, however, return migrants prove to be key innovators who help invigorate local social systems and promote rural development (e.g. Lopreato 1967).

LOCAL COMMUNITY INITIATIVES

In any assessment of locality-based development initiatives, the evaluative framework used must be realistic. On their own, locality-specific forces are minor in scope and impact compared with more encompassing national and international forces. Opportunities for locally initiated development are constrained by the structural impositions of these broader forces. Miracles should not be expected from local inhabitants. Perhaps it is feasible for them to influence direction and speed of flow but they are unable to turn back the tide. Deviations from the main orientation of societal change are feasible, but carving out completely new channels is extremely unlikely (even in instances where social groups have seemingly 'opted out' of mainstream society - as with the Hutterites on the Canadian Prairies - community survival still depends in significant ways on the 'good will' of, and compromises reached with, agents representing the predominant way of life; see Bennett 1969; Flint 1975).

Even for residents of single localities, the fortunes of locally initiated development projects can have important bearings on rural living standards. Farcy (1976), for example, has reported on the income-enhancing effects of a cooperative initiative which provided French farmers with off-season work making 'rustic' objects for sale to tourists. Similarly, both Simon and Gagnon (1967) and Adams (1969) show how economic and demographic decline in mid-western US rural settlements were being reversed (or at least retarded) by the willingness of economic influentials to invest money and take a chance. As Channon (1971:19) reported for the Canadian Prairies:

> ... so frequently the difference between a moribund community and a viable one is that the latter has at least one interested person with ideas and initiative. These ideas have led to the encouragement of other ideas and activity that, in turn, have led to an improvement of the way of life for people in prairie rural communities.

The suggestion that successful local development initiatives tend to have cumulative effects – in that gains in one area elicit improvements in others – has been found in other geographical settings (e.g. Francis 1982). The questions which still have to be asked is how such initiatives emerge and under what circumstances they are successful. The answers to these questions are not easy to provide. Neither in the rural development literature nor in sociology in general have local community initiatives received much attention (Wilkinson 1978). Yet as Francis (1982) reported for Kent (England), there are clear spatial distinctions between localities which are active (or successful) in their initiatives and those that are inactive (or unsuccessful).

A critical distinction exists over the origins of local community development initiatives. In broad outline, two approaches are dominant (Wilkinson 1978). In the first, the initiative comes from the locality, but outsiders are often used as a resource. In the second, as was common under colonial British rule, it is outside agents who seek changes – commonly in only one or at least a restricted range of activities – and characteristically implant a paternalistic hand on the locality which seeks to push change in the direction these 'outsiders' favour. In some instances local residents act independently in promoting improvements. In reality, this occurs in a minority of cases. Dependence on 'outsiders' is important, if only to provide a market for the new opportunities which are being promoted. Commonly, insufficiency of local resources means that outside agents (whether governmental, corporate or otherwise) must be relied upon for financial (and perhaps skilled worker) expertise. If the initiative for change comes from within the locality, however, we are dealing with a quite different animal from that when outside agents provide the first pressures for change. For this reason, this section only considers locally initiated development pushes. Efforts originating beyond the locality are to be examined in the second half of the chapter, when attention is drawn to 'national' pressures on local development paths.

A further distinction that should be drawn in order to separate local development efforts with dissimilar end results is that between entrepreneurial (or elite) initiatives and mass cooperative movements. Analysts have been inclined to see more potential in a small group of entrepreneurs than they have in cooperative movements (e.g. Cloke and Laycock 1981). As we shall see, based on available evidence, there is some justice in this assessment. However, it needs emphasising that comparatively little work has been done on local community initiatives. Certainly, it is more difficult to mobilise the general populace than it is a small group of like-minded entrepreneurs. But, as opposition to such events as school closures show, the general public can be roused (Cleaver 1979). Of course, it is easier to motivate people in a concrete situation, where 'loss' is guaranteed if people do not act (or is at least very likely), than it is in more hypothetical contexts, where it is hoped that improvements will emerge from local cooperation. Yet such cooperative arrangements can prove successful both in a positive and a negative sense. In part this success depends upon the character of the local social system. Thus, MacDonald (1963) showed how rates of outmigration in Italian rural communities were highly dependent upon local social circumstances. In strongholds of working class militancy like Apulia and Centre provinces, social solidarity resulted in low rates of out-migration, whereas in Sicily, where strong repressive measures were taken against working class organisations, heavy rates of population loss occurred. That similarity of circumstance does not provide a sufficient uniting force was evinced by the experience of Alpine localities, where equality (but insufficiency) of farming resources left people with the sense that the root cause of their problems lay within the area (hence encouraging emigration). By contrast, the spectre of large landowners, identified as principal causes for poverty, gave residents in Apulia and Centre provinces a target for their anger and an image of how development could be attained. Dissimilarity of goals and underlying social tensions emanating from a history of local class conflict has also been shown to affect local initiatives in US rural settlements. Thus, in Simon and Gagnon's (1967) 'East Parrish', antagonism and distrust between local miners and entrepreneurs resulted in failure to agree on a development strategy and led to slight use being made of an already established Industrial Fund. Since analyses of development initiatives have paid scant attention to the importance of local social structures, however, it behoves us to use caution in interpreting these results.

There are a large number of case studies which focus on local elites and point to how their lack of initiative led either to economic and social decline or to a failure to attain a locality's 'true' potential (e.g. Lantz 1972; Clark 1974). This was not on account of any deliberate wish not to see socio-economic change (and particularly economic growth), but rather arose on account of an unwillingness to take risks and a failure to see potentials for advancement. When looking at successful local promotions, investigators have tended to emphasise three particular points. First, local elites must believe in the future of the locality sufficiently to commit themselves and be happy to build their children's futures around that place (e.g. Simon and Gagnon 1967). Second, elites must band together so that a corporate risk-taking mentality emerges (e.g. Adams 1969; Gold 1975; Lloyd and Wilkinson 1985). Third, and perhaps most important, these socio-economic elites must be outward looking; that is, they should possess a keen awareness of trends outside the locality and have significant contacts with extra-local institutions (e.g. Adams 1969; Bachtel and Molnar 1980; McGranahan 1984). In addition to these social considerations, the success of local initiatives are linked to the human ecological context of development promotions. Larger settlements, for instance, ceteris paribus have more resources to draw on, so their initiatives start with a better chance of success (Francis 1982; Krannich and Humphrey 1983). Size also has an indirect impact in that it is linked to the social heterogeneity of localities. When formalised into a variety of organisations, with dissimilar interests (what human ecologists have referred to as structural differentiation), this has positive implications for attempts to improve local environments (e.g. Luloff and Wilkinson 1979). To paint a complete picture, however, it would have to be emphasised that, even within a single country, disagreement does exist over the real effectiveness of such local initiatives. Once account is taken of advantages like proximity to cities,* initial employment structure, and particular resource advantages, evidence on the effectiveness of local development pushes varies. Some researchers seem convinced that such efforts have a positive impact (Johansen and Fuguitt 1984; Luloff and

---

* Though care must be taken here, for proximity to cities and position in the urban hierarchy are now seen as much less significant than they were thought to be for most of the 1960s and 1970s (this issue is treated in the next section of this chapter).

Chittenden 1984; McGranahan 1984). This is probably most explicitly proposed in Molotch's (1976) idea of a 'growth machine', which holds that locally-based capitalists must promote local economic growth in order to enhance their profits and maintain a competitive edge over entrepreneurs in other places. In a number of cases, however, Krannich and Humphrey's (1983) investigation of Pennsylvania localities being one example, analysts have concluded that 'outside' events and extra-local institutions are so dominant that growth machine processes affect too small a portion of community life to be significant.

Most especially in more peripheral rural locations which lack any especially advantageous attributes (like tourist appeal or an abundance of oil resources), there is a strong tendency to look for outside help (most especially governmental) in order to promote development. As Matthews (1976) described the situation for present-day outport Newfoundland, people simply lack the expertise, initiative and capital for promoting local development. They look for government assistance because they know of no other way. As we have stressed in Chapter Three, this feature of outport life owes much to a history of isolation and domination by local merchants, who discouraged initiative from the general population. Local social conditions have likewise restricted innovation in many Italian rural settlements. Here it is suspicion of those who put themselves forward as leaders that acts as a restraint. Put simply, the implicit assumption is that such people must be self-seeking, so they should not be wholly trusted (Wade 1971; Davis 1973). Such beliefs undoubtedly restrict opportunities for cooperative action and for the mobilisation of a broad spectrum of local residents for development purposes. Further restrictions on such behaviour arise from the 'free-rider' problem (Olson 1965). In a nutshell, why should you expend effort if there are others who will seek your desired solution without your aid? If the addition of one extra person's efforts makes little difference to the overall effort that is made, yet little change occurs in the likelihood of attaining the desired outcome, is it worthwhile participating? In Olson's (1965) argument, based on ideas of economic rationality, it is not. In rural locations this argument should be less significant, given that the contribution of each person is made more significant by the small size of the total (local) population. In addition, if romanticised images of rural areas being dominated by tight-knit social communities are correct, then communal responses should be easy to cultivate; a threat to one should be seen as a threat to all. In fact, however, reports on rural areas have revealed a strong and fairly widespread reluctance to engage in communal and cooperative ventures with a development focus. This is probably most evident in the behaviour of farmers; largely due to their common adherence to beliefs about the sanctity of farmers' independence (thus, Warner and Heffernan 1967 found that farmers tended not to participate in agricultural organisations if they could obtain benefits without attending meetings – attendance increased when benefits were confined to those who were present). However, this effect is in no sense restricted to this occupational group. Self-interest is integral, and in many cases dominant, in mass support for development efforts. Thus, when Napier and Maurer (1978) reported on a survey of 1,493 families in southwestern Ohio, they observed that community development efforts were favoured most when personal benefits were expected to accrue. Similarly, in a survey of 32 Wisconsin settlements, Buttel and Johnson (1977) found that 73 per cent of socio-economic elites favoured growth centre policies if their own settlement was chosen as a growth point (compared with 40 per cent for the general population), whereas the percentages were 68 and 60 if their locality was not. What should be noted about these figures is: first, that socio-economic elites were always more in favour of growth centre policies – a not surprising result given their control of profit-making institutions and the greater likelihood that they would directly benefit (Molotch 1976); and, second, the direction of the percentages changed according to whether home towns were selected as growth centres (signifying that socio-economic elites put personal benefits before more general local preferences).

It would be misleading to picture the weakness of local development initiatives predominantly in terms of the values of local people. Such movements (as with non-profit cooperative organisations in general) operate in a socio-economic environment which restricts their potential. Undoubtedly, there are cross-national differences in the potency of these restraints. The electoral strength of French farmers has led to at least one case of a government actively encouraging rural cooperative endeavours. In Britain, however, the responses of local governments to local development projects has run the gamut from hostility to indifference to overt encouragement (McNab 1984). President Nixon's 1971 veto of financial support and technical assistance for cooperatives for low-income farmers provides a curt reminder of the active opposition which exists to organisations whose goals run counter to those of dominant power interests. Yet it is precisely these groups that require the greatest local development commitments, for this is one sphere in which people have the potential to structure their lifestyles in a manner more to their choosing (albeit within a broader social system from which they are not immune). Yet, for poorer and more peripheral rural localities, further problems arise in organising promotional efforts. Poverty itself raises significant problems, since it instils a stronger urge for quick results and an impatience with slow progress (see Lipsky 1968 and Dearlove 1974 for a more general discussions of this phenomenon). As Wenger (1980) found in Wales, initial enthusiasm amongst community groups can quickly wane when the complexity of a problem is realised (see also Myers 1976). Energetic leadership is a help, but the poor resources of such localities often means that they cannot hire qualified personnel to assist them (a good example of this is found in farmers' cooperatives, where managers commonly have farming rather than administrative skills; e.g. Gale 1977). All this merely serves to emphasise that the development experiences of localities are subordinate to processes within the national political economy.

NATIONAL EFFECTS ON RURAL LOCALITIES

Rural geographers and agricultural economists have tended to stand at the forefront of theoretical and empirical expositions which de-emphasise the importance of locality specific societal forces. With their emphasis on community and local case studies, anthropologists and rural sociologists have long been accused of leaning the other way; viz. of neglecting the significance of extra-local social processes and structures (Bell and Newby 1971). Nowhere is this difference more notable than in rural development studies. Capturing the flavour of the sociologically-based view, Matthews (1976:48) charged:

> Most planners, particularly those who favour top-down, industrial development, are inclined to regard rural communities as archaic vestiges of a dying way of life. Even those planners who are orientated towards preserving rural communities are prone to condemn a village if it can no longer demonstrate that it is economically viable.

Characteristic of the pro-centralisation view is Brian Berry's (1969:19) argument that:

> If metropolitan development is sustained at high levels, rural-urban differences are progressively eliminated and the space-economy is integrated by outward flows of growth impulses through the urban hierarchy, and the inward migration of labour to cities in a reverse stepwise or ratchet fashion. Troughs of economic backwardness at the inter-metropolitan periphery are eroded, and each

Figure 6.1 Urban gradient hypothesis in four eras

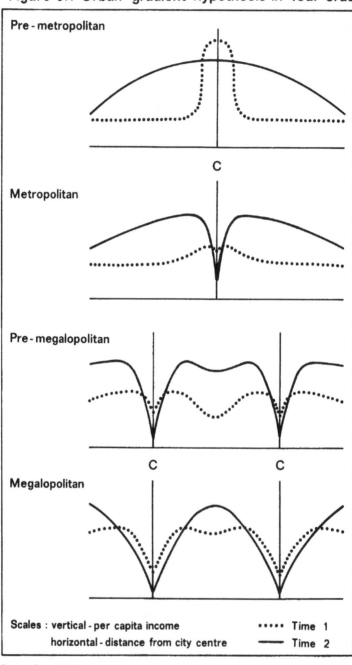

Pre - metropolitan

C

Metropolitan

Pre - megalopolitan

C   C

Megalopolitan

Scales : vertical - per capita income   ••••• Time 1
horizontal - distance from city centre   ——— Time 2

Source : Berry (1971)

then finds itself within the influence fields of
a variety of urban centres of a variety of sizes.
Concentric bands of agricultural organisation and
efficiency around metropolitan centres are
eliminated or reduced in importance and
agricultures also begin to specialise, taking
full advantage of differences in local resource
endowments.

Berry's conception of development paths emerges directly from
both a consensus view of society (Chapter Three) and a
diffusionist image of spatial patterns of development (Chapter
Five). As expressed above, this conception has often been
referred to as the urban gradient model (or hypothesis). This
emphasises that most benefits accrue if public and private
investment is concentrated in a few urban locations. More
peripheral places are effectively ignored because development
benefits will, it is hypothesised, ultimately 'trickle down'
to them (Figure 6.1). As such, this model has provided a
convenient justification for centralisation tendencies which
already existed in government and corporate intentions.

This predisposition toward centralisation is not
absolute. There are contradictions in the working of major
institutions which encourage dispersion at the same time as
centralisation. As O'Connor (1973) observed for capitalist
societies in general, operationally the state has a dual role
of encouraging wealth creation (i.e. capital accumulation) and
maintaining public faith in the existing social order (i.e.
preserving social harmony, which involves appeasing population
elements hurt by the state's primary pro-accumulation goal).
These distinct but inter-related processes help explain why
state institutions encourage population concentration (key
villages), service centralisation and growth promotion
policies (like housing land release) in rural areas, while
also promoting 'remedial' policies (like development boards
and special programmes for less favoured areas) and imposing
some restraints on growth (through environmental legislation
and land-use planning). Because of this dual regard for
accumulation and legitimation, governments require an
ideologically acceptable justification for promoting
centralisation. This can be readily found in the claim that
the concentration of economic activity enhances general
economic growth and so benefits society as a whole. However,
since one by-product of this process is a loss of services,
future job opportunities and perhaps the homes of those in
smaller places, the justification which is required must go
beyond mere wishful thinking or experimentation. It should
rest on a firm theoretical foundation. For spatial dimensions
of the centralisation process (as opposed, say, to the
growth of large organisations), important elements in this

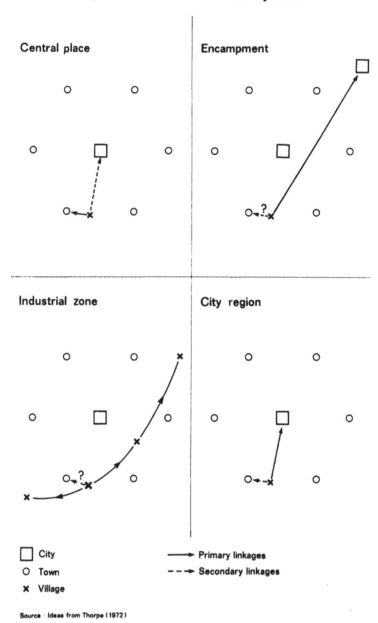

Figure 6.2 Rural settlement systems

Central place

Encampment

Industrial zone

City region

☐ City

○ Town

✕ Village

⟶ Primary linkages

--⟶ Secondary linkages

Source : Ideas from Thorpe (1972)

legitimising process are embodied within the urban gradient hypothesis.

The widespread acceptance of the urban gradient hypothesis in the 1960s and early 1970s was largely based on analysts' assumption that the prevailing pattern of economic growth at that time was unchangeable and that spatial patterns implied spatial causation. Classical theories like Von Thünen's and Alfred Weber's merely added icing to the cake. Not only could the urban gradient hypothesis statistically account for the spatial expressions of economic growth, but its theoretical heritage could be traced back to founding fathers of theoretical approaches to geographical inquiry. Unfortunately, mere statistical association does not make a sound theoretical base. Explanations for causal processes are required. In the urban gradient hypothesis, what explanations are presented seemingly assume that the structures of capitalist economies have barely changed since Von Thunen wrote The Isolated State in 1826. The emergence of large corporations is ignored, as causation is seen to depend upon competition between small-scale producers and consumers. With transport accessibility portrayed as a key dimension in locational behaviour, and metropolitan central cities visualised as points of maximum profitability (assuming small firms were competing to maximise profits), this model inevitably conceptualised settlement systems as oriented around metropolitan central business districts (with reduced transport costs, the growing size of cities and changes in manufacturing processes later favouring single-storey, land-extensive sites in suburban areas). Even ignoring the major impact that the organisational structuring of large-scale corporations has had on the geographical paths of growth-inducing impulses (Chapter Five), this model provides a very simplistic view of development processes. It clearly over-emphasises the importance of a single pattern of inter-settlement linkages (i.e. metropolitan dominance). As Thorpe (1972) has made clear, alternative linkage patterns exist (i.e. settlement systems) which have quite different implications for the geographical spread of development (Figure 6.2). Certainly, there are rural settlements in which growth is primarily induced by linkages with metropolitan centres (whether growth is in population or income terms). This is obvious from innumerable commuter village studies (for one review see Herington 1984). However, as the earlier studies in this tradition illustrated (e.g. Pahl 1965), while primary growth impulses are directed toward metropolitan centres (Figure 6.2), secondary linkages for other activity spheres (like social, recreational and retail activities) are more significant with more localised places. These local settlements might well comprise part of a Christaller-like central place hierarchy. Close to major metropolitan foci,

the effects of intermediate urban places might well be usurped
by the pull of the city. But the more distant a settlement is
from a metropolis, the more powerful the draw of intermediate
centres will be. The significance of this point is especially
evident when it is juxtapositioned alongside the advice of
urban gradient advocates that cities of 250,000 inhabitants
are needed for effective growth promotion (e.g. Berry 1970).
There are numerous studies which have shown that many smaller
places have had very positive growth consequences for their
hinterland areas (e.g. Moseley 1973; Todd 1983; Krakover
1984). This caveat might appear to be somewhat minor, given
that both central place and city-region conceptions of
settlement linkages suggest that proximity and movement up the
urban hierarchy are principal attributes of settlement
systems. This is not a criticism we would challenge. As
numerous investigations of the US Great Plains and Canadian
Prairies signify, central place theory has had as strong a
hold for nonmetropolitan regions as urban gradient ideas have
for metropolitan regions. The emphasis on 'friction of
distance' and 'perfect competition' nevertheless makes both of
these questionable models of inter-settlement linkage.

Thorpe's (1972) notion that certain localities are better
regarded as semi-isolated 'encampments' provides an
alternative to the above spatially deterministic models. By
encampments he is referring to mining settlements, military
camps, fishing outports and tourist centres. These
settlements owe their existence and growth to connections
within an institutional (or ownership) network, not through an
intermeshing of complex socio-economic linkages with proximate
places. Indeed, they might share few connections with nearby
settlements and be largely dependent on international
associations (e.g. see Moore 1976 on the impact of Swedish
tourists on Los Santos in the Canaries). Sharing similarly
'unique' linkage patterns, but this time coordinated into a
unified system, is Thorpe's (1972) 'industrial zone'. The
example he provided for this is that of the Pease family's
enterprises in the north east of England. Here, the
connection between settlements was the dominance of one
family, but each place had its own specific function, and the
fortunes of any one place depended on the enterprise as a
whole. Thus, the family base was Darlington, which provided
the transport focus for family activities, connecting its coal
mines (centred on Bishop Auckland), to its port developments
(in Middlesbrough) and thence to its sponsored holiday
developments (at Saltburn). Possibly such linkage systems are
not that common, yet large landed estates, and the joint
interests of Great Plains settlements with grain elevators in
their branch railways, reveal similarities of condition
(likewise Hudson 1985 has shown how the initial establishment
of North Dakota villages was predicated on railroad initiative

which tied settlement fortunes to branch line profitability). If these 'control related' linkage patterns are insignificant, then city-region and central place propositions should largely account for village growth and decline. Consequently, human ecological variables like distance from nearest metropolitan centre, access to places higher in the central place hierarchy, initial population size and settlement employment composition, should go a long way towards explaining variation in population, income, and employment growth in nonmetropolitan areas. Yet even in environments conducive to the interplay of free-market forces, these variables are now weakly related to local growth rates (albeit the further one goes back over this century, the stronger their effects appear to have been; Groop 1978; Hodge and Qadeer 1980; Taaffe et al 1980).

The spatial (or human ecological) attributes of rural settlements are then poorer predictors of growth and decline than they once were. Many of the reasons for this have already been examined in this book. These include counterurbanisation and the re-emergence of rural industrial-isation forces. To understand both of these, an appreciation of organisational structures is required. For counter-urbanisation the independence of (certain) higher socio-economic groups from organisational ties affords freedom of home location which has aided this process. Amongst lower socio-economic groups, it has been the emergence of large industrial corporations and their utilisation of locational policies to promote profit-making which have provided rural jobs (which both reduce outmigration and foster in-migration). That centralisation processes have continued, and have grown in importance, even as these dispersals have occurred, must also be recognised. As can be seen from the changing availability (private and public) of services in rural settlements, centralisation processes are integral components of the breakdown of central place and urban gradient principles of organisation.

Service Centralisation

Studies from both Britain and North America show that even where rural settlement growth is taking place, the number of retail outlets has declined as the threshold population needed to sustain individual units has increased (Shaw 1976; Johansen and Fuguitt 1984). Indicative of this pattern, a survey of Gloucestershire and Wiltshire discovered that 13 per cent of villages with shops in 1972 had lost them by 1977 (Standing Conference of Rural Community Councils 1978). Similarly, in the Département of Finistère, Ardargh (1982:399) found that over 1,000 small shops had closed between 1954 and 1967. The causes of rural retail decline cannot be divorced from general trends of retail transformation in the West. The dramatic growth of out-of-town and suburban hypermarkets in much of Europe and North America is part and parcel of the increasing average size of firms in capitalist economies. Deriving profits from low margins on huge turnovers (at least until a local monopoly position is achieved; (Cross 1976), the objective of such large-scale developments has been to enhance the market share of giant corporations at the expense of smaller outlets. In this they have been particularly success-ful. This can be seen in the spatial distribution of surviving rural retail outlets. Evidence clearly shows that proximity to large retail units has markedly detrimental effects on small rural shops (e.g. Borich et al 1985; and see Barbier 1972 for a more general view). As a result, the survival of independent rural outlets has shown a clear-cut geographical distribution. Proximity to major urban centres, while enhancing population growth and so market size, actually induces a decline in retail service provision (Johansen and Fuguitt 1979). Being too close to a large centre simply makes the services the city offers too attractive to rural and small-town residents. As a consequence, the viability of 'their own' outlets declines. In addition, the shop in a peripheral rural location obtains some benefits from isolation per se. Obviously this occurs insofar as the greater the distance to the next outlet, the less likely people will by-pass the first. Yet it also arises on account of government policies (in some nations at least), which recognise the social necessity of maintaining services in isolated areas and provide grants for this purpose (see Pacione 1984 for a brief review of the Scandinavian set-up).

The picture painted in the above paragraph might seem a beneficial developmental trend; close to larger settlements, better retail facilities are emerging (in terms, at least, of price, quality and quantity) and these are moving down the urban hierarchy quite rapidly, with small independent sellers surviving in more isolated places. In fact, however, these benefits are very unevenly spread over rural populations. To gain from new urban-oriented outlets requires easy transport mobility. With an increasingly elderly population, set beside a general trend of public transport decline, and higher poverty rates than urban centres, rural populations are commonly not well placed to take advantage of potential benefits. When Pahl (1968) answered his own question, 'Is the mobile society a myth?', with a resounding 'yes', he laid stress on a dimension of rural life that should not be forgotten. The spatial arrangement of rural settlements does enhance the importance of transport mobility, yet there are many for whom easy mobility is a severe problem. Better retail facilities are readily accessible for many rural inhabitants. But for others they are as inaccessible as ever. Further, owing to associated small outlet closures, the rise of large retail concerns has significantly lowered the quality of retail services available to those with transport handicaps. Their problems are not simply induced by large companies, however, for the character of retail outlets has also changed. This is especially seen in areas with tourist appeal, for what stores do continue to operate often change their function to cultivate an 'outside' market (Dawson's finding that craft shops increased by 32 per cent in north Wales between 1970 and 1974 is one numerical illustration of a trend that has seen sharp falls in general grocery stores with a causal, complementary or merely co-existing rise in hand-made pottery, antiques and wood-carving outlets; Dawson 1976). The rise in such specialised establishments is nonetheless spatially selective (being especially evident in tourist areas and in villages close to urban day-trip routes). As studies on both sides of the Atlantic have shown, across rural areas in general there have been sharp declines in the number of retail functions and outlets (e.g. Mackay and Laing 1982; Johansen and Fuguitt 1984).

When examining such centralisation tendencies, we should not expect to find a single causal process is dominant. In the retail sphere, the rise of large companies has been particularly instrumental. Certainly small establishments can gain some scale economies by joining buying groups. But these only go part of the way toward reducing costs. The small size of independent outlets still means that substantial travel distances must be covered for only small deliveries at each establishment (greatly increasing transport costs per sale). In general, stores in rural areas offer commodities for sale at significantly higher prices than urban supermarket competitors (Mackay and Laing 1982 found prices in rural Scotland 8-10 per cent above those in Scottish cities; also Woollett 1982). Lack of choice and older stock are further handicaps for establishments in small centres. As well as the aggressive intrusion of large companies into areas previously dominated by small family stores, changes in retailing have arisen from inadequacies in the independent shopkeeper body. In this, retailing in rural areas reveals many similarities with farming. Large institutions are increasingly dominant, with small family concerns lacking the market power to adapt operating principles to their own advantage.

When British commentators compare retailing or farming with schools, public transport or hospitals, they are prone to pinpoint one critical difference; namely, that the latter institutions are dominated by government control. This is perhaps an important difference, but it has been over-emphasised. For one thing, in the United States, hospitals and medicine in general are predominantly profit-making institutions, yet US service provision changes exhibit the same (or more accurately, very similar) trends as in Britain. Furthermore, in both nations it is difficult to distinguish services in the private and public realms by their change patterns. Just as retailing is moving toward large outlet dominance, so is education. In 1930, for example, there were 149,000 single teacher schools in the US but only 755 survived to 1980 (Hobbs 1982). In similar light, the mid-1950s saw the peak in rural public transport services in Britain. More recent decades have seen sharp declines in route mileage and service frequency (for both bus and rail). As with retailing, there are usually clear-cut expectations of service quality improvements associated with such changes. Thus, the British Government's Plowden Committee emphasised in its 1967 report that schools for those under 11 years should have at least three teachers in order to justify the services and range of teaching experiences suitable for a 'modern' school education. Similarly, while health services are becoming more capital and skill intensive, more based on high technology, and more focused on curative treatment – all of which favour large, urban-based institutions – this does have the potential of producing much higher quality care. As Carlsson et al (1981:120) observed for US doctors:

> Unfortunately, some of the physicians who choose to practice in rural communities are there because of personal inadequacies such as alcoholism, drug addiction, or other mental health problems.

Indeed, they argue that there is sufficient evidence to suggest that small rural settlements (in general) contain the least able practitioners. The dominance of single-doctor practices cannot help here. As with single-teacher schools, these increase pressure on professionals and help isolate them from advances in their occupation. It is this lack of a professional approach to service provision which has been particularly evident in rural areas. As Honey (1983:31) recorded for rural ambulance services:

> All too often in small town America, the most readily available vehicle was the local hearse. This potential conflict of interest seemed not to deter service.

The consequences of such an amateurish approach were often mortifying:

> ... a litany of horror stories has emerged

206

> relating instances of tragedies which could have
> been averted had personnel been a little better
> trained, a two-way radio available, less time
> expended before reaching a victim, etc.
> (Honey 1983:31)

Not surprisingly the US federal government has become more
actively involved in the process of setting standards and
promoting improvements in rural health services (viz., for
ambulances, the 1973 Emergency Medical Services Act).

However, the involvement of 'outside' agents and the
search for better services has commonly resulted in both a
faster urban drift in service provision and an increasingly
bureaucratic mechanism for provision. These, we believe, are
not inevitable consequences. Yet they are by far the most
likely given existing power relations in advanced capitalist
societies. In exactly the same way as new health regulations
forced dairy farmers to purchase new equipment, which could
only be justified financially if herds were large, so too we
commonly find changes in standards for public services exert
upward pressures on the size of the supplying unit.

Based on an impressionistic improved service-quality
argument, the centralisation of services appears to be
warranted. The real situation is not surprisingly more
confused. As Sher and Tompkins (1977) reported for schools,
the problem with arguments favouring school consolidation is
that they lack a sound theoretical grounding. Although our
common sense impression might be that larger schools have
better resources and should improve student educational
attainment, in fact:

> ... educational research has failed to identify a
> single resource or practice that is consistently
> effective in bolstering achievement. Moreover,
> the presumed linkage between school success and
> economic success in later life has been shown to
> be consistently weaker than common sense would
> suggest. (Sher and Tompkins 1977:61)

Those favouring consolidation (in education or health) appear
to approach their task with an almost missionary zeal. Local
community fears over the social consequences of losing their
own school due to centralisation are generally pushed to one
side.* To be fair to consolidation advocates, research has
shown that such fears are commonly over-charged. Any direct
social disruptions emanating from closure tend to be minor
(see Forsythe 1984; Tricker 1984). As for the oft-expressed
fears that the busing of children will weaken their academic
performance (or general socio-psychological well-being),

evidence on this score is mixed. There are those who have found that such problems do arise, more especially in earlier school grades (e.g. Lee 1957; Solstad 1971; Lu and Tweeten 1973), but others discount this (e.g. Dunlop et al 1957; Warner 1973).** Where uncertainty over harmful (or positive) effects is much less marked is in the clash of values which consolidation highlights. As Nash (1980) has pointed out, one of the most central and persistent tensions in rural education comes when demands for schooling with locally relevant knowledge conflict with demands for a nationally relevant teaching disposition. Consolidation marks yet another stage in the subjugation of locality-specific goals in favour of national priorities (the failure to promote rural studies as an academically acceptable school subject and resistance to farm schools provide two further illustrations of opposition to locally relevant education; Nash 1980).

Numerous arguments could wage over the benefits of local as opposed to national foundations for knowledge. We feel it behoves us to point out that there are dangers in locality-specific biases (most especially when local power structures are very elitist; see Chapter Seven). However, it has to be stated that the loss of a school (or hospital) can be an alienating experience for affected residents. Forsythe (1984), for example, found that school closure did generate a feeling of powerlessness and a sense that a locality had been 'attacked'. This feeling arises not simply in the actual process of closure but also in the continuing operations of an institution. This is because preferences for the mode, quantity and quality of service provisions do have locality relevance. This is seen in general terms in evidence that rural residents are less supportive of welfare programmes than other people (even after socio-economic attributes are taken into account; Camasso and Moore 1985). More specifically, it is evinced in residents of open countryside locales preferring

---

* Forsythe (1984) lists the main fears local residents report as: (1) the decline of existing population levels; (2) the loss of teachers who could provide leadership; (3) the closure of a village meeting place; (4) a weakening of social integration as the school brings people together; and (5) a decline in community identity and purpose.

**There is clearer evidence that increases in travel distances consequent upon centralisation do have negative effects in medicine. A variety of studies have shown that out-patient visits (especially for second visits) decline sharply with distance from hospital (e.g. Girt 1973; Haynes and Bentham 1979), and that the likelihood of patients receiving visitors is distance related (e.g. Cross and Turner 1974).

quite different packages of public services from residents of adjoining villages (e.g. Joseph et al 1982), in country voters rejecting better services in order to maintain their existing control over local government (e.g. Browne 1982) and in them purchasing and reopening 'their school' when public agencies close it (Woollett 1982). Consistent with this pattern, studies are increasingly reporting conflicts between recent urban in-migrants and long-time rural residents over the 'inadequacy' of local services (e.g. Hennigh 1978; Green 1982). In developmental terms, therefore, centralisation of services has negative consequences in that it marks a reduced ability to control an important local resource. Given the questionable character of benefits which school consolidations (at least) are purported to bring,* the question of what clear-cut gains do exist must be raised.

Generally, policy-makers justify both school and hospital closures on cost criteria. Some clear indications of higher costs for smaller units do exist. Molyneux (1975), for example, reported that Kesteven's one teacher schools cost 76.3 per cent more to run per pupil than the mean per student costs across the whole country. Of course, closing small schools (or hospitals) does not produce this level of saving. Consolidation requires new capital investments (in central sites), loss of previous social investment (in closed sites) and increased travel costs (to bus children to school or for patient and visitor transport). A large number of studies have examined the cost-savings relationship for school consolidation. Unfortunately,

> With rare exceptions, this body of research is methodologically unsound, with almost every study open to criticisms severe and significant enough to make the findings extremely suspect. The conclusions are, at best, inconclusive, and, at worst, simply incorrect. (Sher and Tompkins 1977:45)

Too commonly transport costs have not been taken into account in these analyses. Yet both the spatial distribution of

---

* As we have argued above, there are reasons for believing that centralisation does bring clearer benefits in the medical sphere. However, this is not the only means through which improvements can be made. The WAMI programme in the northwestern USA provides one alternative, based on improved training and personnel placement (Carlsson et al 1981). Furthermore, it is pertinent to point out that centralisation reinforces trends towards curative medicine at the expense of preventative practices.

students and the size (and spatial arrangement) of school
districts have obvious impacts on cost structures (White and
Tweeten 1973). Of course, where costs can be passed on to
consumers (e.g. many patients and visitors to hospitals
receive no travel assistance and the closure of 'under-used'
rural public telephones usually sees no compensations), then
the savings are more clear-cut. Even here, however, the gains
might be short-term only, since the larger size of consolidated units, and promises of improved services which
frequently accompany closing smaller outlets, exerts pressures
which favour purchasing expensive new pieces of equipment. In
general terms, this all suggests that cost advantages will be
minimal (and as Holland and Beritelle 1975 suggest, for
remoter rural areas one often suspects that they are in fact
negative).

Our conclusion, then, is that the consolidation of public
services constitutes a particular (spatial) manifestation of
the centralisation of power in advanced capitalist societies.
We are not inclined to favour Weaver's (1977) conclusion that
(school) consolidation is permeated with class conflict,
wherein upper classes seek to gain advantages over the working
classes. Our reasoning is that while a national (or aspatial)
cultural norm might well advantage capitalists, there is no
reason why this has to be manifest through consolidation. The
same goal can be attained in other ways. Further, the cost
advantages (i.e. tax cuts) which are likely to emanate from
consolidation are at very best only uncertain. With question
marks also raised over the issue of service quality
improvements (as in education), the benefits capitalists could
foresee which would make them favour consolidation per se are
unclear. For the service professions, however, there are
clear advantages from working in a larger organisation (more
and better paid promotion spots for one), especially when this
has superior facilities (leading to improved occupational
satisfaction; on these points see Dunleavy 1985). That these
professional goals meet with the approval of higher class
elements seems clear. After all, such organisations carry in-
built biases which discriminate against the working classes.
In their very operations, large bureaucracies intrinsically
work in favour of higher social classes (Sjoberg et al 1966).
Both elected and appointed officials in these institutions
also tend to be drawn disproportionately from higher-income
groups. At the same time, the professionalist goals which are
emphasised in these larger units provide a more accepting
market for the 'high tech' products of large corporations (as
these promote a high status image for a profession and
consequently aid income enhancement potential - see Heller
1979 on these last points in a health care context). Two
further aspects of discrimination relate to the resources
available to social groups. Put simply, those with higher

incomes are better placed to find alternatives to a consolidated unit and, if they choose to use it, are in a better position to do so. Thus, the wealthier and well connected are better able to refuse to send their children to newly centralised schools, because they can opt for private education, or they can even purchase the old school and reopen it (noticeably Woollett's 1982 list of parishes which have done this in England is biased towards the Southeast). Then, if they decide to use the new facility, the fact that it is farther from their home is less troublesome than it is for low income villagers, since they can afford their own transport, whereas others must rely on a decaying public transport system. As Parsons (1980:7) observed: 'The two-car household is now more a reflection of genuine need in the countryside, than of relative affluence'.

It is at this point that we can see a critical difference between government-provided and private sector services. What governments must always bear in mind is that the people these centralisation policies discriminate against are voters. While there is undoubtedly leeway – in that votes are not automatically lost for every 'negative' government act – there is still good reason for elected representatives to be cautious of objections to their policies. In addition to which, not all bureaucrats accept the goals their profession promotes, or some at least see mitigating local circumstances to warn against implementation. For these (and certainly other) reasons, a whole host of government policies exist which seek to tone down the harsher aspects of prevailing centralist trends. Like 'welfare' payments for small farmers, these are a minor ebb which floats over the surface of a dominant tidal movement (i.e. in agriculture, payments aiding large-scale farm units). Nevertheless, they do exist. In character they come in numerous forms. For education, this usually includes cooperative arrangements across a number of small schools (sharing specialist teachers and equipment, perhaps having one head teacher for three or four schools which now operate as one unit and providing trained advisers to assist teachers). These projects have had varying degrees of success (see Bowen 1973; Boulter and Crispin 1978; Woollett 1982). In this they are no different from the large number of 'experimental' public transport services (e.g. Carpenter 1973; Hertfordshire County Council 1975; Blunt 1976). Both share a common problem, in that they seek to provide a quality rural service within a socio-economic environment which favours large urban-based institutions. Furthermore, they usually require high levels of funding. Yet these demands for money are made in an organisational environment that is increasingly favouring the overthrow of social criteria for action in favour of economic efficiency. In Britain, for example, local governments are being squeezed by current national government

demands for reduced spending. With costs rising rapidly,
grant cuts and ever more restrictions on local tax-gathering,
local councils are increasingly pushed toward abandoning
'social' programmes. Where lower income groups comprise a
small proportion of the electorate, such cuts have often
already been made (the notoriously low subsidies given to
public transport services by many Conservative controlled
English local councils is indicative of this). Whether or not
the long-term viability of such programmes will become a
critical issue, however, will partly depend on other
governmental centralisation schemes. For a long time now,
governments in many different geographical settings have
recognised the discriminatory effects of centralisation
processes and have sought to alleviate these through settlement centralisation policies.

Settlement Centralisation

Centralisation trends are seen in two particular planning
strategies for settlement systems. The first of these is
found in areas of rapid (or at least consistent) population
growth. In Britain this is usually referred to as key settlement policy. Its aim is to channel growth into a small number
of selected places. The second is more common in economically
depressed areas. Here, through the mechanism of encouraging
expansion in a limited number of centres (i.e. growth centre
policy) it is hoped to enhance the economic vitality of all
places within a region. In some areas, these two kinds of
policy have been associated with schemes to 'close down'
smaller settlements. This seemingly represents a somewhat
different centralisation goal, but the aims of these three
policy instruments are in fact difficult to disentangle at
times. All of them are really components of a general attempt
to restructure a settlement system.

This can be seen in Britain, where the introduction of
key settlement policies in the 1950s was intended as the
'... planning panacea for the physical and economic ills of
the countryside' (Cloke 1983:168). Not only was this policy
to help meet the rising demand of the metropolitan middle
class for rural housing (and a resulting need for low-cost
housing for the indigenous rural working class), but it was
also to help halt depopulation in remoter areas (in this guise
it was really a growth centre policy, with a
characteristically British hope that growth would somehow
occur without being comprehensively, coherently and actively
promoted). Since that time key settlement policy has been at
the forefront of planning responses to rural development. In
idealistic terms:

> ... planners perceived a logical progression from
> the identification of existing rural centres to
> the continuing support of these centres as the
> basic focus for investment in rural areas. In
> effect, many planners were attempting to build up
> certain key settlements into the ideal central-
> village model whereby additional service
> provision in one central location would benefit a
> wide rural hinterland. (Cloke 1979:42-3)

In practice, key settlement policy has been justified on the
grounds that it is cost-effective. Too commonly, simplistic
ideas taken from central place theory (which models the flow
of expenditures up the settlement hierarchy) have been relied
on to justify the assumption that benefits from centralising
resources (and growth) will trickle down the settlement
hierarchy. In this regard key settlement ideas are no
different from their growth centre counterparts. Both in fact
suffer from a weak (central place inclined) theoretical base
(Moseley 1973b; Cloke 1979), so that the identification of
appropriate centres to designate for special status has lacked
coherence. Frequently, simplistic measures like population
size have been used, though there is little agreement over
what is an appropriate size. For growth centres, for example,
Berry (1970) has suggested that 250,000 inhabitants are
required for effective growth promotion. Yet as Morrill
(1973) convincingly informed us, size alone is hardly an
adequate guarantee of economic stability (e.g. Seattle or
Pittsburgh). Moreover, other researchers have shown that
places of 25,-50,000 can have significant development inducing
impacts (e.g. Moseley 1973) and, if the conditions are right,
even places as small as 1,-5,000 can have beneficial impacts
(e.g. Grafton 1980). Not unexpectedly, with such looseness in
theoretical back-up, the field was open for policy-makers to
use almost any criteria they favoured. It would be startling
if we were unable to report that politicians consequently did
not utilise key settlement and growth centre designations to
favour their own supporters or attract votes in marginal
constituencies (Ryan 1970; Cloke 1979). In addition, lack of
clarity over what properties a settlement should have if its
growth is to be promoted has led to some farcical
designations. Probably the most widely quoted example of this
is Devon County Council's key settlement scheme. Here, half
the designated key villages could take no further growth
anyway, since they had already reached full capacity in their
water or sewerage systems, with no plans for new investment to
increase capacity (Cloke 1983).

The fact that certain designated places were favoured by
key village or growth centre status nevertheless meant that
where you lived came to have a growing significance as a

determinant of service quality. Key settlements were more likely to see service improvements and were less inclined to experience service cuts. For those in non-designated centres, the promise was of benefits through association. This was most visibly seen in expectations about the trickle-down effects of growth centres, In fact, however, these trickle-down effects tended to be minimal. Quite apart from attracting less employment growth than was commonly expected (e.g. Cloke and Shaw 1983), growth centre institutions provided few spin-offs for other local endeavours. Instead, growth centre institutions appeared to purchase their inputs from larger places or within their own organisations (i.e. other branches). Instead of trickle-down, the process was one of trickle-up (e.g. Moseley 1973a). In study after study, the picture painted revealed that growth centre benefits were largely restricted to short-range commuter zones (e.g. Moseley 1973; Todd 1983). Further, within localities benefits were very uneven. Through their ownership of local retail and commercial outlets, those of higher income saw a disproportionate share of government-infused funds passing through their hands (e.g. Mellor and Ironside 1974). Meanwhile low-income residents living close to (but not in) the designated centre (whether growth or key) often received few benefits, since their poor transport mobility restricted their chances of garnering some of the extra funds flowing through their locality (Mellor and Ironside 1974; Grafton 1980). In fact, their overall position in this regard was severely weakened in the medium-term. The fact that growth was being restricted made non-designated settlements even more attractive living places. Through the resulting gentrification process (Parsons 1980), this has not simply forced up house prices but has also reduced the stock of village residents who required public transport (thereby threatening the mobility of lower-income groups).

Implicit within the above commentary is the assumption that some economic growth (or at least stability) existed. Accompanying the assumption that 'viable' settlements should be the focus of employment, population growth and service provision, has been the antithetical assumption that 'non-viable' villages in the lower tiers of a hypothetical settlement hierarchy should decline. Planned decline policies have either been adjunct to concentration policies or have comprised policies in their own right. The difference, as Cloke observed (1983:93) is '... the latter's presumption against local need development in small villages as compared to the former's more lenient attitude in this respect' (see also Gilder 1984). Planned decline policies are rooted in the notion of a settlement's economic viability. As such, they have been widely criticised (e.g. McLaughlin 1976; Matthews 1976; Gilder 1984). Cohen's (1977:6) observations provide a wide-ranging commentary:

> The small town itself is demoted to a purely
> economic entity in a regional hierarchy of
> communities allowed to function until reaching
> economic obsolescence whereupon it is abandoned.
> Blithe ignorance of the small town's social,
> political, historical and cultural significance
> allows some planners to call for the outright
> elimination rather than revitalisation of such
> 'inefficient' economic units.

In Britain, the most infamous, and well referenced, example of
planned decline was Durham's 'D' village policy (Barr 1969;
Blowers 1972). Of 357 rural settlements in the County of
Durham in 1954, 114 were listed in the 'D' category. This was
where no investment should occur, where unoccupied dwellings
should be pulled down and where services should be limited to
those necessary to maintain only the life of the remaining
property. Although this policy was terminated in 1964, with
only 8 of the 114 villages cleared, the 'D' village policy
attained a notoriety in the literature of settlement planning
(Cumberland's very similar policy of the same era caused much
less comment). Durham's policy was nevertheless quite tame
compared with the settlement restructuring efforts of other
governments. One of the most effective of these was
Newfoundland's resettlement programme. Existing in dissimilar
form (and with varying objectives) across the 1953-75 period,
this programme was formally a voluntary scheme for aiding
families to move from isolated outports to sites with better
physical infrastructure or job-generating potential. Once set
in motion, however, the programme developed an inner motion of
its own, as officials realised its potential for reducing the
number of outports which were clamouring for public services
(including basic facilities like roads or water and sewerage
systems). Through deliberately starving some places of
services, confusion over the need to evacuate (and the job
potential at reception centres) and generous grants, the
scheme was able to remove 261 settlements from the province's
total (Hoggart 1979a). The bad feeling provoked by the
programme ultimately proved its undoing. Increasingly people
resisted government overtures and pressures to resettle
(Matthews 1976); this movement constituted a key element in
the 1972 Progressive Conservative electoral victory over the
provincial Liberal administration which had ruled continuously
since 1949.

With rural resettlement schemes existing in areas as
different as Greenland and the Canadian Prairies, it is
pertinent to ask if some inner driving force pushes governments into centralisation promotion. On the one hand, this

question can be answered in the affirmative. Examining documents from both Durham and Newfoundland reveals that there was a strong element of altruism in these policies (as in key village and growth centre policies). Policy-makers were conscious of the poor services and slight job opportunities which residents in remoter villages were faced with. We are convinced that behind these policies there was a genuine desire to see improvements in living conditions for remoter rural dwellers (albeit we also shudder at the heavy-handed, paternalistic approach which was adopted to promote this aim). On the other hand, it must be recognised that capitalist economies are predisposed toward favouring centralisation policies. With increased demands for business and household services being made on governments, with changes in technology making such improvements ever more expensive, and with the innovatory urge within capitalist markets being predisposed towards serving large numbers (i.e. there is little profit in helping small groups of people), there is strong pressure towards centralisation. This is especially so given that it is governments (rather than capitalists) who are usually 'blamed' for such societal ills as unemployment, inadequate public services and high taxation. Some feel for the strength of this centralising imperative can be obtained from academic commentaries on the minimum population size for a 'viable' village. Thus, whereas in 1948 Dudley Stamp was recommending a figure of 500, by 1962 Michael Chisholm had raised this to 1,200, with the analysts in the early 1970s leaning more toward a 5,000 figure (Chisholm 1962; Green 1971).

Built into this centralising movement, there is the complication of internal resistances. It would warm the hearts of key village and growth centre advocates to see processes of change which emphasised a hierarchical order in settlement structures, but this has not been happening. Even in the 1960s, Hart et al (1968) were warning us that rural settlements (in the US Midwest) had become specialised service outlets. What they found was that settlements tended to have much higher per capita expenditures on single retail service functions than the state average, but usually had below the state average for others (they examined nine service functions). This indicated that specialisation was occurring. One village was achieving the status of a significant eating and drinking spot, its neighbour was a major automobile sales centre, while the next-but-one place provided above average grocery facilities. Rather than a hierarchical order, what was emerging was a dispersed (but integrated) settlement system. Since that time, researchers have identified the same pattern in a variety of locational settings (e.g. Weekley 1977; Dahms 1980; though see Johansen and Fuguitt 1984 who question this trend). What is more, this pattern of complex settlement inter-linkages can also be seen in production spheres like agriculture (Hoggart 1979) and in inhabitants' social behaviours (MacGregor 1972).* In a sense, all this represents is a local version of the breakdown in hierarchical settlement orders, which at the national level is a product of the growth of large corporations and their promotion of intra-organisational linkages (see Massey 1984; Markusen 1985). At the local level, the specialisation process emerges partly from entrepreneurial success (in a market that can only sustain one grocery outlet, the most entrepreneurial enterprise should survive - assuming comparative ease of travel for many consumers - and this operation will not inevitably be located in the largest settlement). In addition, specialisation emerges from conflicts over social change (which differ in intensity across settlements). Aspects of such resistance have already been considered in the 'Traditionally Rural' section of this chapter. Significantly, however, recent decades have seen new conflicts arising in rural locales. This occurs when an 'outside' (nationally-oriented) force makes its mark through urbanite in-migration.

'OUTSIDER' ANTI-GROWTH MOVEMENTS

As a sentiment and basis for policy-making, anti-growth has become increasingly popular in developed nations over recent decades. At heart, anti-growth sentiments hold that economic expansion and population growth, along with the land-use and landscape changes associated with them, are no longer pre-requisites for social betterment. If allowed to continue unchecked, economic expansionism is believed to lead to the despoilation of the environment. It will thereby threaten the enjoyment of those improved qualities of life that are the fruits of past economic progress. Although not restricted to rural localities, anti-growth lobbies have become particularly strong within the countryside. Here, their 'targets' are

---

* The reader might wish to ponder on this point, for it suggests that government policies favouring centralisation into key villages in fact neglect what is happening to service distributions through private sector actions (it might for instance be that the focusing of public transport links on key villages is inappropriate since many private services are not ending up in these places). As Owen's (1984) research has suggested, settlement restructuring into a hierarchical order might be appropriate if an area is being planned afresh (though Dutch planners found their 'ideal' system did not work on Zuider Zee polders), but when a settlement structure already exists, such concentration policies are not the most efficient settlement organisation.

creeping urbanisation, industrialisation, landscape change, the laying-waste of 'unique' habitats for flora and fauna and the destruction of symbols of local (or national) heritage. In their efforts to protect rural areas from 'unwanted' growth, anti-growth lobbies have gained considerable strength. This is not only seen in battles with individual pro-growth agents, but also arises in creating a context of restraint and conservation within statutory planning processes.

Concern for the environment is not new in either Europe or the United States. However, the antecedents of the current anti-growth movement are relatively recent. Whereas earlier preservationist feelings developed largely around the concerns of a few wealthy philanthropists, mass support is evident for recent anti-growth pressures. This broader appeal derives from basic shifts in popular attitudes away from concerns of economic stability, peace and security toward those of cultural identity and aesthetic indulgence (Inglehart 1977). These new concerns are founded upon both national and personal economic security. An underlying sense of 'well-being' has led to people questioning the precepts of growth arguments. Increasingly, calls are made for the slowing down or halting of economic expansionism (Schumacher 1973). To place this in the context of Maslow's (1970) hierarchy of needs (Chapter Two), it appears that many people are satisfied with the fulfilment of their more basic (physiological and safety) needs, which has 'freed' them to show greater concern for aesthetic needs associated with environmental quality. Of course, the incidence of such anti-growth sentiments does vary. It is particularly evident in richer areas and, significantly but not exclusively, amongst the middle and upper classes (e.g. Buller 1983). In addition, as Lowe and Goyder (1983) pointed out, there is an episodic strength in environmental (or anti-growth) concerns, associated with a stage in world business cycles towards the end of periods of sustained economic expansion (Lowe and Goyder 1983:25). Ultimately, this temporal relationship is paradoxical, for environmental concern (and a concomitant anti-growth feeling) is in a sense a product of economic progress. The transition to post-materialist values arises partly because the achievements of economic growth free people from an over-riding search to fulfil more basic social needs. Yet the environmental consequences of those achievements have directly stimulated anti-growth sentiments and pressures to slow down further growth. Resulting from this initial paradox is the central issue of benefit allocation. For those who have been less able to reap the benefits of previous economic growth, anti-growth policies threaten to widen the gap between those with post-materialist aspirations and those who continue to struggle for economic security. In rural areas, the issue of distributional inequity takes on a particular significance.

It creates conflict within localities and mars the appearance of social consensus that conservationists (and rural residents in general) have sought to promote.

The history of the anti-growth movement has been investigated by others from many different perspectives (Johnson 1974; Lowe and Goyder 1983). Analysts have noted that anti-growth forms but one part of a wider movement that embraces 'issues' like political decentralisation, the control of population growth, wildlife protection, building conservation and landscape preservation. The anti-growth lobby considered in this chapter is essentially a locality-based lobby of citizen groups seeking to protect the existing built environment and countryside from 'unwarranted' change. Although these local groups form part of a national movement, because they are primarily concerned with individual localities, the notion of a single anti-growth lobby is not sustainable. It is the 'commitment to locality' (Lowe and Goyder 1983:29) that is of prime interest.

Local anti-growth groups are a post-1945 phenomenon of growing significance both in Europe and the United States (Lowe and Goyder 1983). In Britain, popular opposition to urban sprawl, and concern for its economic and social effects, date back as far as the nineteenth century (Kennet 1972). The construction of 862,500 new houses in England during the inter-war years (Sheail 1981:25), and the associated growth of urban sprawl and ribbon developments, led to a second burst of anti-growth sentiments. In both these time periods, however, environmental concern in Britain was expressed in national endeavours. After 1945, this concern became more visible in the proliferation of local anti-growth lobbies (Barker 1976; Lowe 1977). The collective size of this movement is difficult to assess, since precise enumeration of both the number of local groups and their memberships has proved elusive. Some indication of its magnitude is given by Lowe et al's (1980) suggestion that a million British people belong to these local organisations. Many of these people will be members of well established groups. Thus, the Civic Trust, which acts as a federative body for established organisations, had 1,250 affiliated local groups in 1975. Yet the majority of local societies are independent. This is illustrated by the organisations which exist in West Sussex, where local anti-growth groups are so numerous that an estimated four per cent of the population are members (Buller 1983:108). Here, 116 individual groups were identified in 1981, of which only 52 were affiliates or members of national environmental bodies (Buller 1983:106). In West Sussex, as elsewhere, such organisations have become one of the '... strongest land use lobbies in many parts of the country' (Fujishin 1975:43).

In Britain, the strength of anti-growth lobbies in the countryside, if not even their very existence, has been consequent upon the evolving socio-political composition of rural areas and the changing circumstances of land-use decision-making. Economic prosperity has been fundamental for both of these. This is not simply an account of 'freeing' people from more basic needs. It also arises from improved transport mobility, enhanced choice in residential location and increased leisure time. Together these have made countryside habitats more attractive homes for those who have reaped the benefits of economic growth. This particularly applies for the (former urban-based) upper middle class. These beneficiaries now comprise a major element in both urban-rural in-migrant flows and local anti-growth movements. Their very 'newness' to country locales nevertheless makes them somewhat suspect 'ruralites'. Consequently, to legitimise their involvement in rural social life and, as important, to protect the status and character of their new-found rurality, these newcomers have been particularly active in promoting local anti-growth groups. They reveal their true 'rural' heritage by being more 'concerned' than most 'indigenous' residents about the 'threats' of expansion to the built environment, new (predominantly agricultural) land-uses and land utilisation practices which they believe are incompatible with the ideals of 'traditional' rural areas. Not surprisingly, local anti-growth lobbies tend to be dominated by people who have migrated into rural areas. However, strictly it is not accurate to refer to all these people as 'newcomers'. Many activists have now been rural residents for over 20 years (Buller and Lowe 1982; Little 1984). Nevertheless, few local lobbyists have either lived or worked within their locality all their lives. Farmers, and members of the rural working class, are poorly represented amongst anti-growth lobbies. Being comprised largely of an adventitious, professional and mobile middle class, the anti-growth lobby has no long-established role in rural areas.

The basic concern of local anti-growth lobbies is the protection of the residential and recreational environment through resistance to change. Thus, one West Sussex group summarised its aims as:

> To preserve and enhance the essential atmosphere of the parish and to improve its appearance and amenity and to stimulate public consciousness and appreciation of its beauty, history and chara-cter. (Buller 1983:113)

Such organisations differ from other components of the wider environmental movement since their aims revolve around representing and defending the special interests of residents of a single locality. However, local anti-growth lobbies have become proficient in relating their predominantly parochial concerns to wider environmental issues. As farmers have found to their cost, this apparent coincidence of private and public interests is advantageous. As Gregory (1976) suggests, it is hardly surprising that owners of large detached properties in potential greenbelt areas are among the most vociferous advocates of protectionist designations. Property owners cannot fail to be aware of the increases in property value that accrue once restraint designations are set (Shaw and Willis 1982). In arguing their case, however, these property owners can draw on the undoubted public benefits that arise from having areas of green space between urban centres. The recreational and aesthetic value of such open spaces can benefit all social groups (Countryside Review Committee 1977). It would be both uncharitable and incorrect to believe that a sense of altruism was not intrinsic to the aims of local anti-growth lobbies. But it would be a distortion to ignore the heavy hand of self-interest. This is illustrated in an 'anywhere but here' protectionist ethos. This is well illustrated in locational conflicts in which a number of alternatives are feasible sites for 'unwanted' facilities (e.g. Gregory 1976; Buller 1983). The political effects of moves to push 'detrimental' change towards 'other localities' have been to confound strategic planning intentions in both the UK (Herington 1984) and the USA (Fuguitt et al 1979). This sentiment has also inspired doubts over the genuineness of the anti-growth lobby concerns for the environment (e.g. Gregory 1976).

Although Herington's (1984:152) cynical observation, that local lobbies emerge '... when specific development proposals threaten the property values of local residents', might go too far (see Barker 1976; Lowe and Goyder 1983:99), group formation does nonetheless tend to be linked to perceived threats to the local environment (rather than to a general environmental concern - i.e. the originating catalyst has usually been a negative, 'stop it', rather than a positive, 'let us enhance it'; see Buller 1983). This is carried into the everyday life of these organisations in that their raison d'être is to act as a watchdog over individual development projects and secure either formal presumptions against growth or protectionist designations within official land-use policies.

The political strength of the local anti-growth lobby has been contingent upon a number of factors. Obviously there is some significance to the over-representation of in-migrant professionals within this movement (Barker 1976). Frequently these people have achieved higher levels of education than indigenous residents, as well as possessing more economic resources and 'contacts' (both of which can prove to be valuable attributes in political conflicts). Their 'expertise' and sense of responsible participation – being cooperative rather than confrontational, informal rather than formal – have also been in their favour. They <u>are</u> able to 'talk the same language' as planning officers. Additionally, they can project their opposition to private development within widely acceptable idioms; their case is presented as favouring the public good rather than private profit. These factors have enabled local groups to become important contact points between localities and planning agencies. To this end, links with local 'notables' and holders of formal positions of power are important resources for local anti-growth lobbies. This is especially so since anti-growth groups usually lack the ability to directly threaten policy-makers with sanctions (viz. economic penalities). Thus, in many rural localities, anti-growth lobbies have incorporated members of traditional elites into their fold (Eversley 1974; Buller and Lowe 1982). Not surprisingly, then, the relationship between British anti-growth lobbyists and local planning agencies has been described as informal and close (McDonald 1969; Lowe 1977). In part, this has come about through concerted lobbying, but it also emerges from a communality of social status, educational background and sentiment concerning rural areas (Cohen 1975a; Gamston 1975). As rural gentrification enlargens the anti-growth proportion of rural electorates, local government priorities and outputs either come to reflect anti-growth sentiments (Buchanan 1982) or, if initially pro-growth, can see priority changes in the face of public pressure (e.g. Blowers 1980). Thus, in an overt way, the anti-growth lobby seeks influence which is dependent upon a harmony of values, rather than relying on an ability to wield sanctional power.

Nevertheless, despite their vitality and active involvement in land-use decision processes, commentators are by no means united in their assessments of the political strengths of anti-growth lobbies. For Herington (1984:158), for example, '... it would be misleading to suggest that the anti-growth lobby has had a substantive impact on the outer city in the last decade'. As research we have conducted on West Sussex has shown, abundant examples can be presented to reveal how anti-growth pressures were associated with the defeat of proposals to extend the built environment in rural locales, yet it is still very difficult to show that these groups actually determined that growth would not occur (e.g. Buller 1983). A critical problem is that there is a strong ideological compatibility between opponents of growth and policy-makers making land-use decisions. Partly, this convergence of interest can be traced back to romanticised images of the countryside which are endemic in England (e.g.

Keith 1974; Youngs 1985). Additionally, it owes much to the
distaste shown towards 'money-making', and an associated
preference for 'leisure' (and particularly country-based
leisure), amongst the English upper classes (Tylecote 1982).
Together these value dispositions provide a 'natural' orientation towards no-growth. Of course, policy-makers might be
pleased to receive the support of anti-growth lobbyists; if
only to ease their task in explaining away a decision to
refuse 'development' permission. Even so, there is little to
make us believe that such pressures should usually be
effective in their own right. The norms of English local
authorities veer much more towards ignoring public pressures
than they do towards accepting them (Hoggart 1984). Yet it is
likely that we do not need to look for active opposition to
growth proposals in order to see anti-growth sentiments in
action. Our research in West Sussex clearly showed that both
the likelihood of 'development' proposals being made and their
chances of being accepted were influenced by the social
composition of local populations (Buller and Hoggart 1986).
Most probably because stronger opposition was expected there,
areas of higher social standing had fewer proposals for growth
than otherwise similar areas of lower status. In some
localities, especially given the propensity for anti-growth
groups to form once 'threats' to the local environment have
already been made, the existence of anti-growth groups is a
measure of weakness rather than of strength. Had the populations of such localities seemed more powerful to growth prop-
onents (even without acting), it is less likely that a growth
proposal would have been made (Buller and Hoggart 1986).

In these circumstances assessing the precise impact of an
anti-growth lobby is difficult. Interpreting the impact of
anti-growth policies is more clear-cut. There is little doubt
that the primary beneficiaries are wealthier rural residents.
To rephrase this, it is not difficult to show that these
policies discriminate against the rural working class (Newby
1980a). In terms of environmental quality, for instance, it
is property-owners in the 'nicer' parts of villages (or in
'nicer' villages) who see most benefit (Barker 1976). As Lowe
and Goyder (1983:100) have pointed out, there is a significant
environmental implication here:

> The danger is that certain areas, inhabited by
> the poor and the deprived, and already suffering
> from environmental degradation and dereliction,
> come to be regarded as environmental sinks where
> all sorts of non-conforming and noxious land-uses
> can be sited without provoking effective
> opposition.

On the housing front, the consistent opposition of local

lobbies to new housing often culminates in raising the price of the existing housing stock, while promoting only low-density, high design-standard units to be built without opposition. This affects not only indigenous residents, but also aspiring ruralites of the future, who find those who have got there before them have 'pulled up the drawbridge'. Public housing in rural areas is already a dwindling resource (Shucksmith 1981), particularly following the concentration policies embodied in key settlement planning (Cloke 1979; Phillips and Williams 1982). Indeed, for anti-growth lobbies, council housing frequently represents the twin evils of poorly designed, high-density residential units and unlooked for social change. Although 'local need' policies are commonly contained within development plans, market forces often dictate that some or all of these will fall into the hands of affluent 'outsiders'. In such a high-priced local housing market, it comes as little surprise to find that residential caravans are commonly an alternative form of accommodation (Dunn et al 1981; Phillips and Williams 1982). However, policies which alleviate local housing problems by promoting caravan sites have also been the focus of anti-growth lobby opposition (Buller and Lowe 1982:39). To counteract this, it is rare that anti-growth groups seek to mitigate the effects of restraint policies upon the housing problems of the less economically endowed. A further handicap for the rural working class which emanates from anti-growth sentiments comes from opposition to industrial expansion. Through their anti-growth stance, lobbies help deny the emergence of alternative employment opportunities and frustrate attempts by manufacturers to decentralise their operations. As for housing, in fact, whereas anti-growth lobbies might be successful in opposing schemes within a single locality, the end-product is usually the transferring of that project to an area where opposition is weaker (Goodman 1972). Paradoxically, at a time of economic recession, one outcome of placing restraints on further industrial expansion is that it has given manufacturers more power. Their hand has been strengthened by being able to play the sanctional threat of moving their enterprises elsewhere. The proposal of the London Brick Company to rebuild one of its major brickworks in the Marston Vale, Bedfordshire, provides one example of this (Blowers 1983). Although environmental groups offered qualified support for the Bedfordshire development, they wanted enforced landscaping and pollution control measures. The principal effect of a lengthy period of opposition was to delay the eventual decision until a period when economic recession had taken hold and altered the circumstances of the debate. Threatened with the total withdrawal of the company, employment rather than the environment became the major concern of the county council. Opposition collapsed and planning permission was granted without the environmental controls sought by the plant's opponents. Although the victory of the Brickworks Company was 'pyrrhic' (Blowers 1983:406), in that the final result led to a net loss of jobs, it does demonstrate the limitations of the anti-growth argument when faced with the principle of employment generation. This Bedfordshire example is neither unique nor typical. Anti-growth lobbies have succeeded elsewhere in preventing major industrial developments, as for example at Belvoir (Arnold and Cole 1982), yet have failed in others, most notably at Stanstead. Often of major importance to the outcome of such issues are features beyond the reach of anti-growth groups. Clearly, the ability of anti-growth lobbies to influence policy-makers is restricted by the emergence of more politically sensitive issues (like job creation or job retention). Hence, there is a clear spatial expression to the influence of anti-growth sentiments. In particular, as Fliegel et al (1981) observed for the USA, no-growth movements are more characteristic of (or more intense in) areas closer to metropolitan centres and other rural places of economic vitality.

Another more subtle effect of English anti-growth lobbies has resulted from their domination of participatory processes with local governments. This has increasingly allowed them to be seen as the legitimate representative of their locality. Numerical dominance does not guarantee desired goals, but, where public responses to invitations to participate in policy-making are poor (Walmsley 1982), particularly in working class areas (Batley 1972), it comes as no surprise that groups which do participate are accorded a voice which is disproportionate to their numbers. Significantly, anti-growth lobbies are increasingly sought as the 'legitimate' representatives of their locality (Buller 1983; Vernon 1981). This has twin effects upon the ability of contrary interests to affect policy. First, the close, informal links that often come to develop between anti-growth lobbyists and policy-makers cannot be duplicated (most especially if social backgrounds are not shared). Working class rural residents thereby rely more on elected representatives and a minority of pressure groups, like Shelter, whose aims are to help the deprived (Newby 1980a). Second, by presenting their viewpoint – which probably are majority views in areas where rural gentrification has long been established – as 'the public interest', anti-growth lobbies strengthen their argument by helping promote a consensus over land-use policy. Dissent is easily identifiable and too readily dismissable.

Today, the view that rural anti-growth lobbies are comprised of the '... ever present ancient establishment, the landed aristocracy, the products of Oxford and Cambridge, the landowners, the officer class and behind them, the hangers on'

(Eversley 1974:14) is no longer a valid one. Indeed, the conservation and preservation lobby acts more as a challenge than as a support to traditional elites. The recent entry of agriculture and farm organisation into the arena of conservation politics has increased the ideological distance between landowners and conservationists (Shoard 1980; Lowe et al 1986). Nevertheless, both groups share a uniting bond, In that they belong to the property owning classes (Newby et al 1978:276). As such both are favoured by socio-economic stability within rural localities. Although much of the justification for stability rests in the belief that the public interest is served by restraint, it has been largely the property owning classes who have benefited from this. The fact that, as Allison observed, '... there is no coherent working class opposition to conservation' (1975:108), nor is there any overt party political role adopted by restraint interests (Lowe and Goyder 1983), should not blind us to the distributional consequences of anti-growth.

In the United States, it would be difficult to conclude differently (Frieden 1979), although the precise manifestations of the processes involved are somewhat different. In the main, this is due to the weakness of the US land-use planning system. It also owes something to basic differences in attitudes towards economic growth (compared with Britain that is). Whereas in Britain anti-growth groups are often formed to influence planning agencies to stop 'unwanted' construction proposals, in the USA it is often necessary to first of all introduce a planning system. The pro-free-market sentiments that pervade US public life have greatly restricted the occurrence and strength of land-use planning controls. Analysts considering places as far-flung as Arkansas (Dailey and Campbell 1980), Kentucky (Garovich 1982) and Vermont (Olgyay 1982) have reported that only as the countryside landscape was becoming 'blighted' – through uncontrolled construction – did rural residents generate pressures to introduce planning controls. Even then, the more 'open' political context of US local government policy-making has led to controls which lack bite. The continuation of land-use controls is, for example, frequently open to referendum votes (consistent with the picture in Britain, Medler and Mushkatel 1979 reported that Oregon's Measure 10 was retained after a vote in which support for the pro-land-use-control side was characteristically from upper income and faster growing localities). Along quite different lines, but similar in its effect of weakening the supervisory role of planners, local policy-makers are more open to direct business pressure. As Gottdiener (1977) reported for New Jersey, this can even take the form of companies (or individuals) essentially 'buying' planning permission (the resulting income then swelling the coffers of controlling political parties). More important than such (probably somewhat rare) practices are the ambivalent attitudes which exist towards land-use control.

These are partly evinced in the failure of many localities to adopt effective land-use controls, but they are also evident in situations where controls are sought. Strong (1975) for example has shown how agreements to restrict future urban expansion can break down as area residents see personal benefits accruing from land sales for housing construction. This idea is associated with Fliegel et al's (1981) finding that recent rural in-migrants to the Midwest do not 'pull up the gang plank' once settling in a new area. They are conscious of the benefits which growth brings in increased business profits and an enhanced tax base (and so hopefully in reduced taxes). These pro-growth sentiments are undoubtedly part and parcel of dominant US value-codes. The emphasis given to personal gain and economic growth (Molotch 1976), and the concomitant de-emphasising of general societal benefits, stand in sharp contrast with the British situation (even though there is substantial abuse of this sentiment in the UK). It is nevertheless worth noting that traces of a rural no-growth mentality do exist in the United States. Partly this can be traced to the rise of environmental concerns in the 1960s, associated as it was with in-migration to particular rural locales (e.g. in Colorado and Oregon) by middle class professionals. Additionally, and more recently, this sentiment probably owes much to the widely-publicised economic, social and psychological problems which have emerged in more peripheral settlements which have experienced rapid population growth (Albrecht 1978; Freudenberg 1982; England and Albrecht 1984). Certainly, analysts have increasingly found support for a questioning of the growth ethic. In places this seemingly owes much to personal gain (Frieden 1979), much as it does in many parts of Britain, but there is also a developing commitment to the preservation of the countryside's character, even when this leads to lost opportunities for personal gain (e.g. Alanen and Smith 1976; Browne 1982).

PERSPECTIVE

The organisation of this chapter has revolved around two basic divides; that between pro-growth and anti-growth sentiments and that which separates local from broader (or 'national') forces of change. We believe that these constitute critical dimensions in social processes which determine the developmental fortunes and potentialities of individual rural localities. Our treatment of the four-fold division which reflects these basic categories (i.e. local anti-growth, local pro-growth, national pro-growth and 'national' anti-growth) has not been equal. That this was so is not inevitable, but it does symbolise the fact that these dimensions cannot be understood without reference to societal power structures, which determine the relative force of social processes aligned with each of these dimensions. Without any doubt, emerging from the basic structural arrangements and imperatives of capitalist societies, the strongest impulse has been national in origin and pro-growth in flavour. Trying to 'buck the system' via local pro-growth movements must at times seem like blowing a feather eastward in the face of a westward gale. Pressures of centralisation are dominant. This is seen not simply in service provision, but also in settlement form. It is nonetheless necessary to be aware that there is no direct correspondence between social form and spatial structure. There are cross-currents between these two. Thus, while there is centralisation in economic organisations, this has been accompanied by a decentralisation of economic activity (as seen in the flow of manufacturing plants into more peripheral locations). Our argument is that social groups have been differentially placed with regard to drawing advantages from these counter-currents. Where working class residents would have benefited from such decentralisations, the vested interest of local elites has at times deprived them of the opportunity. This is seen not simply under 'traditional' power structures (local anti-growth) but also from the actions of ex-urbanite in-migrants ('national' anti-growth). In both instances it is those who already have more resources who benefit most from this process. These benefits come from more 'sanitised' environments for urban in-migrants and, frequently, lower wage payments for 'traditional' elites. For the working classes, it means further deprivations – poorer services, lower incomes, higher costs and less accessible housing. Not that local elites could (consistently) outweigh the force of national institutions and processes. However, in a socio-economic environment in which national agents have numerous alternative locations to chose from, and in which many of their imprints are in any event effectively nation-wide, local resistances can be accommodated or circumvented. The main pressures for change are national in origin and scope, but their precise local manifestations are significantly conditioned by local power networks and social structures.

Chapter Seven

DEVELOPMENT WITHIN LOCALITIES

The scales at which rural development has so far been considered mostly lie beyond the experiential reference points of individuals. International, national and inter-local development forces structure opportunities for people's betterment. They also condition the intensity, duration and breadth of socio-economic changes. Yet these forces are not experienced directly by individuals. Rural inhabitants' primary contacts with such structuring processes are mediated through locality-specific effects. President Reagan's strictures on the necessity for large increases in defence spending impinge on the consciousness of US rural inhabitants through images of communist containment, national regeneration and preserving 'freedom'. Yet they are experienced by individuals in the form of lower (relative) rural incomes, since defence expenditures are focused on cities, with social welfare and agricultural programmes suffering reduced financial commitments in order to fund military expansionism (Buttel 1980). The aim in this chapter is to focus explicitly on these experiences of development. Of course, each of the previous three chapters has examined one aspect of that experience. But these have concentrated more on what conditions the path of socio-economic change in localities per se. Here, how these processes have an uneven impact within localities will attract our attention.

Most usually, investigations of the distribution of socio-economic benefits within localities have taken Max Weber's (1948) class, power and status as the organisational medium for their analysis (e.g. Wild 1974). Many studies have devoted themselves to just one of these, either implicitly or explicitly contending that a single dimension of social stratification is pre-eminent. Other analysts have sought to reinterpret the results of earlier studies, emphasising that the initial authors misconceived the primary dimension of social division. As yet, however, researchers have not provided a convincing alternative to the class, power and status schema. To accept this as a categorisation through which the distributional consequences of developmental processes can be appreciated does nevertheless require explanation. For one thing, it would be incorrect to believe that rural residents necessarily see these dimensions as critical to their everyday lives. This is especially so for social class. As West (1945:115) reported for one US locality: 'Many, if not most, Plainvillers completely deny the existence of class in their community'. With a multitude of analysts recording the same result,* both the interdependence of social stratification dimensions and the frequent divergences of popular imagery and underlying causality must be recognised. How we have approached the problems of disentangling social stratification dimensions is by emphasising the specifically rural focus of our investigation. The first component of the chapter draws attention to social class divisions. This section presents no general analysis of social class or even of locality-specific manifestations of class relations. Rather it questions whether a specifically rural basis for class divisions can be identified; and investigates distributional elements emerging from class structures which are grounded in rural localities. This is followed by a section on local power structures. Again, the intention is to identify the character of power networks which carry explicitly rural articulations. Then, in the third section (on social status), a separation has been made to distinguish the issues examined from those in the previous two parts of the chapter. While both class and power standings have status implications, for this section of the chapter we have interpreted social status as an attribute of individuals (not of their possessions or according to people's role in social processes). The question asked here is how the physiological features of birth (i.e. gender and race) become ingrained into the distribution of societal benefits within rural locales.

---

* Which does not mean that this is a general condition in rural areas. Douglass's (1984:13) observation that, for social class divisions in Agnone (southern Italy), '... the chasm is so great in the minds of so many that it practically assumes caste-like overtones' has been duplicated in array of studies (e.g. Cutileiro 1971; Duncan 1973; Wild 1974). While the existence of social class divides does vary across rural locales (compare Williams 1956 with Williams 1963), it is evident that, even where class divisions are critical, rural residents regularly deny their existence (e.g. West 1945; Vidich and Bensman 1968).

SOCIAL CLASS

For both Karl Marx and Max Weber, social class referred to the division between capitalists, who owned means of production (like factories or land), and the proletariat, whose members did not. Various gradations can be recognised in this distinction (as between 'little capital' and 'big capital'), but the essential divide is binary. This fundamental division has nevertheless been played down in studies of rural areas. This is most evident in investigations of the 1950s and 1960s, but still finds its way into more recent analyses. A probable cause for this, as evinced in British studies, was the character of the investigations themselves. Throughout the 1950s and 1960s, research on rural Britain (and the USA, for that matter) was dominated by socio-anthropological studies of small, peripheral 'highland communities' (e.g. Rees 1950; Williams 1956, 1963). With the exception of Littlejohn's (1963) Westrigg, these places were characterised by family farming, an absence of hired labour, strong kinship ties and tight-knit (introspective) social communities. With the rural-urban continuum as an underlying conceptualisation, these studies appeared to confirm the idealisation of 'rural' as a bastion of resistance to the incursion of 'urban' ideas and practices. Seemingly included amongst these 'urban' practices were the organisation of work along class lines and the development of class consciousness. In a British context, the major break with this conception of rural social structure came through the publications of Howard Newby (and his associates). Prior to Newby's work, studies of rural social organisation in lowland Britain were dominated by considerations of the impact of recent in-migrants (largely urban commuters) on village social life (e.g. Pahl 1965; Ambrose 1974). Newby launched rural studies into a new era (and in sociological terms a new respectability as well) by showing that class divisions were firmly entrenched in East Anglian farming areas. This evidence did more than help break down notions of rural social uniqueness (which studies of commuter villages had already started), for it also directed analysts' attention to the applicability of general theoretical ideas to rural contexts. In a sense, the opening had been made for drawing rural studies away from their almost isolationist stance into the mainstream of social science theorising. These changes have occurred at the same time as researchers have begun to emphasise the somewhat distinctive character of rural class relations. In this, the key point of differentiation rests in land ownership (Stinchcombe 1961; Urry 1984).

The significance of land ownership derives from the key position that agriculture plays in rural economies. Obviously we must be conscious of the fact that farming employs many fewer works today than in the past. Recognition must also be given to the enhanced role of manufacturing, government and service sectors in rural economic endeavours. However, in spite of the relative decline in agricultural activities, it is through land ownership patterns that we see specifically rural manifestations of class conflict. The very existence of giant farms in the West, coupled with the questionable legality of their establishment, had earlier drawn US attention to the significance of land control in determining the distribution of socio-economic benefits. As McWilliams (1939:21) recorded for California:

> The ownership patterns established by force and fraud in the decade from 1860 to 1870 have become fixed; the social structure of the state is, in large part, based on these patterns. (see also Chapter Sixteen in Encel 1970 which paints a similar picture for Australia.)

In a British context, for much of the post-1945 era, social researchers have largely ignored the role of large landowners, tending instead to concentrate on areas dominated by family farms. Perhaps the mistaken supposition that the 'decline' of the landed aristocracy, along with a sharp reduction in the number of employees on farms (with a proportionate increase in the number of family farms), lulled analysts into assuming away the importance of large land holders. The scares of the early 1970s over the seemingly large-scale movement of financial companies into farmland ownership, combined with the work of Newby and associates (e.g. Newby et al 1978), have had the effect of dispelling any such misconceptions. This is a step in the right direction, for it is clear that large-scale land holders do impose severe restraints on the social well-being of their employees and perhaps even of the entire localities in which they are based.

Starting with workers themselves, popular conceptions of agricultural labourers implicitly see them fulfilling an occupational role that is soon to become extinct. Caught between the anvil of rising input prices and the hammer of declining product receipts, farmers are 'forced' to reduce factor costs. A key means of achieving this is to squeeze out labourers, replacing them with new machinery. Evidence provides support for this view. For example, surveys of those leaving farm labourer jobs have found that redundancy arising from replacement by machinery is a major reason for job loss (e.g. Drudy 1978). However, it would be far too simplistic to take this surface impression as the underlying causal mechanism. This is most visibly evident, and most closely documented, for the US West. Here, researchers have confirmed that landowners frequently employ mechanisation primarily to 'discipline' workers rather than to reduce costs. Thus, for as long as the federal government allowed large inflows of (low cost) Mexican workers to assist on southwestern holdings, farmers showed little inclination to mechanise. With the termination of Public Law 78, the growing strength of (and broad public support for) agricultural labour unions, and with government moves to tone down the shabby and exploitative conditions under which these labourers were employed,* farmers responded by seeking mechanisation to reduce their workforce numbers. As Price (1983) found in California farmer interviews, machinery was frequently turned to as a means of harming the development of unionisation. In a sense, mechanisation was a means of 'disciplining' the agricultural labour force. If unions were pushed for, farmers would respond with redundancy and replacement machines (Friedland 1980). In California at least, mechanisation was a highly visible symbol of class conflict. Given workers lose jobs in these circumstances in order that owners can reap higher rewards, mechanisation is inevitably integral to class conflict. But this does not have to be openly so. That California farmers were open (to researchers) in revealing the 'anti-worker' character of their mechanisation efforts, does not mean that this is how they projected mechanisation decisions to their workers (though to be an effective 'disciplining' agent, this likely was the case in California). Outside observers, without detailed knowledge of farmers' decision criteria, could easily be led into believing that farmers were merely responding to a cost-price squeeze. That class conflict is a fundamental dimension of farming today is nevertheless seen in those occasional instances when employer-employee disagreements erupt into confrontation.

One clear example of this occurred in the infamous Raymondville farmworker strike. Amongst Mexican workers in this Texas community, a successful strike against a single farmer (who unilaterally reduced payments for onion harvesting) led to pressure for union recognition. For local entrepreneurs, owner-worker conflict ending in the minor success of reversing a somewhat disreputable act was one

---

* To give some indication of this working environment, by 1983 only six US states allowed agricultural workers to hold unemployment insurance (Price 1983), while only 13 states made provision for compensation for work-related injuries equivalent to that available for other workers. In all, 24 states allowed no such compensations (Congressional Quarterly Inc 1984).

thing, to officially sanction trade unions was quite another. This could disrupt the dominant power structure in the locality (and perhaps by example even beyond it). In response, a concerted bourgeois offensive was waged against the agricultural labourers. From across the locality, business leaders and their spouses crossed picket lines and became labourers on the strike-hit farm enterprise (ensuring production was not stopped). Police with riot equipment escorted the strike breakers, and provided further assistance to growers by arresting strike leaders. 'Outsiders' were introduced to help harvest the crop and to disrupt strikers' meetings. Had this simply been a local dispute, the use of the police as an arm of the local farmers' 'defence force' could be interpreted as another example of local business-professional elites using publicly-employed personnel for their private gain (such business-professional dominance being common in rural local governments in the United States; e.g. Vidich and Bensman 1968). In truth, however, this interpretation will not do. As Jennings (1980) has made clear in his examination of the Raymondville strike, this was not simply a local controversy. Large agribusiness concerns became involved. Political intervention went up to the state level (including the use of state police troopers). Outside trade union help was called upon. This was no local incident, but one instance of a society-wide class struggle that found expression (and for capitalists, at least, had to be stopped for fear of contagion) at a local level.

That 'outside' trade unions were interested and involved in the Raymondville conflict is indicative of recent changes in US farm labour conditions. As many commentators have pointed out, a long-time intrinsic feature of relations between US trade unions and farm labourers has been neglect by mainly white trade unionists for the plight of non-white agricultural workers (e.g. Baker 1976; Goldfarb 1981; Majka and Majka 1982). In this regard, the experiences of Britain and the United States are quite different. Undoubtedly, what strengthened the hand of US growers was their ability to promote and provoke racist sentiments amongst the working class. This they could accomplish because a variety of cheap labour sources could be drawn on to feed the agricultural production machine. As the Majkas have pointed out:

> When labour unrest and organisation became widespread and the [currently dominant ethnic or racial] group no longer was the source of low-wage, powerless labour, the large agricultural landowners attempted to undermine its dominant position in the agricultural labour force and hire another group to undercut the organised. (Majka and Majka 1982:5)

Thus, as we move from the mid-nineteenth century to the present day, we find southwestern agricultural labour dominated by the Chinese, then the Japanese, then Filipinos, followed by a short-lived burst of poverty stricken 'Dust Bowl' migrants (in the 1930s), culminating in the more recent Mexican dominance (Congressional Quarterly Inc 1984). For a variety of reasons, including their becoming too well-organised (as with the Japanese) or their working too hard (for too little) and threatening white workers' jobs outside agriculture (the Chinese), these groups have superseded one another. Goldfarb's (1981) appropriately titled <u>Caste of Despair</u> signifies the isolation of these different groups of workers from mainstream labour organisations, as well as informing us of their relative socio-economic standing.

For British agricultural labourers, there are certainly similarities with the US situation, but there are also significant differences. Thus, British agricultural workers do stand in a low socio-economic position. Admittedly this is not of the same magnitude as their US counterparts, but on a relative income-earning scale the two groups sit in comparable positions. In addition, the relationship between agricultural workers and the mainstream British trade union movement has some notable similarities with the USA; albeit, in Britain their separation is less intense. No major differences of race affect the British scene. Yet, on account of the small units of agricultural production, their dispersed nature and, hence, difficulties of organising collective action, agricultural labourers have never been major contributors to trade union pressures for improved living conditions (Newby 1977). In a sense, they have followed on the coat-tails of industrial trade unionists' actions. But, by not being integral to these struggles, the benefits they have derived have come in a diluted form. In good measure, this situation can be tied back to the character of capital-labour relations in British agriculture. The key similarity between the UK and the USA in this regard is that these two groups are in conflict over the monetary benefits derived from farming. A critical difference is that, whereas in the United States this is seen through exploitative, de-humanising working conditions, in Britain it is characterised by almost passive and friendly social relationships; namely, by paternalistic capitalism.

There are key differences between the UK and the USA which help explain the dissimilar manifestations of capital-labour relations. Apart from questions of racial composition, there is also the impact of farm size. British farms with agricultural labourers tend to be smaller than their US counterparts and employ fewer workers (frequently only one). In the workplace, British farm labourers have more usually been treated as members of a farmer's household (e.g. Rees 1950; Williams 1956). Give and take is the order of the day. Some researchers have concluded that employers who try to impose industrial style management-labour relations have great difficulty hiring workers (e.g. Williams 1956). Of course, in areas dominated by larger farms, like East Anglia, the workplace model does take on stronger overtones of a manufacturing plant. Yet, as Newby's (1977) investigations made clear, work units are still comparatively small, and something of a 'family' atmosphere generally persists. A consequence of this social environment is that worker pressure for improved income or employment conditions is difficult to muster. There is the fact that, on smaller farms, strike action by agricultural labourers can be circumvented by the use of family labour and by many non-urgent jobs being put off for a time. Yet there is also the existence of a strong social bond between worker and employer, on account of the 'family' atmosphere which these small, semi-isolated work units engender. Both the sense of semi-isolation and feelings of work unit camaraderie are further aided by the increasing dependence of labourers on housing tied to their place of employment (over half the male agricultural labourers in Britain currently live in tied accommodation, e.g. Gasson 1975). Studies vary in the support they find for such housing amongst agricultural workers, but there is clearly some concern about this housing style (Newby 1977, for example, reported that 49.8 per cent of his sample of hired workers wished for changes in tied cottage conditions and 29.6 per cent wanted their abolition). This is hardly surprising given the clear restraints tied accommodation can place on both obtaining other jobs and, indirectly through this, garnering better pay from an existing employer. If change of job means a change of home as well, this weakens a worker's bargaining position (for instance, in his survey of Tiverton Rural District, Fletcher 1969 found that 90 per cent of agricultural labourers who wished to leave farming could not do so on account of their housing problems). This is especially so since farmer-dominated local governments have been extremely poor providers of alternative low-income housing facilities (Rose et al 1976; Larkin 1979).* Here, the influence of large-scale farmers stretches beyond the boundaries of their own production units into the locality as a whole. What the relationship of large landowners to the broader local

---

* Since 1976 farm labourers' security of tenure has improved, but farmers can still require a labourer to vacate tied accommodation if the home is essential for farm operations. With few workers' homes per farm and little alternative accommodation locally, this is usually the case. In practice, then, security of housing tenure is still weak.

community actually is, is the subject of some controversy. For a number of analysts, a critical element in this relationship is that the impact of large-scale operators is negative for general social well-being.

The most well-known proposition of this kind is the Goldschmidt hypothesis. Analysing two California settlements in the 1940s, Goldschmidt concluded that the quality of life (income, level of living, social and physical amenities, and involvement in local political processes) was higher where family farms were dominant than where large-scale enterprises were the norm (Goldschmidt 1978). Dependent upon ideological perspective, Goldschmidt's study was either damned (as Hayes and Olmstead 1984 have noted, largely without considering the 'facts' he produced) or critically acclaimed (by those on the Left and those opposed to the demise of family farming). Since Goldschmidt's work was completed, a number of investigations have lent support to his basic contentions, albeit not in a wholly convincing manner. Goldschmidt (1978a) himself showed that there was a strong positive relationship at the state level between the percentage of the population in the lowest income groups and the dominance of agricultural production by large enterprises. However, as this was at a state level, it runs into innumerable interpretive problems associated with ecological correlation fallacies and the intervening effects of other income-generating economic processes (his result also contradicts that of Harris and Gilbert 1982 who attempted to analyse both direct and indirect farm size effects at a state level). Support for the claim that commercial facilities are poorer where large farms are plentiful has also been upheld in other research projects (e.g. Bible 1978); though others are less certain and report that they found no such relationship (Korsching 1984 qualified his conclusion by noting that his Iowa study area did not have extremes in farm size composition on account of family farms being prevalent in the state). In principle, the logic of Goldschmidt's case vis-à-vis commercial facilities is difficult to deny. Large farms tend toward less intensive production practices and so both produce less per acre (requiring fewer farm-related facilities) and employ fewer people than smaller farms covering the equivalent geographical area (thereby needing less household retail amenities). There is nevertheless a danger in equating these two occurrences in a causal manner. Farms are increasingly selling their produce through contracts with agricultural processors (this process affecting 40 per cent of UK output and 20 per cent of that in the United States even by the early 1970s; Perelman 1977; Hart 1978). Of necessity contracting usually leads to a by-passing of local farm input suppliers and local farm output traders (e.g. Dalton 1971; White and Watts 1977). Centralisation of both agricultural processing and the provision of farm inputs is occurring as large corporations increasingly come to dominate these components of agriculture. The result is a more dispersed pattern of farm linkages; the decline in farmers' reliance on local trade centres is not inevitably a function of farm size, for it is happening anyway, even where family farms are dominant (e.g. Hoggart 1979). Certainly, commentators have found that larger farm operations are more closely aligned with all of contract farming, direct purchases from factories and direct sales to processors or large, non-local wholesalers (e.g. Bible 1978). This, however, might be largely on account of smaller-scale farmers being less aware of (or merely more resistant to) trends which need to be followed to stay competitive in agriculture (see Gasson 1969). As we have stressed in earlier chapters, farmers respond to large corporate pressures. They do not mould them. We remain to be convinced that it is large farms per se that affect local amenities. More likely, it seems, 'outside' corporate pressures are simply more advanced in areas of large farms.

This does not mean that large farm dominance does not have detrimental effects. Reviewing studies on the topic up to the early 1970s, for example, Heffernan (1972) concluded that there was clear evidence that farm workers are much less involved in local social and political activities where corporate agrarian structures hold sway. Similarly, Heady and Sonka (1974) have concluded that gross incomes are higher where farms are smaller in scale than where they are larger; although average farm earnings would be close to the poverty level under a small farm regime (requiring additional off-farm work for a 'respectable' salary). This suggests that the alternatives to large farm dominance are not that rosy. Multiple job holding amongst farmers is clearly linked to distance from major metropolitan centres and, to a lesser extent, proximity to medium-sized towns (e.g Fuguitt 1959; Mage 1974). So is the mean average size of farms. Hence, there is the danger of circularity in arguments aligned with the Goldschmidt hypothesis. Larger farm acreages tend to be found in more sparsely populated places anyway, so poorer community facilities should be expected (more recent criticisms of Goldschmidt's initial study have likewise argued that the two settlements he contrasted provided an invalid basis for comparison, since the corporate dominated locality was more recently established and had had insufficient time to develop its 'true' infrastructure; Hayes and Olmstead 1984). On the other hand, if the size of farms is measured by gross output levels and 'large farms' actually have small acreages, then income generation must be tied to intensive production practices (like dairying), which leave comparatively little time for 'outside' political and social involvements. Variations in product lines (associated with how the farm is considered 'large') and location (for population density, alternative job opportunities and pressures for residential land sales) combine to induce complex relationships between farm size and Goldschmidt's ideas. To date, research on this hypothesis has been insufficiently detailed to truly assess its validity. Certainly, we know that larger farm enterprises are linked to lower farm worker wages, but whether this effect carries into the general character of localities has yet to be established. This issue however is an important one. This is especially so in theoretical terms, since theorists have recently sought to emphasise how land ownership has a critical role in rural social stratification (Stinchcombe 1961; Nelson 1979; Newby 1983).

Recognition of the importance land ownership has as a social structuring mechanism has in our view come too late. The horse had not exactly bolted before the door was shut, but as researchers gained satisfaction from seeing one door shut, the horse was strolling out of another exit. The importance of land owning has declined as an instrument of rural social stratification. Yes, land ownership is important in determining class relations within the workplace (or the scale of land ownership is anyway). Yes, very large rural 'estates' – like those of highland Scotland or the Duchy of Cornwall – still have important effects on local social units (e.g. Kendall 1963; Bird 1982). Yet the impact of these large estates has declined significantly over the last century. This is especially seen in land reform movements which breakup large holdings (for southern Italy for instance see King 1973). It has also occurred as a 'natural' process, tied to inefficiencies of operation, inheritance taxes or taking advantage of more profitable returns through land sales (Clark 1981, for example, found that between 1872 and 1970 there was a 49.3 per cent decline in the number of estates larger than 20,200 hectares in the Highlands of Scotland). There is no doubt that average farm sizes in general have been increasing, so that 'land power' should be concentrated into fewer hands. As McEwen (1977) has pointed out, however, farmland ownership is increasingly a meaningless concept in social class (or social power) terms, since farmers are increasingly at the mercy of oligopolistic firms which dictate input and output conditions (this is far from a new idea, for Veblen 1923 pointed to the same occurrence more than 60 years ago). As we have argued in earlier chapters, the agricultural sector (and with it much of the 'power of land ownership') has been usurped by industrial and financial capital. This is not simply seen in farmers' input-output interactions, but even in land ownership itself. Thus, both the surge in British farmland purchases by major financial institutions in the early 1970s (Munton 1977) and the growth of foreign farmland ownership in the USA (Vogeler 1981) owe more to rates of investment return than to any peculiar interest in rural class relations (the price of agricultural land in the United States rose by 900 per cent between 1950 and 1979, compared with a general inflation rate of 320 per cent; Healy 1980:75). All too frequently, for larger landowners, interest in land as a production unit (and so as the base for rural class structure) is incidental. Thus, the Tenneco Corporation became a major landowner (and farmer) in the United States largely to protect access to the under-surface mineral resources it wished to exploit later (Cortz 1978). With 44 per cent of US farm and ranch land now owned by non-farmers (Healy 1980) and with national governments (like France) actively promoting non-farm land ownership (to reduce farmers' capital investment requirements), the trend is towards agriculture being subjected not only to 'outside' control of its inputs and outputs but also of its major means of production (viz. land). This hardly forms an acceptable background for proposing that rural areas have distinctive local class structures.*

SOCIAL POWER

Analysts have traditionally distinguished power relations in rural and urban areas along a rural-urban continuum. In theoretical positions, like Rokkan's (1970) on voting behaviour, the whole of socio-political affairs in rural environments is characterised as being 'retarded' compared with their urban counterparts. Illustrating this point is Rokkan's (1970) conception of urban areas rejecting the old-style voting method of selection by candidate's home location or personal reputation, in favour of voting on the basis of

---

* Two points need to be made about this conclusion. First, we are not claiming that distinctive local class structures do not exist in rural areas. Urry (1981) makes the general point in this regard, and there are innumerable investigations from widely differing rural environments which show that the class system of particular rural localities is not the same as that in the nation as a whole (e.g. West 1945; Pitt-Rivers 1960; Emmett 1964; Christian 1972). Rather, our contention is that land ownership cannot provide a valid distinguishing criterion for identifying distinctly rural class structures. Second, while we have contended that land ownership is not critical to the farm production process, this does not mean that land ownership per se is not critical. As Barlow (1986) has argued, through controlling the magnitude of land released for housing or other constructions (and by imposing conditions on the use of that land), landowners can have major impacts on rural development.

social class. Remoter rural areas, by contrast, are portrayed as only slowly and reluctantly forgoing this electoral style. The underlying image of rural socio-political activity presented here is of a population whose ideas are slow changing, locality-bounded and respectful of 'traditional' leadership structures. Immediately objections can be raised over this image. Repeated incidents of agrarian protest against prevailing societal (economic) norms, the quite distinctive political party attachments of US farm organisations (the National Farmers' Union being tied to liberal Democrats, while the American Farm Bureau Federation is aligned with conservative Republicans), and the fact that the first 'socialist' party to attain provincial or state elected office in North America had its roots in rural areas (the CCF in Saskatchewan; Lipset 1950), belie this simplistic idea. Yet they do not discredit its basic premiss. Agrarian participation in political affairs is a very self-interested form of involvement. Periods of intense farmer political participation are associated with 'trouble' periods in the farm sector. With present-day government sorties to reduce farm production and prices, for example, even North America's pro-private enterprise, anti-government intervention, cattle ranchers have become active political lobbyists (Stokes 1985). Such self-interested political involvement leads to short spurts of activity followed by long lulls (exceptions being where other interests coalesce with rural protests – as with the 'anti-central Canada' and provincial rights elements of socialism on the Prairies). In truth, there is considerable evidence that rural areas are characterised by 'traditional' socio-political cultures. However, there is nothing particularly 'rural' about this. Similar political structures exist in many central cities. Most particularly, this is seen in the ethnic-based, non-ideological political machines of North American cities (e.g. Guterbock 1980). These have their counterparts in the formal political machines which exist in a few rural areas (e.g. Sacks 1976) and in the traditional support which is bestowed on the most prestigious local families in many others (e.g. Johnson 1972; Dyer 1978). The similarity of these situations does not arise on account of equivalent habitats. Rather it emerges from comparable power structures (and their resulting socio-cultural norms). To understand how rural environments are commonly associated with traditional political stances, we must first appreciate that, by their very nature, rural localities are small in scale. This provides a forum in which a concentrated (i.e. elite) power structure can more easily emerge. Over time, through constant efforts to bend people to the will of the elite, such socio-political environments can produce all the hallmarks of an unquestioning, respectful-of-authority political culture.

The apparent self-reinforcing stability that many take to

be the hallmark of rural society in reality masks an underlying structure of power. The precise character of that power structure is nonetheless open to considerable variation. Recent emphasis on the importance of land ownership in rural social stratification has focused attention on the role of large corporations, landed estates and large-scale farmers. The key component here is inequitable access to means of production. Although most usually expressed as if rural was equatable with agricultural (hence land ownership disting-uishes capitalists from the proletariat), this is an incorrect inference. Similar class-based power structures exist equally well in rural areas with no agricultural endeavours (see Perlin 1971 or Cohen 1975 for descriptions of class-based power structures in the fishing outports of Newfoundland). Rather than referring to a landowner-dominated power structure, therefore, we prefer to call the power relations specified in this class-based system 'traditional'. Most characteristic of such traditional power structures are the landed estates of England. While the incidence of such large land holdings has declined, new class-based power systems have arisen to take their place (i.e. neo-traditionalist structures). Yet class-based systems represent only one kind of rural power network. A much less centralised (or elitist) pattern, wherein social class does not overtly provide the basis of the power divide, can be recognised in rural idealist power structures. These have similarities with traditional power relations, insofar as both are usually assumed to be dominated by tranquil social relations. In this regard, they stand in marked contrast with conflictual power structures, which comprise our third category of rural power network. Increasingly analysts are identifying and describing the dimensions of this model of rural power relations. It is already a significant element of rural social structures and looks likely to become more so. As with the other two power structure categories, these conflictual structures are associated with quite specific distributional consequences for socio-economic change.

Given its status as the 'traditional' rural power structure (in a European context at least), the class-based model assumes the most logical starting position for our analysis. In reality, the political supremacy of the landed gentry has ceased to be a major component of rural power structures in all but a few (predominantly remoter) rural areas. This 'traditional' element of class-based structures has withered away under the influence of death duties, inefficient management and altered goals. In areas like the Highlands of Scotland, however, the laird commonly continues to occupy the roles of political leader, major employer, custodian of economic welfare and trustee of traditional values. As Dyer (1978:40) has reported:

> Even as late as 1974 it was still possible to see
> all [local government] councillors as leaders in
> the way their nineteenth century predecessors
> were: gentlemen meeting together to discuss the
> business of the county.

There is something characteristically British about such
systems of 'benevolent paternalism'. Not that the traditional
model is restricted to Britain. The early squatters who came
to establish themselves as Australia's largest landowners have
pursued a quest for 'aristocratic' standing (Encel 1970). An
integral component in this process has been copying the
behavioural norms of the British landed gentry (Wild 1974).
The desire for a rigid, unchanging elitist power structure,
based on rural economic endeavours and espousing a
paternalistic intent, has also been present ever since
Europeans first settled the southern United States (Nicholls
1969). Fundamental to the continued existence of such
traditional power structures is the general populace's
acceptance of the legitimacy of elite dominance. As Newby
(1977:49) expressed it:

> The beauty of traditional authority was that it
> was the most stable of all forms of legitimation,
> since deference was granted both to the tradition
> itself and to the person holding the position of
> authority.

Although some tensions have long existed between landowners,
tenants and workers (e.g. Charlesworth 1983), the dominant
attribute of this social order is the lack of opposition to
its continued existence. Rarely, as in parts of southern
Italy (e.g. MacDonald 1963), did workers organise themselves
to oppose local aristocrats. When this did occur, the social
order was most likely already breaking down or was undergoing
transformation. Thus, to preserve the traditional social
order, mass support could not be assumed; landed elites had to
work at preserving it. The gentry was expected to 'care' for
its wards (to reciprocate mass acceptance of social inequities
which favoured the elite). Conflict amongst elites was also
to be avoided, since an aura of consensus and unity helped
preserve the legitimacy of the existing order (Dyer 1978).
Failure to meet these two basic requirements could readily
induce pressures for change in existing power relations.

There are many reasons for the decline in traditional
power structures. In part, landed elites themselves decided
that this should be so. Seeing chances for higher monetary
returns, many deliberately dismantled existing patterns of
mutual obligation (in the case of highland Scotland by
forcibly evicting much of the population; Prebble 1963; also

Bird 1982). At times, it has been 'external' influences which have wrought the change. In the southern USA, for instance, an influx of new industry seeking cheap labour, while facilitated by some members of the local socio-economic elite (e.g. Cobb 1982; Molnar and Lawson 1984) introduced a new element into the apex of existing local power structures. The passing of national social welfare legislation (pensions, unemployment insurance and health programmes) has likewise weakened the need for local inhabitants to rely on elites for aid in times of distress (although the more localist orientation of US local governments has meant that this particular change has made fewer inroads here than in most of Europe).

What has come to occupy the attention of many rural analysts since the publication of Howard Newby's major East Anglian studies (Newby 1977; Newby et al 1978) is the manner in which these traditional power structures have lingered on (in somewhat differing form) with large-scale family farms at the apex. In Britain, these farmers largely trace their roots not to the gentry of Victorian times, but to tenant farmers of that era. That they occupy important local positions of power is readily apparent. The Maud Report for example found that 35 per cent of all rural district councillors in England and Wales were farmers (Moss and Parker 1967; see also Broady 1980), which needs interpreting in the context of Newby et al's (1978) finding that 73 per cent of farmer councillors hired labour (42 per cent employing at least five) compared with 32 per cent (8 per cent) amongst the general farm population (see also Buller and Lowe 1982). In general, in lowland agricultural areas, farmers and landowners are well represented in local political institutions. This they achieve not simply through elected office but also through holding influential positions in local political parties (Johnson 1972; Newby et al 1978) and from their close liaison with governmental agencies (like local water boards; see Saunders 1984). Despite the multiplicity of their power bases, leaders in neo-traditional power structures are less dominant than elites were in traditional structures. Principally, this is because the top rung of the local power structure is now held by a particular social class (say large-scale farmers). In true traditional structures this position was in the hands of a single family or at most a few families. So far as we can ascertain from the limited empirical evidence which is available, conflict amongst elites in neo-traditional structures is comparatively rare. However, with a larger membership (in the elite ranks) the potential for disagreements obviously increases. Hence, even if conflict is not present on the surface, its potential for emergence should restrict the range of actions which elites are prepared to support (in other words, with a larger group of 'significant' power-holders, the chances of disagreement over any proposed action is greater than with a smaller group). Hence, <u>positive</u> action should not be expected to be a hallmark of neo-traditional power structures (though it <u>could</u> be of traditional ones). In fact, this is not that critical. In rural development terms, what is a fundamental element of neo-traditional and (most) traditional power structures is that elites seek to block change. This we saw in the previous chapter in our discussion of opposition to industrial and residential developments amongst large-scale farmers. What is critical in traditionalist power structures is the elite's attempt to restrict opportunities for self-advancement (except amongst themselves), while promoting the belief that the existing social order is the only acceptable one.

The situation in the other two rural power structure types is quite different. In the first of these, initial appearances might suggest a strong similarity. This 'rural idealist' power network is characterised by relative stability in socio-economic systems. It is also characteristically associated with dominance by local capitalists. However, the structure of power is more diffuse and the ability of local leaders to direct socio-economic change is weaker. First, we should ask why local capitalists do not provide leadership? In answering this question, differences in social class structure must be noted. Almost certainly these are localities in which inequalities in economic standing are smaller than those under neo-traditional structures. No hard and fast lines can be drawn to demarcate where one power network should begin and another end. Yet it is notable that where socio-economic frameworks veer toward neo-traditionalism, but where 'rural idealist' structures are dominant, local capitalists tend either not to depend on controlling local socio-political events for their economic rewards or else they are 'isolated' from mainstream socio-political events. The former of these is evinced in Bealey <u>et al</u>'s (1965) observation that the local gentry were <u>not</u> concerned with local events in rural Staffordshire, while Burnett (1951:114) provides us with an example of the latter from Hanna, Alberta:

> That only the socially superior class are active participants in the town's associations decreases the stability of the town and its contribution to the country region around. The business and professional people are probably able enough, but they are too busy, too fun-loving, too removed from and too disliked by the rest of the town to give wise and constructive leadership.

Even where an economic elite exists, under rural idealist

power structures the effective population from which that power structure emerges is more egalitarian than that found in neo-traditional sectors.

Five main features characterise rural idealist networks. First, as with more traditional structures, there is an attempt to personalise political activity and local social affairs. Political parties, for instances, are disliked and the politics of personality (or family heritage) is stressed (Grant 1977). Bureaucratic rules and regulations are likewise shunned. Those who exert power emphasise that such rules and regulations do not take account of particular family (or personal) problems. They are damned as 'urban' and impersonal; as intrusions into a personalised 'community' (hence analysts have found that the idea that the allocation of public housing to people should be based on a points system - that is, according to an 'objective' list of housing needs - has commonly been rejected in English rural areas in favour of allowing councillors - who 'know' the applicants' true 'worth' - to select tenants; Newby et al 1978; Dunn et al 1981). Second, largely on account of a more egalitarian resource distribution, turnover rates can be fairly high amongst those who exert most influence on local affairs (i.e. there is no long-term power elite). Indicative of this is Bokemeier and Tait's (1980) findings for two small towns in the United States, where around two-thirds of the power 'elite' in 1962 did not occupy a similar role in 1973 (with just over three-quarters of the 1973 'elite' not holding similar positions in 1962). Third, obviously associated with power actor change-over, there is commonly a reluctance to take up formal positions of authority. Occupying such positions can be a time-consuming business. This is especially so when 'leaders' are pressured into a host of secondary (often unwanted and alien) activities, merely because failure to accept these positions can lead to the accusation of being a selfish power-grabber who wants to take and not give (see Vidich and Bensman 1968). In effect, returns from position-holding are slight. Little extra influence in local affairs is derived and financial rewards are generally minimal (if not negative). Moreover, personal losses are potentially high, for decisions might have to be made that will offend friends and neighbours. Not surprisingly, analysts have found many people unwilling to take on formal positions of authority in these circumstances (e.g. Williams 1963). Alternatively, strategies are evolved for minimising intra-local conflicts. Thus, in 'Pentrediwaith', 'outsiders' ('strangers' or upper class figure-heads) were 'invited' to occupy key organisational positions. These people were then blamed for destroying village unity when 'hard' decisions had to be made (Frankenberg 1957). This brings us onto the fourth aspect of rural idealist structures; namely that they are characterised by attempts to preserve intra-local social harmony. As Banfield and Wilson (1967:25) expressed it:

> Nothing in small town politics is as important ... as the preservation of peace and harmony and the maintenance of easy personal relationships ... To put the matter more generally, the function of politics in the small town is less to resolve issues than, by suppressing them, to enable people to get along with each other while living together in very close contact.

There is abundant evidence to support this view (for rural idealist power networks at least), but one numerical example is sufficient to illustrate its potency. The example selected here comes from Wild's (1974) investigation of the Australian locality 'Bradstow'. Here, between April 1967 and September 1968 a total of 1099 formal decisions were taken by the municipal council. Of these, 1087 (all bar 12) were unanimous. This is hardly the kind of ratio one expects from a government seeking to address the problems that any locality inevitably faces over a 17 month period. So we come to our fifth and last point. This has been most vividly expanded on in Vidich and Bensman's (1968) Small Town in Mass Society. In brief, the message is that, as local 'leaders' do not make the difficult decisions that must be made to direct change in a locality, the end-result is a lack of local autonomy. Change in the locality is left to be directed by outside forces. In more traditional power structures there is both the will and the political savvy to block many external impulses. But in rural idealist surroundings, the power 'elite' has none of the experience, the drive, the public support or the political connections to avert unwanted intrusions.

The designation 'rural idealist' for this power structure emerges for two main reasons. First, because this type of locality appears to live under the illusion that it controls its own future. Thus, when Small Town in Mass Society was first published, 'Springdale' residents were outraged at Vidich and Bensman's portrayal of their self-imposed inability to address local problems (although in recent years a growing awareness of local powerlessness in the face of outside interventions has emerged in many rural locations; e.g. over such issues as local school closures - see Forsythe 1984). There was also uproar over Vidich and Bensman's descriptions of the differences in social rank which existed in the town (Vidich and Bensman 1968). This stems from the second component of rural idealism; the claim that such localities are 'classless'. In fact, the operations of rural idealist power structures have very marked, class-related distributional consequences. Although there is a less formalised

elite in such localities, there are still social discriminations. A small group does not control the main organs of
socio-political life. But there are obvious restrictions on
who is acceptable as a position-holder in local organisations.
Predominantly, position holding goes to the middle class and,
most particularly, to owners of small businesses and to
professionals. Fearing innovation in case it raises conflict,
position-holders find their main role in ensuring that local
taxes are kept low (Vidich and Bensman 1968; Claval 1983).
This is in keeping with the prevailing ideal for small
business enterprises (albeit it is frequently justified by
reference to the plight of the elderly on fixed incomes). It
also has the effect of reducing self-advancement opportunities
and service supports which are available for lower income
inhabitants (who cannot privately afford the services local
governments might otherwise provide). As a further element of
rural idealism, the reverence which is bestowed on non-
partisan politics, and on local officials' personal
involvement in service provision, disguises the fact that
lower-income residents have little opportunity to have their
demands heard. No political group represents their interests
in elections. In the general course of organisational
operations, they also find their needs are easily discounted.
In terms of obtaining public housing, for instance, Connell
(1978:117-8) found moralistic overtones were strong in
assessments of whether applicants in Guildford Rural District
were successful. By making criteria like, whether a
householder's existing home was 'clean' or whether the
applicant was 'nicely spoken', the service allocation
procedure implicitly incorporated goals of social control.
This 'social service' was not operated simply for 'need'
satisfaction. In this situation, the local power structure -
more egalitarian than 'traditional' arrangements - is actually
quite claustrophobic for low-income inhabitants. It requires
outside involvement for change to occur.

Such 'outside' interventions are characteristic of our
third power structure category - the conflictual model. Most
commonly this kind of power structure has been associated with
localities experiencing a large influx of new in-migrants.
The characteristic picture which emerges from the literature
has been described by Newby (1977:330):

> Quite quickly, then, a new social division may
> arise - between on the one hand the close-knit
> locals, who form the rump of the old occupational
> community and, on the other, the newcomers. The
> former occupational community then tends to
> retreat in upon itself and become what might be
> called an encapsulated community, since the
> locals now form a community within a community -

a separate and dense network encapsulated within
the total local social system.

The theme that counterurbanisation pressures are including a
bifurcation in the social structures of rural localities is
now well established on both sides of the Atlantic (e.g.
Dobriner 1963; Pahl 1965; Ambrose 1974; Sinclair and Westhues
1974; Wild 1983):

> ... as Jeffrey Bernard put it a few years ago in
> the New Statesman: 'The place [commuter-belt
> Surrey] is divided into two halves. At one end
> it's lavender, medium sherry, The Daily Tele-
> graph and what a wonderful job the Conservatives
> are doing, at the other end it's baked beans,
> bicycle clips and what a lousy job Sir Alf
> Ramsey's doing'. (Cooper 1980:241)

The importance of such divisions comes in the manner in which
it alters existing power arrangements. The imagery of Surrey
which Jilly Cooper adhered to (above) is one in which new,
in-coming middle class residents were grafted onto an already
subservient working class population. The previous upper
class elite did not conflict with the newcomers. Rather, the
clamour for middle class homes in the countryside was used by
the upper classes (to reap large sums from land sales), who
then removed themselves to avoid the emerging middle class
invasion. As Cooper (1980:240) made clear:

> ... the real upper classes - except for one or
> two like the Earl of Onslow whose family have
> lived near Guildford for centuries - wouldn't be
> seen dead in Surrey.

In such instances, the change in power that heralded in middle
class supremacy was not conflictual. Rather the power
structure was eased out of a traditionalist mould into a rural
idealist mould. Where existing power-holders stay on in the
face of new influxes, however, the scene is set for a clash of
values:

> To the villager, Old Harbour represents
> continuity between the generations, stability
> instead of the city's chaos, and a place of
> permanence in a universe of bewildering change.
> The suburbanite sees in the village a weekend
> away from the advertising agency or the pilot's
> compartment. He [sic] experiences Old Harbour as
> a series of isolated, fragmented, unconnected
> social situations. Old Harbour is the family, a
> cocktail party, a bathing beach, a movie, a

supermarket, a country club, a school, a church, a PTA meeting. It is a one-acre wooded retreat from all the drive, bureaucracy, and anxiety of the city. But the weekend is enough for the necessary physical and psychological repairs; it's back to the city on Monday. (Dobriner 1963:135)

The reason for including this long quotation is not because it represents the universal character of recent in-migrant behaviour. On the contrary, many studies have shown that many (even ex-urban) migrants are actively seeking to become firmly entrenched in existing rural social structures (e.g. Forsythe 1980; Ploch 1980). However, Dobriner's description does reveal the kind of gulf that exists between many long-time residents and newcomers. Even when the will is there to become part of the existing social order, these value differences can lead to a seemingly unbridgeable bifurcation of local community interests (e.g. Forsythe 1980).

Attempts by long-standing ruralites to maintain a sense of locality which is distinct from that of village newcomers (whose mobility permits them to maintain more extensive social networks), have often accelerated encapsulation and entrenched segregation on both sides of the fence. In rural Surrey, where affluent newcomers outnumber the traditional working class almost to the point of the latter's extinction, polarisation is extreme. The main form of contact between the two social groups comes from the latter's role as gardeners or cleaners for the former (Connell 1978). Forsythe's (1980) example of the Orkney community of 'Stormay' illustrates in considerable detail the nature of the newcomer – longstanding resident divide. It particularly reveals the distrust and misunderstanding that kindles conflicts. In 'Stormay', the ability of ex-urbanites to dominate and even 'capture' local voluntary organisations is aided by their more 'forward' and authoritative style of communication (which contrasts with the local custom of not engaging in self-promotion). Certainly, the differences recorded by Forsythe were aligned with basic dissimilarities in socio-cultural background, but much more than this was involved. Although Forsythe draws the distinction between the ex-urbanite who migrates to the remote Orkneys and those studied by Pahl (1965) and Ambrose (1974), whose destinations were villages within the commuting distance of major urban centres, the conflicts surrounding these two groups are essentially the same. This observation is confirmed by similar conflictual situations which have been identified in the USA (Carlsson et al 1981:16). In all of these instances, the potential for class-based divisions to provoke conflict must be recognised. As traditional rural activities, roles and ideologies are taken up and transmogrified by newcomers, the sense of isolation and fatalism of the local working class can increase. When, with their newly acquired rurality, newcomers seek to impose their view of what does and what does not constitute acceptable activities in the countryside, fatalism can turn to exasperation. An example of this, in the form of recreational gentrification, is provided by Elson et al (1986) in their review of the recreational conflicts associated with off-road motorcycling. Such activities have long provided an important component of the recreational and sporting opportunities of members of the rural working class. In many areas of Britain, off-road motorcycling has been taking place regularly for over 50 years. Increasingly, however, local objections, almost entirely from newcomers to the rural scene, have forced a large-scale reduction in these activities. As Forsythe (1980) reported for the Orkneys, newcomers seek to live out an idealised rural lifestyle that they believe does (or at least should) exist. Activities they believe are non-rural are resisted, even if rural residents have long practised them. The resulting clashes in ideology reveal not only the variety of rural images that exist, but also the supremacy of the newcomer vision over that of long-standing rural interests.

In the face of such 'provocations' from new in-migrants,* local residents often withdraw from socio-political affairs (Popplestone 1967; Newby 1977; Forsythe 1980). However, whether this does or does not occur will depend upon the character of the existing power structure and the stake existing powerholders have in its continuance. Of particular relevance here is the provision of public services. Most especially for education, but also in other spheres, much evidence can be brought forward to show that new in-migrants commonly expect a higher service standard than long-time residents have been accustomed to (Summers et al 1976; Hennigh 1978; Price and Clay 1980; Green 1982). In part this picture emerges on account of differences in the socio-demography of new residents and long-time residents (in-migrants are generally of higher socio-economic standing - leading to expectations of higher service standards; and they are commonly younger - which increases demands for expensive services like education and recreation). In part, it simply arises because they have become used to better provision standards in their former, more urban habitats. In any event, the end-result is commonly conflict over who shall direct the provision of services and future patterns of change in the locality (e.g. Hennigh 1978).

---

* By 'new in-migrants' we are excluding return migrants or those with long-time family ties within a locality.

There are no easier answers to the question, what is the outcome of such conflicts? The picture we have painted above is a very simplistic one. Conflict has been presented as an interaction between two elements distinguished by length of residence. In truth, however, conflictual power structures arise in a variety of ways. The black civil rights movement induced many local conflicts in the US South which led to changes in power networks (Molnar and Lawson 1984). Likewise challenges to the existing order have come unexpectedly from younger entrepreneurs in many localities. Admittedly, these young hopefuls do not wish to see any marked changes in how societal benefits are distributed in their locality (except that they want more), but they do seek to usurp the 'old guard's role' and instil more dynamism into socio-economic affairs (e.g. Gold 1975). Even where the conflict is between long-term and short-term residents, outcomes are difficult to predict. The basis of conflict which has been outlined above emerges from dissatisfaction with existing service provisions. As described in the last chapter, however, conflicts drawn along similar lines of social division also emerge when newer inhabitants wanted less services provided (largely where these services were housing related). It might be tempting to see such conflicts solely in class terms, but this would be a mistake. New middle class migrants' opposition to additional housing undoubtedly works against the working class. But their demands for improved education facilities would benefit working class children (you do not usually see newcomers arguing for service improvements that specifically help the working class – like welfare – rather, these benefits arise as spin-offs from middle class gains). Here the incidence of conflict is more usually between two middle class groups (i.e. newcomers and 'old' business elites). Even the word 'conflict' might appear to sit with some discomfort over this power structure category. Some of these 'conflicts' are visited by violence, but many others take the form of quiet discussions and covert manipulations. In effect, conflictual power structures are those in which power relations are in flux. Social researchers currently know too little about the consequences (or even processes) of change in these circumstances. This is an area that needs a stronger research commitment, as do the elements of social status divisions in rural locales.

SOCIAL STATUS

Having earlier noted the criticism that rural studies previously placed too heavy an emphasis on the importance of social status, the claim of the last sentence requires some justification. This comes from our particular treatment of social status. Studies conducted in the 1950s and 1960s, have been criticised in three ways: first, the bias towards agrarian settlements with roughly equal-sized, family farm holdings (which, hardly surprisingly, often led to the conclusion that class was unimportant in rural locales); second, the almost myopic local frame of reference which was dominant in rural investigations (this ignored broader power structures which moulded the character of local social systems - that status often appeared to be important was because the critical features of the local social order, which were determined extra-locally, were not examined); and, third, the seeming reluctance of researchers to pinpoint the importance of class structure even when it was implicit within investigators' reports. The framework we have used in this text avoids these problems. It still leaves us with instances in which social status is an important dimension in rural development. By status we are referring to prestige. But prestige differences are intricately woven into the fabric of class and power structures. The way we have sought to extricate ourselves from this complexity is to treat class more explicitly with reference to income generation processes, to analyse power networks as a locality-wide phenomenon and to focus on status as an attribute of individual people. Social status differences, in the context examined here, are power-related, but they cannot be reduced to class structure and they prompt different (though related) power networks to those of locality-wide power structures. Viewed in this manner, we have distinguished status dimensions by their being attributes of individuals per se (not of social processes or material possessions); attributes that is which make a significant imprint on the distribution of societal benefits within rural localities. Most significant in this regard are the social status differences which differentiate men and women and which separate white and non-white rural inhabitants.

The position of women in rural areas has become a topical issue for research and debate in the last decade (Little 1986). Research has thrown considerable light upon the importance of traditional stereotyping of gender roles, and has helped highlight an insufficiently explored dimension of rural social structure. Probably three main images persist in the stereotyping of gender roles. The first of these is of women as supportive wives and mothers:

> In agriculture, the father is the head of the household and the chef d'éntreprise at the same time. The mother is queen of the hearth and the hearth is the centre of attraction and the convergence of the farm and family. (Anon, quoted in Franklin 1969:73)

The second image is that of women as defenders of family and local community norms (as Worster 1979 reported for the Dust Bowl states in the 1930s, the family was the main institutional force against instability - i.e. outmigration - excessive individualism and 'outside' influence, with family concerns being primarily allocated to women to organise and control). Thirdly, there is the image of women as a cheap labour pool. This is seen: (1) on family farms, where wives (and daughters) are called upon both as labourers in times of peak demand (Morkeberg 1978) and to undertake repetitive, unskilled tasks (Gasson 1981); and (2) as employees, since the notion that women work for 'pin money' is used as justification for their pre-eminence amongst low-paid, part-time employees (93.3 per cent of part-time retail workers in Powys are women, for instance; Watkin 1978) and their increasing share of the workforce in low-income occupations (like agricultural labourers - Gasson 1981 for England and Douglass 1984 for Italy). None of these stereotypes is restricted to rural inhabitants, but each of them possesses a peculiar manifestation in rural locales (most especially on account of the importance of farming).

To appreciate how these stereotypes have substantial effects on women's lifestyles and well-being in rural locations we must recognise how their behavioural options have been manipulated through the male domination of society. Predominantly, women have been treated as if they are 'factors' in a male production process (Haney 1983; Sachs 1983). Their joint roles as custodians of family and community norms occupy key positions in the perpetuation of a male-oriented social reproduction process. Then, in the production sphere, they have been called on to bolster the activities of their male partners rather than pursuing goals of their own. To give one example:

> ... as one respondent in the Farmers' Weekly survey put it, 'Farmers' wives have fewer career opportunities because their husbands expect them to be consistently available in case of emergencies'. (Gasson 1984:221)

As Golde (1975) made clear in his West German study, men who are unmarried, and cannot obtain female help, find it next to impossible to continue farming for long. Not surprisingly, given male domination, this 'need' has been translated into norms for female behaviour. In the early part of this century, for example, it was not uncommon to find the expectation that young rural women would either obtain work on a farm or else become a school teacher. Any ventures into nearby towns to obtain alternative work (like being a waitress) were frowned upon, and often carried the assumption

that the woman had 'turned' to prostitution (West 1945).
Clearly, this acted as a restraint on women's desire to break
away from a local social system; even though this system was
directing them into a subservient position, as farm wives.
Change in societal norms in general (partly consequent upon
improved access to 'outside' information in rural
settlements), and reductions in the numbers of farms (and
hence 'demand' for 'farm wives'), has changed this situation
in recent years. Younger women now show a marked reluctance
to become subservient farm chattels (e.g. Brody 1973; Douglass
1976). Outmigration amongst women is common (most especially
for those with higher educational levels; Fulton 1975). But
those who stay, still find the hardships of rural life bear
most heavily on rural women (and this applies in a variety of
contexts; see Burnett 1951; Albrecht 1978; Momsen 1983). To
appreciate more fully this facet of rural living, we will
return to the three stereotypes listed above and explore their
components in more detail.

A variety of studies has emphasised that households in
rural areas are characterised by a highly segregated division
of labour, organised along lines of gender (Dunne 1980;
Stebbing 1982). The role of supportive wife and earth-mother
is embodied in the attitudes of Stebbing's (1982) 'country-
women' (who comprised 65 per cent of her Kent sample).
Proclaiming no distinctions on class or age grounds, these
women:

... encapsulate the very essence of the rural and
domestic idyll – the calm well ordered home as a
refuge from the whirl of meaningless urban
activity, presided over by a woman in touch with
nature and life's real values. (Stebbing
1982:202)

Similarly, although less adhered to by younger women, a
dominant female job stereotype which is held by both men and
women in rural areas of the United States is that of
'homemaker' (Flora and Johnson 1978; Dunne 1980). In
reality, the role of 'homemaker' goes hand in hand with the
persistence of the family as a basic social unit.
Particularly in areas of family farming, and where income
levels are low, the family achieves a significance far beyond
that provided by its roles for partnership or child-rearing.
As Fitchen (1981:122) observed, in poorer areas:

Due to the collapse of the local rural community
and the failure of the urban community to become
a social substitute, the family in the rural
depressed areas has been forced to take on extra
functions. Because of the social marginality,

the multigenerationally poor families must provide for their members most of the social and psychological functions that more affluent members of the community satisfy through a variety of secondary relationships and groups. By default, the family is the only group in which poverty stricken rural people regularly participate on a sustained basis.

Underlying the practice of women dominating family-centred life is a critical power relationship in which women are subordinate to men. Of 'traditional' working class women, for example, Strathern (1984:187) declared:

These women are kin keepers, highly competent in family matters, keeping kin networks open, and they rather than their husbands tend to give voice to social aspirations and to see possibilities for mobility not open to men. Marriage is crucial for women in opening up choice and domains of decision making where they are principal actors ... There is some evidence that these women see themselves in key boundary positions.

This duality of position - the power house of family affairs, but dependent upon a male's social position for broader status - is perhaps changing as many women recognise the inequity in these gender work-divisions (hence many male farmers are finding an increasing female reluctance to become farm wives; Golde 1975; also, more generally, Dunne 1980). In truth, however, this appreciation is not well developed (as evinced in Stebbing's 1982 finding that, even in comparatively wealthy and gentrified Kent, 81.8 per cent of respondents accepted 'traditional' gender-based divisions of work and responsibility). Furthermore, by continuing to act as zealous guardians of local social mores, and in seeking to maintain (or secure) local social stability, women frequently help perpetuate the disadvantages of their existing situation.

For women, the role of family maintainer spills over into the function of local 'community' preserver (Davidoff et al 1976). As Strathern (1984:188) has put it '... the habits of daily life are status indicators, crystallised in household management. In this sense, the style of life of a local community is a domestic style of life writ large'. With children commonly forming a focus of their social endeavours (Bouquet 1981), women have traditionally taken the leading part in organising local social organisations and activities (Fitchen 1981; Stebbing 1982; Haney 1983; Sachs 1983). What we wish to emphasise is that these joint roles of home and local community 'organisers' are critical underpinnings of production (or, more generally, income-generating) processes. Without stability in the home and in local social systems, the capacity for earning a living would be severely reduced. Further, the continuance of existing social and productive mechanisms intimately depends upon these joint roles. Young people must be socialised into both family and locality norms, if existing social arrangements are to be reproduced. The notion present in much rural writing, that women play background, supportive roles in rural development processes, we therefore reject. That men dominate the production process (less so now than in the past) has, through male-domination itself, been projected as evidence that men's work is more important than women's. Our point is that social reproduction is just as important as production (or income earning). They are effectively opposite sides of the same coin. Moreover, as we shall see below, men have both relied on women to ease their production-related tasks and, at the same time, have restricted female opportunities for advancement in the production sphere.

Even in the production sphere, women take on multiple roles with a reproduction (or socialisation) component. As Gasson (1980:166) reported, '... the role of wives in producing and rearing successors and in socialising them to accept that role is crucial to the survival of most farm businesses'. In effect, in farm households women fulfil three (alternative) ideal roles - as a worker who lives on a farm, as a working farm woman, and as a woman farmer. Where individual farm women fit into this schema depends upon their own socio-economic status, the size of the farm enterprise and prior socialisation. At the bottom of the pile in socio-economic terms is the woman farmer. Evidence from a variety of nations clearly shows that farms run by women tend to be amongst the smallest in size (First-Dilic 1978; Gasson 1981). Further, even when formally qualified for agricultural work (e.g. by completing technical agricultural courses), women are less able to obtain work relevant to their training than equally qualified men (Gasson 1981). Even with the help of a rung up the ladder - as provided by land ownership - women farmers still find their progress restricted by male prejudices (as seen in the difficulties experienced in seeking to increase farm sizes or in obtaining bank loans; e.g. Salamon and Keim 1979). To circumvent these prejudices often requires the surrendering of decision-making power to male partners. For working farm women, opportunities are likewise restricted by male domination. For example, as Sachs (1983) has pointed out, up to 1981, farm inheritance tax laws in the United States discriminated against women, so that even when they did inherit, the farm often had to be turned over to a man. Although social practices do vary considerably, there is a very marked tendency for farm heirs to be male. In farm operations itself, the sting of prejudice is also felt. Gasson (1981), for example, found that 40 per cent of women in a Farmers' Weekly survey blamed male conservatism for the poor way they were treated on farms (the figure rose to 59 per cent for hired labour). In work tasks, women have predominantly been 'relegated' to repetitive, unskilled tasks: 'The woman's place, it seems, is in the byre or barn rather than out in the field' (Gasson 1981:5). Their decision-making roles are highly restricted. Usually administrative tasks are found to lie in the 'woman's realm', but not production or purchasing decisions (Salamon and Keim 1979; Coughenour and Swanson 1983; Buttel and Gillespie 1984; Fassinger and Schwarzweller 1984). Of particular note for the future, it should also be reported that women appear to have a smaller decision role the larger the farm (e.g. Sharp et al 1986). What is more, there appears to be no justification for assuming that this pattern is a product of capitalist, social or production relationships. Studies reveal that very similar male-female farm work divisions exist under communist regimes (e.g. First-Dilic 1978). As Bokemeier and Tickamyer (1985) reported for the United States, even with a 34 per cent increase in female agricultural activity in the 1970s, the job rewards women receive (income, prestige, stability of employment) have not improved.

As made clear throughout this text, agriculture is a 'pressured' economic sector. Perhaps as a result, not much can be expected of it as a vehicle for improving female work conditions. Yet evidence does not suggest that other economic sectors provide a much better employment base. Too frequently, non-farm employment for women offers little more than relatively low pay and little call for occupational skills (Dunne 1980); albeit, as Bryant and Perkins (1985) report, off-farm work does at least provide more opportunity for interaction with friends, neighbours and kin, which have important social-psychological benefits in their own right. As a variety of researchers have found, rural female employment tends to be in competitive sectors of the economy, where employers constantly seek to keep wages low (Bradley 1984; Deseran et al 1984; Horan and Tolbert 1984). That women take such low paying jobs is frequently a result of economic (and to some extent socio-psychological) necessity. This is shown in studies of variation in female activity rates. Here, it is not the characteristics of females or of their households that determines employment rates. Rather it is the openings offered by local labour markets. High rates of female (non-household) work participation are directly linked to the opportunities for such work within a locality (e.g. Moseley and Darby 1978; West 1981; Bokemeier and Tickamyer 1985). Improved transport mobility (i.e. car availability) also has a positive effect, as does the availability of such services as day-care centres (Morkeberg 1978; West 1981; Stebbing 1982).* Each of these affects the earning ability of women, but they in no sense account for the lower income women derive from their work.

In highlighting contrasts between male and female work conditions in rural areas, a critical point that needs emphasising is wage differentials. Many researchers have catalogued the fact that women are not less educated than their male counterparts (often the reverse is the case; e.g. Dunne 1980; Gasson 1984). Indeed, this situation is reflected in the higher incidence of white-collar work amongst women working off-farm compared with the dominance of blue-collar work for men (e.g. Coughenour and Swanson 1983). Yet female wage rates (for equivalent work commitments) are significantly lower than those of men. Undoubtedly, a reason for this is that men are more prevalent in the monopoly sector of the economy (where employers find it easier to pass on wage increases to their customers), whereas women occupy a substantial part of competitive economic sectors (Horan and Tolbert 1984). The dualistic structure of labour markets helps explain why large wage differentials exist. It does not answer the question, why do women end up in the poorer paying jobs. For this we need to examine how gender discrimination is integrated into structural economic divides.

The ability of women to enter local labour markets on an equal footing is partly a function of the persistence of traditional expectations about 'a woman's place' (Flora and Johnson 1978). As Morkeberg (1978:104) pointed out for residents of farms: 'In most cases employment outside the farm has given the farmer's wife a double work load'. Maintaining the traditional role of family 'organiser' greatly restricts work opportunities and weakens any sense of urgency over repairing the poor working conditions women face (such work often being seen as a part-time or incidental household role):

More significant deterrents to women taking or keeping jobs have to do with attitudes and

---

* As a pointer to the future, it is appropriate to note that the centralisation of employment and service facilities into 'key villages' or 'growth centres' has important negative implications for female employment. This is especially so since rural women often have restricted automobile availability (Moseley 1979) and usually fulfil the role of family child-minder (picking up children from school for example can be very tying and leads to some rural mothers seeing themselves as glorified taxi drivers - Momsen 1983).

> beliefs about women's roles, and with the relationship between husbands and wives. Some husbands openly disapprove of or actually prohibit their wives from working. Some wives report that their husbands would be jealous of the competition, particularly if the husband is out of work or earning very little. A number of women also report that their husbands believe a woman should be at home with her children. (Fitchen 1981:70)

Although this might seem to boil down to an intra-family disagreement, traditional social mores designate the husband as the ultimate family decision-maker. In the end, '... the men appear to have the positional advantage of power, either in the domestic group or the workplace, to discriminate against the access of women to equal labour market opportunities' (Bradley 1984:83).

Socio-economic stability in rural areas has served to perpetuate traditional stereotyping of gender roles. Yet change per se has not necessarily been a panacea for female disadvantage (i.e. change is not unidirectional). Constraints on opportunities and choices continue to exist and in some measure are intensified by current trends in society. The trend towards higher rates of divorce and single-parent families, for instance, has an accompanying consequence of female headed households being more likely to live in substandard housing, with incomes below the poverty-line (e.g. Marantz et al 1976). As sole head of the household, such women can see their living standards slipping away as public and private services are centralised into larger places (making them more inaccessible). Meanwhile, new job opportunities for women, although increasing the chances of employment, offer little in career prospects or salary. The fact that around 40 per cent of the British workforce is now female, and that this percentage has increased consistently over the last 20 years, should not lull us into believing that labour markets are uniformly open to female penetration. Women are still highly restricted in their employment prospects. Part-time, low-wage, unskilled jobs with little opportunity for self-advancement are still the norm. This applies in both the city and countryside, but it is more marked in its incidence in rural locales. This is because of the strength of traditionalist values (induced partly through the 'necessities' of family farm operations), as well as emerging from the particular paths which current socio-economic trends are following.

The existence of similar conditions in city and countryside environments, but with greater intensity in rural areas, applies to household poverty in general (in 1981, for example, the rural USA had 32 per cent of the nation's population but 40 per cent of the people living below the poverty line; La Follette 1982). As Cullingford and Openshaw (1982) found in their empirical analysis of northeast England, there are no fundamentally different dimensions to socio-economic conditions between these two physical environments. Indeed, Cullingford and Openshaw recorded only one socio-economic dimension that was distinctly rural. This was equivalent to deprived areas in major cities, except for the high incidence of unfurnished rented accommodation and the low level of unemployment. The combination of these two distinguishing variables is very suggestive of low-wage employment associated with tied housing. Given trends towards more low-wage manufacturing jobs, aligned with an existing emphasis on jobs in the competitive economic sector, perhaps this 'low-wage but employed' dimension will become an enhanced component of rural socio-economic environments. Already we have seen that this occupational position is linked to female employment. What we also need to recognise is its associations with race.

In truth, race is a dimension of rural employment patterns that has been little explored. Partly, this is on account of its uneven appearance as a significant issue. In Britain and most of Europe, for example, the bulk of minority racial groups are urban dwellers. It is mainly in the United States that the racial element of rural labour markets appears significant. Even here some care must be taken in assessing its importance. As Flynt (1979) has pointed out, in parts of the United States, the rural white population is differentiated from its black (or hispanic) counterpart only by reason of colour. Poverty of white and black coexist. There is certainly evidence of racial discrimination. As an analysis of six rural housing markets in different parts of the US found, homes in black submarkets sell for significantly higher prices than equivalent housing in white submarkets (Marantz et al 1976). Likewise, industry moving into the southern US has acquiesced to local 'customs', so that blacks have been denied equal employment or promotion prospects (Greenberg 1981). Similarly, through a white elite's 'capturing' of the organs of the Tennessee Valley Authority, policies have been pursued with a clear anti-tenant, anti-black direction (Selznick 1949; McDonald and Muldowny 1982). Whether these examples are revealing deliberately racist practices or merely identify racially-specific outcomes (which are by-products of other processes) is a moot point. In the Tennessee Valley, for example, the lack of help given to black agricultural colleges can be seen as part of a broader opposition to institutions or programmes that aid small-scale farmers (Selznick 1949). Similarly, while the proportion of

US farmland in black ownership is now around 40 per cent of its 1910 figure, this must be placed alongside a general trend for smaller farms to be bought-out by larger concerns. Even the fact that the appallingly treated agricultural labourers of the United States are overwhelmingly from racial minorities is not necessarily evidence of racism. The white 'Dust Bowl' migrants of the 1930s worked in conditions which, for the time, were also quite appalling. Growers have been concerned with finding a population group that has little political clout, which they can exploit with few worries over repercussions. Of course, racism undoubtedly helps farmers in this regard (as it does industrialists; Greenberg 1981). But racism is not the root cause of major socio-economic disparities in rural areas. Racism intensifies such disparities and racialist arguments have commonly been employed to justify the policies of the dominant white elite (as with the 'theft' of vast tracts of land from the North American Indians). With an insignificant population from minority racial groups, however, countries in Europe reveal similar problems and inequities in their rural areas as does the United States. Racial differences perhaps channel the wheels of change more in favour of specific population groups than would otherwise be the case. Yet they do not determine the general direction that they follow. Race is an important element of rural development (in the USA at least), but it is not a major determinant of broad change patterns.

PERSPECTIVE

In a chapter that examines intra-local divisions in social standing, it is perhaps to be expected that sections will be given over to rural poverty and the particular problems of the rural elderly. This has not been the case in this chapter. We would not wish to claim that these are not population groups that merit attention. Neither would we argue that there are not issues specific to rural areas that justify distinguishing the rural dimension of these social groups as distinctive topics. Yet, in deciding on material to include in a text, selectivity is inevitably an important consideration. In our view, the particular problems and issues of these social groups carry tags which are insufficiently unique to merit them being accorded status as specifically 'rural' entities. In other words, while recognising certain differences in rural-urban behaviour patterns, we accept the arguments put forward by many rural researchers that the situation of the rural elderly is basically similar to that of the urban elderly (e.g. Wenger 1981). Further, the dispersed locational pattern of the rural elderly makes governmental actions to alleviate their plight both seem less pressing and intensifies problems of service delivery. In this, issues 'surrounding' the elderly are little different from those affecting the poor. Basic underlying poverty-causing processes are eqivalent in rural and urban locales even though their manifestations and policy consequences are somewhat different. In the main, these processes operate at an international, national and inter-local level. Movements of capital produce locationally-specific patterns which primarily determine variations in the incidence of poverty. This is not to say that locality-specific causal processes do not have a role to play. But these processes are fuelled by the social class, power and status considerations we have examined in this chapter. It is these that provide the peculiarly rural input into locality-based causation. This does not mean that the topics we have covered exhaust the range of causally significant, but locality-bound, social processes with a rural overtone. Of particular interest for geographers, for instance, are the social divisions and conflicts that have been reported in a wide array of studies between village residents and those living in the open countryside (e.g. Burnett 1951; Pitt-Rivers 1960; Davis 1973; Oxley 1974; Bax 1976). Our decision not to include such divisions was not based on their insignificance. Rather, it reflected our view that these divisions were less important than others as determinants of rural development trends.

The material examined in this chapter does not cover issues which are unique to rural locales. The intention has nevertheless been to focus on social divisions whose manifes-tations do carry peculiarly rural markers. In terms of social class structure, despite our belief that it provides an insufficient basis for identifying a uniquely rural system of class-based social stratification, land ownership clearly provides a specifically rural structure to social divisions. For community power structures, the three power network types we identified are certainly not restricted to rural areas. However, by the very character of rural localities - most especially in terms of their overall size (and the implications this has for preserving socially harmonious populations) - the incidence and consequences of such power structures carry significant rural overtones. Similarly, gender and racial divisions possess specifically rural manifestations (although for race this often appears more as an enhanced incidence of general patterns; as Jim Hightower 1973:13 observed, 'If whites in rural America are in trouble, blacks are facing disaster'). What each of the class, power and status stratification divisions reveal is a peculiarly rural-based, locality-specific set of causal processes. These not only help determine the quantity and quality of change within rural localities, but also play a major part in

263

channelling benefits arising from such changes (or the lack of them) toward specific social groups (and away from others). These locality-specifc causal processes have major implications for the ways rural residents actually experience development processes and outputs. At present, however, these specifically local manifestations are among the most poorly theorised aspects of development processes. This is so in general terms, but is even more strongly felt when researchers seek causal models to explain inter-local variations in the incidence of locality-specific processes.

Chapter Eight

SOCIAL RESEARCH AND RURAL DEVELOPMENT STUDIES

Conclusions about the relevance and validity of social science theories for studies of rural development are inevitably conditioned by our understanding of the purpose of theory and the nature of rural living. It is critical that we are clear about our standing with regard to these issues, for they go a long way toward pre-defining the scope of what is seen to be relevant for rural-based investigations. We can illustrate this by contrasting the approach favoured in this text with that adopted by Andrew Gilg in his recent An Introduction to Rural Geography. Gilg's prescriptions for future rural investigations share some common threads with our own; most especially in his belief that too great a concern has been expressed over establishing a sound conceptual basis for the concept 'rural':

> It is probably better to argue that the need for a conceptual framework is probably a passing academic fashion, and that the best concept for rural geography should be a spatial [i.e. ecological] one, in which rural geography is seen as the study of extensive land uses or activities in which the friction of distance plays a major role. (Gilg 1985:172)

Obviously, from the approach adopted in this book, we are happy to agree with the first part of this statement. Yet we are not sympathetic with the second. For one thing, if the friction of distance demarcates rural activities, then urban journey-to-work or shopping patterns are presumably 'rural'. It would of course be easy to exclude such activities, merely by introducing a population density component into one's definition of 'rural'. This still would not bring this conceptualisation into line with our perspective, for the very notion that the friction of distance can be used to distinguish issues of rural relevance runs counter to our philosophy. As shown in Chapter Four, transnational corporations can have very significant impacts on rural development, yet the friction of distance is largely irrelevant for most of their activities (as seen in the production network for Ford's J-car; Figure 4.4). By restricting attention to activities for which the friction of distance is critical, we effectively restrict our interest to studies of individuals (or families) within localities. Further, as telecommunications improve and transport mobility increases (assuming both do), rural geographers will see themselves faced with a diminishing range of activities for which the friction of distance is important (already, using this definition, urban-rural shifts in manufacturing should not be investigated by rural analysts, since these relocations are predicated upon the weakening of communication and transportation restraints). In our view at least, such a delineation is therefore unacceptable. In reality, though, this definition, and its implied orientation in favour of studying individuals within localities, finds its roots quite clearly in Gilg's beliefs about the purpose of rural studies (or, more exactly, rural geographical studies).

The differences between our approach and that of Gilg lie not simply in dissimilar specifications of 'legitimate' subject matters. They also exist in our understanding of the basic role of academic investigations. As Gilg (1985:173) outlined the situation:

> ... the future for rural geography should be an applied one, where it integrates its own research, relates this to the real behavioural world and to policy formulation, and thus attempts to produce a rural environment that is not only physically attractive but also a lively and prosperous place to live.

While this is not a view we accept personally, we in no sense wish to be critical of it. The belief that rural geography will (or at least should) find its future in applied research is a quite valid one. It is nonetheless a view that rests on very clear, but unstated, value-dispositions. Essentially, it assumes a consensus (or integration) view of social organisation, since it rests on the idea that agreement can be reached over what constitutes a 'physically attractive', 'lively' and 'prosperous' environment. It also carries within it heavy overtones of liberalism. There is an underlying acceptance of the view that governments principally are there to serve the people; that they listen to suggestions which serve to promote the 'common good', that they act on those suggestions and, in doing so, are neither self-interested nor promote biases in the distribution of socio-economic benefits. This is a status quo image of society that runs counter to the conflict-based conceptualisation which has underpinned the approach in this book. Of course, this does not mean that our approach is different from Gilg's simply because we wish to see change, whereas he does not. Gilg's (1985) emphasis on applied research signifies that he accepted (perhaps even welcomed) change (and saw research as a means of selecting the most 'beneficial' change paths). For us, this suggestion does not go far enough. It implies that researchers should primarily restrict their attentions to identifying the most appropriate means of 'tinkering' with existing socio-economic conditions in order to weaken the impress of 'malevolent' trends. This fails to recognise that the processes which brought about current maldistributions or malpractices (that 'applied researchers' want to help change) are inherent in policy procedures which have to be relied on to alleviate the problems researchers have analysed. There is a certain 'Clockwork Orange' overtone to this. It is somewhat equivalent to claiming that you can eradicate crime by employing criminals in the police force and appointing villains to juries. Quite simply, if the vices you wish to eradicate are structured into the 'healing process', there is little chance that they will be removed. Certainly, the specific problems which applied researchers analyse might be circumvented, but they will generally manifest themselves in other mutations. They will not go away unless dominant policy orientations and socio-economic processes change (even then, as Pareto suggested, an alternative political economy would throw up new societal problems).

The role Gilg proposed for rural analysts would effectively make them data gatherers who focus on 'problems' in the rural environment. Since the objective behind this suggestion is to make rural geography (in particular) a more applied subject, certain other implications must be recognised. For one thing, for research to be 'applied', the results of these studies must be applied (i.e. converted into governmental or corporate policies). Realistically, this means that the topics examined would need to serve the interests of societal elites. Further, if analysts' recommendations are to be applied, then the conclusions they reach must broadly coincide with (or only deviate marginally from) the preferences of existing elites. These comments do not constitute an appeal against applied research. What they warn us against is an over-reliance on applied research. Ultimately, to put all of our eggs into an applied research basket can only lead in the direction of an unquestioning sub-discipline. So far as academic respectability is concerned, this would be self-defeating. Very possibly, the tone of Gilg's appeal in favour of applied research owes much to academic 'fashion'. Certainly, the reality of New Right policies under Prime Minister Thatcher and President Reagan have pushed academic research more towards an applied mould. Some steps in this direction were undoubtedly justified. Academic social scientists have traditionally been too far removed from the problems of the public at large. However, we have the strong impression that New Right preferences for applied research owe less to an intention of helping the public at large than they do to a desire to cut back on 'pure' research; given that 'pure' research has very commonly produced results which have been an embarrassment to New Right advocates. Let us by all means engage in applied research, but let this not detract from more fundamental pure research. As academic investigators, we must identify how problems arise and not simply offer suggestions on how these can be toned down at the margins.

To be able to do this requires abstract theorising. This will involve investigating not simply how social structures emerge and are perpetuated, but also how processes of change are generated, thwarted, alter over time and distribute their benefits. Readers will have their own views on the main weaknesses of this book. Perhaps it is too long-winded in places (or overall); there are omissions and there is undoubtedly some unevenness in coverage. Out of the various problems the authors themselves see, the most serious is the weak development of abstract theorising. In places there has been too much generalising and insufficient abstracting. To illustrate this, we can refer back to our commentary on the impact of urban in-migrants on rural localities, in which social disruption was the theme we emphasised. In this, we were generalising across the evidence available to us. We are in fact aware of a number of research projects in which such in-migrants have fitted into their new locality with ease, and where long-term residents have evaluated their local social contributions as beneficial (e.g. Anderson and Anderson 1964; Hill 1980). As White (1974) has observed, it is often these newcomers who resurrect old customs in villages and breathe new life into local social affairs. The beginnings of abstract theorising are contained in a schema like that presented by Forsythe (1983; Figure 5.1), wherein dissimilarities in the character of new entrants is seen to impact on future social relations. This goes some way towards explaining why the same phenomenon (urban in-migration) produces different outcomes across localities, but we do not think it goes far enough. We are convinced that the character of localities themselves, and the particular time-frame for migrant inflows, are also critical. As yet, however, we can neither find an available schema, nor can we satisfy ourselves over a new theoretical model, which could explain these various outcomes.

The problem of reverting to generalisation in place of

theoretical abstraction is a critical one. The example given above merely reports on 'deviations' about a general trend across localities impacted by a specific process.   In truth, however, the very fabric of, and dominant development trends within, rural localities are affected by processes which cannot be explained adequately through generalisation.   Over the 1970s, for instance, rural localities were marked by a counterurbanisation process that carried population and employment into rural locales.   That this has widely been held to constitute a marked change with past trends is unquestionable. Yet doubts must be raised over whether this change was really a break with the past.   As Ballard and Fuguitt (1985) have reported, similar counterurbanisation trends were evident in the depression of the 1930s. Furthermore, in line with Stillwell's (1982) warning that reversals in patterns of regional growth characterise periods of economic depression, Lichter et al (1985) reported that nonmetropolitan growth rates have recently waned in the United States. It seems that metropolitan population growth is again moving into the ascendancy.   Place this observation in the context of some rural in-migrants being returnees seeking to avoid urban unemployment (e.g. White 1983), with around 90 per cent of new jobs in the nonmetropolitan USA being for women (Carlsson et al 1981), and the picture which emerges is one of a rural environment which has not strengthened its socio-economic position. Rather, it has been 'used' as a convenient geographical setting for capital restructuring.

A pressing need is for theorising on rural development to gain from an understanding of the critical role of class relations in structuring the options available to rural localities.   At the international (Chapter Four) and the national levels (Chapter Five), these relationships have an overriding influence on rural fortunes; this impact being converted in obvious form into events within localities (Chapter Six).   This is not to say that independent, alternative influences are not available.   State institutions offer the most visible and significant agents at the national and international levels in this regard.   Yet it has to be stressed that the state in capitalist societies must ultimately underwrite the logic of capital accumulation, if only for self-preservation (Block 1980; Offe and Ronge 1982). Hence, although imposing deviations of a specific kind, the main thrust of state interventions inevitably smooths the path for development patterns dominated by capital accumulation impulses.   This obviously leads us toward agreeing with Barry Commoner's comment that: 'The giant corporations have made a monopoly out of rural America [or Britain for that matter]' (Perelman 1977:vii). Yet our acceptance of this view is couched in guarded terms. In particular, it should be emphasised that, just as capital accumulation processes are altered by state actions, locality-specific social processes are somewhat autonomous. The same universal processes will not find uniform expression across localities and should not be expected to do so. Loss of autonomy is only partial and is matched by locally inspired adaptation processes (e.g. Bennett 1969). There is no reason why the strengthening of ties between local and national spheres should lead to a weakening of local influences (Richards 1984). As Saunders et al (1978:60) argued for state agencies:

> ... not withstanding the increasing state intervention in the everday life of the citizens of all advanced industrial societies, in rural areas the local political process has retained a good deal of its influence.

Again, however, we must be aware of the dangers of generalisation. Vincent (1978:7) for example is quite correct in his assertion that: 'Some areas, classes, communities, can be subordinate but of central importance to the state, others are subordinate and irrelevant'. How much autonomy localities are 'allowed' or wrest for themselves is variable. It is determined by local, national and temporal conditions and by particular social processes. In other words, developmental processes exist within dynamic (local and national) structures which interact and impose on precise actions to produce a myriad observable outcomes.

In theoretical terms, the logic of the above comments favours the development of causal models which go behind surface impressions to inquire into the underlying processes and structures (at local and national levels) which interact to produce given end-products. To argue this view undoubtedly closes off some theoretical ideas. Perroux (1983), for example, is quite accurate in his observation that the very concept of an underlying societal structure is abhorrent to most economists. Our emphasis on process and structure, even granted that structures are composed of 'regulated' processes, puts us at odds with the dominant view in economics. This stresses that economies are undifferentiated entities within which agents freely move under the impress of price conditioning. This note of discord is not problematical. It is inevitable that researchers cut out certain theoretical perspectives from their work. Hopefully, it is obvious from the tone of this text that '... there can be no theory of rural society without a theory of society tout court' (Newby 1980:9). Our particular guiding ideas have revolved around notions of the dynamism of capital accumulation, the significance of structures in determining development processes and the superior utility of conflict models of societal organisation (over consensus models that is). Each of these raises particular problems for theorising. In particular, social scientists have (to date, at least) shown themselves to be poorly equipped to develop dynamic theoretical models. Too much causal modelling in social science has been static in design. Further, there is an urgent need for social researchers to promote more understanding of the dimensions and limits of structural determination. At this time, we find ourselves having to agree with critics who find structuralist theorising too 'contentless' and excessively removed from actual human behaviour (e.g. Whitt 1984). Indeed, some structural theorists seem to be too inclined towards explanatory modes which are anti-empiricist, in that they effectively seek no empirical reference points for their theoretical ideas (Saunders 1982; Whitt 1984). Ultimately, this opens social science to the dangers of anti-humanism.

While at present it seems very loose in its conception and implementation, work within the broadly-defined structuration mould seems to offer some hope for advancement in this regard (e.g. Gregory and Urry 1985). Whether or not this is the way forward or merely one step on the way toward stronger theoretical understanding is as yet difficult to assess. Certainly, there have been some notable contributions within this emerging 'tradition', which do have rural development implications (e.g. Urry 1981, 1984; Massey 1984; Markusen 1985). However, built within the fabric of studies in this research framework, there still appears to be too strong an emphasis on economic determinism. Indeed, too often analysts fall into promoting explanations based on economic reductionism. It is in this spirit that we would commend the work of researchers like Markusen, Massey and Urry. Exemplifying their more 'balanced' approach, Massey (1984) has emphasised that the personal qualities of Ian MacGregor or Henry Ford are important, but only in the context of a particular historical setting for their industries. These entrepreneurs could not alter the general historical trend within their industries single-handed. The logic of economic forces did not make them act in particular ways, but it did channel their behaviour in particular directions. The idea that deviations about broad trends can be significantly determined by non-economic forces (including personality) is, we believe, a critical one. Indeed, we would extend this idea to incorporate the causal importance of social status. We raise the specific issue of status at this juncture deliberately. There has undoubtedly been a growing trend toward disclaiming the importance of social status in causal structures. Indeed, by the way it is seemingly dismissed in an off-hand manner, one is led to suspect that (for some analysts) to pinpoint the causal significance of social status is to verge on sacrilege. This we find regrettable. As we have stressed, class relations (and national political power) are the dominant causal imperatives in structuring broad societal organisation. Yet the deviations brought about by status considerations can also have significant long-term effects. The change in Britain's position from leader of the industrial world to poor relation within the European Community cannot be adequately explained (for us) without drawing attention to the long-term complications which arose from British capitalists adopting the ideals of a status system drawn from the landed aristocracy. This did not fundamentally influence Britain's economy in the short-term, but it did significantly alter its relative national economic standing in the long-term (and hence the standing of British rural localities).

What this comment leads onto is the contention that recognition must be given to associations between geographical scale and the roles of alternative explanatory modes. At the international scale, the heavy impress of capital accumulation processes must be acknowleded. There are certainly national political considerations which have important bearing at this scale, but (as shown by the actions of transnational corporations) economic imperatives are inclined to outweigh these. Within nations, a stronger political (or ideological) component is evident. This is readily seen if we compare 'anti-profit' policies at a national and international level. Thus, governments have tended to drag their feet over implementing international environmental legislation, even when they were surprisingly responsive to demands from their own nationals for stricter regulations over environmental standards (cf. Britain's reluctance to act over acid rain problems on continental Europe). Viewed in its simplest light, votes could have been lost through not responding to intra-national environmentalist pressures, whereas the electorate is less likely to view a loss of national income in order to alleviate problems in other nations with favour. As we move down in geographical scale to the locality, the importance of non-economic causal processes becomes even more evident. This does not mean that local social processes overturn the structural impositions of processes at higher geographical tiers (which are more heavily influenced by capital accumulation pressures). It does nevertheless emphasise that the experiences of development trends for local residents are potentially conditioned in very substantial ways by a wider spectrum of causal forces. This is not difficult to appreciate. For one thing, localities are more easily dominated by single economic activities than nations (or regions). It is also more likely that localities will be composed of people of similar social standing (as in villages dominated by new commuter-oriented residential developments). Added to this, by the very nature of their size, there is a greater chance that rural localities will be dominated by a small cohesive elite than that more extensive geographical areas will be so dominated. For each of these reasons, the probability that alternative messages will flow through power networks, in some places favouring economic criteria for action, in others leaning toward ideological or status considerations, will be greater. Where this places us in terms of an overall explanatory framwework is to emphasise the importance of power structures and power struggles in accounting for development trends. It is power that provides the common causal link across geographical scales. This is a point many structuralist theoreticians do not seem to grasp. In emphasising the dominance of material imperatives (i.e. capital accumulation), structuralist theories give insufficient attention to the fact that material relations are power relations. Indeed, through utilising power largely as a non-materialist concept, and then (rightly) emphasising the dominant position of material interests, such theorising too easily slips into economic reductionism; in that it discounts all bar materialist objectives as irrelevant. Our perspective contrasts with this view in that we believe power structure research holds the key to linking structural imperatives with actual causal processes (cf. Whitt 1984). Where rural (and indeed urban) researchers still fall a long way short is in identifying the alternative bases of power networks. These vary across both geographical locations and through various geographical scales. Recognition of the dominance of material interests at an international (and national) scale is a step in the right direction. So are attempts to link local and national class structures. But they are only a beginning. There is still a long way to go.

BIBLIOGRAPHY

Abate,Y.,1976, 'Foreign aid, UN voting behaviour and alliances', East Lansing, PhD Thesis, Michigan State University.

Adams,B.,1969,The small trade centre: processes and perceptions of growth and decline, in R.M.French(ed),THE COMMUNITY,Itasca,IL,F.E.Peacock,471-84.

Adams,H.,1975,PRISON OF GRASS: CANADA FROM THE NATIVE POINT OF VIEW,Toronto,New Press.

Agnew,J.A.,1982,Sociologising the geographical imagination: spatial concepts in the world system perspective, POLITICAL GEOGRAPHY QUARTERLY,1,159-66.

Aiken,C.S.,1971,The fragmented neoplantation: a new type of farm operation in the Southeast, SOUTHEASTERN GEOGRAPHER, 11(1),43-51.

Alanen,A.R. and K.E.Smith,1976, Local development concerns within America's Appalachian region, in V.M. Gorkham (ed)REGIONAL GEOGRAPHY,Moscow,International Geographical Union Conference Proceedings,Section 8,75-78.

Albrecht,S.L.,1978, Socio-cultural factors and energy resource development in rural areas in the West, JOURNAL OF ENVIRONMENTAL MANAGEMENT,7,73-90.

Alexander,D.,1974,Development and dependence in Newfoundland 1880-1970,ACADIENSIS,4(1),3-31.

Alford,R.R.,1975, Paradigms of relations between state and society, in L.N. Lindberg et al (eds), STRESS AND CONTRADICTION IN MODERN CAPITALISM, Lexington,MA,D.C. Heath,145-60.

Allison,L.,1975,ENVIRONMENTAL PLANNING,London,Allen and Unwin.

Ambrose,P.J.,1974, THE QUIET REVOLUTION: SOCIAL CHANGE IN A SUSSEX VILLAGE 1871-1971,London,Chatto and Windus.

Ambrose, P.J. and B. Colenutt, 1975, THE PROPERTY MACHINE, Harmondsworth,Penguin.

Anderson,B.,1983,IMAGINED COMMUNITIES: REFLECTIONS ON THE ORIGINS AND SPREAD OF NATIONALISM,London,Verso.

Anderson,R.T. and B.G. Anderson,1964,THE VANISHING VILLAGE: A DANISH MARITIME COMMUNITY, Seattle, University of Washington Press.

Anderson, R.T. and D.L. Barkley,1982, Rural manufacturers' characteristics and probability of plant closings, GROWTH AND CHANGE,13(1),2-8.

Andrlik,E.,1981, The farmers and the state: agricultural interests in West German politics, WEST EUROPEAN POLITICS,4,104–19.

Arbuthnott,H. and G.Edwards,1979, A COMMON MAN'S GUIDE TO THE COMMON MARKET,London,Macmillan.

Archer,J.C.,1983,The geography of federal fiscal politics in the United States of America,ENVIRONMENT AND PLANNING C: GOVERNMENT AND POLICY,1,377–400.

Ardagh, J.,1982, FRANCE IN THE EIGHTIES,Harmondsworth,Penguin.

Arnold, P. and I. Cole, 1982, Political motives of the Belvoir truce,TOWN AND COUNTRY PLANNING,51,207–8.

Astrow, A.,1983, ZIMBABWE: A REVOLUTION THAT LOST ITS WAY ?, London,Zed Press.

Averyt,W.F.,1977,AGROPOLITICS IN THE EUROPEAN COMMUNITY,New York,Praeger.

Ayubi, S., R. Bissell, N. Korsah, and L. Lerner,1982, ECONOMIC SANCTIONS IN US FOREIGN POLICY, Philadelphia, Foreign Policy Research Institute.

Bachtel, D. and J. Molnar, 1980, Black and white leader perspectives on rural industrialisation, RURAL SOCIOLOGY, 45,663–80.

Baird, P. and E. McCaughan,1979, BEYOND THE BORDER: MEXICO AND THE US TODAY, New York, North American Congress on Latin America.

Baker, G.L.,1976,The invisible workers: labour organisation on American farms, in R. Merrill, (ed) RADICAL AGRICULTURE, New York,Harper and Row,143–67.

Ball, A.G., 1974, The nature of rural development, in PRIORITIES IN RURAL DEVELOPMENT, Guelph, Ontario, University of Guelph,Ontario Agricultural College,1–12.

Ballard, P. and G.V. Fuguitt, 1985, The changing small town settlement structure in the United States, 1900–1980, RURAL SOCIOLOGY,50,99–113.

Banaji, J.,1976, Summary of selected parts of Kautsky's 'The Agrarian Question', ECONOMY AND SOCIETY, 5, 2–49.

Banfield, E.C. and J.Q. Wilson,1967, CITY POLITICS, Cambridge, MA,Harvard University Press.

Barbier,B.,1972, Le role des petites villes en milieu montagnard, BULLETIN DE L'ASSOCIATION DE GEOGRAPHES FRANCAIS,400/1,295–8.

Barker,A.,1976,THE LOCAL AMENITY MOVEMENT,London, Civic Trust.

Barker,R.G., L.S. Barker, and D.M. Ragle,1967, The churches of Midwest, Kansas and Yoredale, Yorkshire: their contributions to the environment of towns, in W.J. Gore, and L. Hodapp (eds), CHANGE IN THE SMALL COMMUNITY,New York,Friendship Press,155–89.

Barlow,J.,1986, Landowners, property ownership and the rural locality, INTERNATIONAL JOURNAL OF URBAN AND REGIONAL RESEARCH,10,309–29.

Barnes, P., 1978, The great American land grab, in R.D. Rodefeld et al (eds), CHANGE IN RURAL AMERICA, St Louis,C.V.Mosby,137-42.

Barr, J.,1969, Durham's murdered villages,NEW SOCIETY,3 April, 523-525.

Barrett,S. and P.Healey,1985,Priorities for research in land policy, in S.Barrett and P.Healey(eds),LAND POLICY: PROBLEMS AND ALTERNATIVES,Aldershot,Gower,347-67.

Batley, R., 1972, An explanation of non-participation in planning, POLICY AND POLITICS,1(2),95-114.

Bax, M., 1976, HARPSTRINGS AND CONFESSIONS: MACHINE-STYLE POLITICS IN THE IRISH REPUBLIC,Assen,Van Gorcum.

Bealer, R., F. Willits and W. Kuvlesky,1965, The meaning of 'rurality' in American Society,RURAL SOCIOLOGY,30,255-66.

Bealey,F.,J.Blondel and W.McCann,1965, CONSTITUENCY POLITICS: A STUDY OF NEWCASTLE-UNDER-LYME, London, Faber and Faber.

Bechtel,R.B.,1970,A behavioural comparison of urban and small town environments, in J.Archea and C.Eastman (eds),EDRA TWO: PROCEEDINGS OF THE SECOND ANNUAL ENVIRONMENTAL DESIGN RESEARCH ASSOCIATION CONFERENCE, Pittsburgh, Carnegie-Mellon University,347-53.

Beenstock,M.,1983,THE WORLD ECONOMY IN TRANSITION,London,Allen and Unwin (second edition).

Bell, C.and H. Newby,1971, COMMUNITY STUDIES, London,Allen and Unwin.

Bennett,J.W.,1969, NORTHERN PLAINSMEN: ADAPTIVE STRATEGY AND AGRARIAN LIFE, Chicago,Aldine.

Berger,S.,1972, PEASANTS AGAINST POLITICS: RURAL ORGANISATION IN BRITTANY 1911-67, Cambridge,MA, Harvard University Press.

Bernard,J.,1973,THE SOCIOLOGY OF COMMUNITY, Glenview,IL,Scott, Foresman.

Bernier,B.,1976, The penetration of capitalism in Quebec agriculture, CANADIAN REVIEW OF SOCIOLOGY AND ANTHROPOLOGY,13,422-34.

Berry,B.J.L.,1969,'Growth centres and their potentials in the Upper Great Lakes region', Washington DC, Upper Great Lakes Regional Commission.

Berry,B.J.L.,1970,Commuting patterns: labour market participation and regional potential,GROWTH AND CHANGE,1(4),3-10.

Berry,B.J.L.,1971,'Megalopolitan confluence zones: new growth centres in the United States', Athens, Athens Centre for Ekistics Research Report 10.

Berry,B.,1973, GROWTH CENTRES IN THE AMERICAN URBAN SYSTEM, Cambridge,MA,Ballinger.

Bertrand, A.L.,1972, Definitions and strategies of rural development,SOCIOLOGIA RURALIS,12,233-51.

Bible,A.,1978,Impact of corporation farming on small business, in R.Rodefeld et al (eds), CHANGE IN RURAL AMERICA, St Louis, C.V.Mosby,205-16.

Biggs,T.,1976, 'Political economy of East-West trade: the case
    of the 1972-73 US - Russian wheat deal', Berkeley, PhD
    Thesis, University of California.
Birch,D.L.,1979,'The job generation process', Cambridge,MA,MIT
    Programme on Neighbourhood and Regional Change Working
    Paper.
Bird,E.,1982, The impact of private estate ownership on social
    development in a Scottish rural community, SOCIOLOGIA
    RURALIS,22,43-55.
Blackbourn,A.,1982,The impact of multinational corporations on
    the spatial organisation of developed nations, in
    M.J. Taylor, and N.J. Thrift (eds), THE GEOGRAPHY OF
    MULTINATIONALS,London,Croom Helm,147-57.
Blackwood, L.G. and E. Carpenter,1978, The importance of anti-
    urbanism in determining residential preferences and
    migration,RURAL SOCIOLOGY,43,31-47.
Blake,D.E.,1976, LIP and partnership: an analysis of the local
    initiatives programme,CANADIAN PUBLIC POLICY,2,17-32.
Block, F.,1980, Beyond relative autonomy: state managers as
    historical subjects,in R.Miliband and J.Saville (eds) THE
    SOCIALIST REGISTER 1980,London,Merlin,227-42.
Bloomgarden, K.F.,1983,' Patterns of participation and
    leadership in the American community:a case study of
    cotton farmers',New York,PhD Thesis,Columbia University.
Bloomquist, L.E., and G.F. Summers, 1982, Organisation of
    production and community income distributions,AMERICAN
    SOCIOLOGICAL REVIEW,47,325-38.
Blowers,A.,1972,The declining villages of County Durham, in A.
    Blowers (ed), SOCIAL GEOGRAPHY, Bletchley,Open University
    Press,Unit 12,part 2.
Blowers,A.,1980,THE LIMITS OF POWER:THE POLITICS OF LOCAL
    PLANNING POLICY,Oxford,Pergamon.
Blowers,A.,1983,Master of fate or victim of circumstance - the
    exercise of corporate power in environmental policy-
    making,POLICY AND POLITICS,11,393-415.
Blowers, A., P. Braham and J. Woolacott,1976,THE IMPORTANCE OF
    SOCIAL INEQUALITY,Milton Keynes,Open University Press.
Blum, J.,1971,The European village as community:origins and
    functions,AGRICULTURAL HISTORY,45,157-78.
Blunt, J., 1976, Do-it-yourself village bus, COUNTRYMAN, 81
    (Summer),86-90.
Body, R., 1983, AGRICULTURE:THE TRIUMPH AND THE SHAME, London,
    Temple Smith.
Boehlje, M.D.,1985,'An assessment of alternative policy
    responses to financial stress in agriculture',Ithaca,NY,
    Cornell University Agricultural Experiment Station
    AER 85-2.
Bokemeier, J.L. and J.L. Tait,1980, Women as power actors: a
    comparative study of rural communities,RURAL SOCIOLOGY,
    45,238-55.

Bokemeier,J.L. and A.R.Tickamyer,1985,Labour force experiences of nonmetropolitan women,RURAL SOCIOLOGY,50,51-73.

Bolton, R.E.,1966, DEFENCE PURCHASES AND REGIONAL GROWTH, Washington,DC,Brookings Institution.

Bonnen, J.T.,1972,The effect of taxes and government spending on inequality,in R.C.Edwards et al (eds),THE CAPITALIST SYSTEM,Englewood Cliffs,NJ,Prentice-Hall,238-43.

Bontron, J-C. and N.Mathieu,1982,Agriculture and rural change: the case of France, in G.Enyedi and I. Volgyes (eds), THE EFFECT OF MODERN AGRICULTURE ON RURAL DEVELOPMENT,New York,Pergamon,23-33.

Borich, T.O., J.R. Steward and H. Hatle,1985,The impact of a regional mall on rural main street, RURAL SOCIOLOGIST,5, 6-9.

Bornschier, V. and C. Chase-Dunn, 1985, TRANSNATIONAL CORPORATIONS AND UNDERDEVELOPMENT,New York,Praeger.

Boulter, H. and A. Crispin,1978,Rural disadvantage:the differential allocation of resources to small rural primary schools,DURHAM RESEARCH REVIEW,8(41),7-17.

Bouquet, M.P., 1981, 'The sexual division of labour:the farm household in a Devon parish', Cambridge, PhD Thesis, University of Cambridge.

Bowen,F.W.,1973,The Cambridge village college:a cultural centre for rural life,ASPECTS OF EDUCATION,17,98-110.

Bowers, J.K. and P. Cheshire,1983,AGRICULTURE, THE COUNTRYSIDE AND LAND USE: AN ECONOMIC CRITIQUE,London,Methuen.

Bowler, I.R.,1985, AGRICULTURE UNDER THE COMMON AGRICULTURAL POLICY,Manchester,Manchester University Press.

Bradley, T.,1984, Segmentation in local labour markets, in T, Bradley and P. Lowe (eds), LOCALITY AND RURALITY,Norwich, Geo Books,65-90.

Bradley,T. and P.Lowe,1984, Introduction:locality,rurality and social theory,in T. Bradley, and P. Lowe (eds), LOCALITY AND RURALITY,Norwich,Geo Books,1-23.

Brandes, S.,1976,The impact of emigration on a Castilian mountain village,in J.B. Aceves and W.A. Douglass (eds), THE THE CHANGING FACES OF RURAL SPAIN, Cambridge,MA, Schenkman,1-16.

Breckenfeld, G., 1977 Business loves the sunbelt (and vice versa),FORTUNE,95(6),132-46.

Brewis, T.N., 1969, REGIONAL ECONOMIC POLICIES IN CANADA, Toronto,Macmillan.

Britton, J.N.H.,1980, Industrial dependence and technological underdevelopment:Canadian consequences of foreign direct investment,REGIONAL STUDIES,14,181-99.

Broady, M.,1980, Rural regeneration: a note on the mid-Wales case,CAMBRIA,7(1),79-85.

Brody,H.,1973, INISHKILLANE: CHANGE AND DECLINE IN THE WEST OF IRELAND,London,Allen Lane.

Browett,J.,1980,Into the cul de sac of the dependency paradigm with A.G.Frank,in R.Peet (ed), AN INTRODUCTION TO MARXIST THEORIES OF DEVELOPMENT,Canberra,Australian National University, Department of Human Geography Publication HG/14,95-112.

Brown, R.H., 1967, The Upsala, Minnesota, community: a case study in rural dynamics, ANNALS OF THE ASSOCIATION OF AMERICAN GEOGRAPHERS,57,267-300.

Brown, R.M.,1981, Southern violence versus the civil rights movements,1954-68, in M. Black and J.S. Reed (eds),PERSP-ECTIVES ON THE AMERICAN SOUTH, VOLUME ONE,New York,Gordon and Breach,49-69.

Browne, W.P.,1982, Political values in a changing rural community, in W.P. Browne and D.F. Hadwiger (eds), RURAL POLICY PROBLEMS, Lexington,MA,D.C.Heath,61-71.

Bryan, F.M.,1981, POLITICS IN THE RURAL STATES: PEOPLE,PARTIES AND PROCESSES,Boulder,CO,Westview.

Bryant, C.D. and K.B. Perkins,1985, The poultry processing worker, in C.D. Bryant et al (eds),THE RURAL WORK FORCE, South Hadley,MA,Bergin and Garvey,23-42.

Bryant, C.R. and L.H. Russwurm, 1979, The impact of non-farm development on agriculture,PLAN CANADA,19,122-39.

Brym, R.J., 1978, Regional social structure and agrarian radicalism in Canada, CANADIAN REVIEW OF SOCIOLOGY AND ANTHROPOLOGY,15,339-351.

Buchanan, S.,1982,Power and planning in rural areas:prepartion of the Suffolk county structure plan, in M.J. Moseley (ed), POWER, PLANNING AND PEOPLE IN RURAL EAST ANGLIA, Norwich, University of East Anglia, Centre for East Anglian Studies,1-20.

Buller, H.J.,1983, 'Amenity societies and landscape conser-vation', London,PhD Thesis, King's College, University of London.

Buller, H. and K.Hoggart,1986,Nondecision-making and community power:the case of residential development control in rural areas,PROGRESS IN PLANNING,25,133-203.

Buller, H. and P. Lowe, 1982, Politics and class in rural preservation:A study of the Suffolk Preservation Society, in M.J.Moseley (ed), POWER, PLANNING AND PEOPLE IN RURAL EAST ANGLIA, Norwich, University of East Anglia, Centre for East Anglian Studies,21-41.

Bunce, M.F.,1973,Farm consolidation and enlargement in Ontario and its relevance to rural development,AREA,5,13-16.

Burbank, G., 1971, Agrarian radicals and their opponents: political conflict in southern Oklahoma 1910-24, JOURNAL OF AMERICAN HISTORY,58,5-23.

Burnett, J.R.,1951, NEXT YEAR COUNTRY: A STUDY OF RURAL SOCIAL ORGANISATION IN ALBERTA,Toronto, University of Toronto Press.

Burns, L.S., 1977, The location of the headquarters of industrial companies,URBAN STUDIES,14,211-14.

Buttel, F.H.,1980, The 'new cold war' and rural America, RURAL
    SOCIOLOGICAL SOCIETY NEWSLINE,8(2),59–68.
Buttel, F.H., 1982, Farm structure and rural development, in
    W.P.Browne and D.F.Hadwiger (eds), RURAL POLICY PROBLEMS,
    Lexington,MA,D.C.Heath,213–35.
Buttel, F.H.,1982a,The political economy of part-time farming,
    GEOJOURNAL,6,293–300.
Buttel, F.H. and W.L. Flinn,1977,The interdependence of rural
    and urban environmental problems in advanced capitalist
    societies,SOCIOLOGIA RURALIS,17,255–281.
Buttel,F.H. and G.W.Gillespie,1984,The sexual division of farm
    household labour,RURAL SOCIOLOGY,49,183–209.
Buttel, F.H. and D.E. Johnson, 1977, Support for liberal
    development policies among community elites and non-
    elites in a rural region of Wisconsin, LAND ECONOMICS 53,
    455–467.
Caird, J.B. and H.A.Moisley,1961,Leadership and innovation in
    the crofting communities of the Outer Hebrides,
    SOCIOLOGICAL REVIEW,9,85–102.
Camasso, M.J. and D.E. Moore,1985,Rurality and the residualist
    social welfare response,RURAL SOCIOLOGY,50,397–408.
Canada,Royal Commission on Farm Machinery,1969,SPECIAL REPORT
    ON PRICES OF TRACTORS AND COMBINES IN CANADA AND OTHER
    COUNTRIES,Ottawa,Queen's Printer.
Carlsson,J.E.,L.Lassey and W.R.Lassey,1981, RURAL SOCIETY AND
    ENVIRONMENT IN AMERICA,New York,McGraw-Hill.
Carney, J., R. Hudson, G. Ive and J. Lewis, 1975, Regional
    underdevelopment in late capitalism:a study of the North
    East of England, in M.Harloe (ed), PROCEEDINGS OF THE
    CONFERENCE ON URBAN CHANGE AND CONFLICT,London,Centre
    for Environmental Studies Conference Paper 14,136–63.
Carpenter, E.H.,1980,Retention of metropolitan-to-nonmetropol-
    itan labour force migrants, in D.L. Brown and J.M.
    Wardwell (eds), NEW DIRECTIONS IN URBAN-RURAL MIGRATION,
    New York,Academic Press.
Carpenter, T.C.,1973, Postbuses in Scotland, COACHING JOURNAL,
    41,(November),44–7.
Carter,I.,1974,The Highlands of Scotland as an underdeveloped
    region, in E. DeKadt and G. Williams (eds), SOCIOLOGY AND
    DEVELOPMENT,London,Tavistock,279–311.
Casanova, P.G.,1965,Internal colonialism and national develop-
    ment,STUDIES IN COMPARATIVE INTERNATIONAL DEVELOPMENT,1,
    27–37.
Castles, S. and G. Kosack, 1974, How the trade unions try to
    control and integrate immigrant workers in the German
    Federal Republic,RACE,15,497–514.
Cathie, J. 1982, THE POLITICAL ECONOMY OF FOOD AID, Aldershot,
    Gower.
Catton,B.,1963,TERRIBLE SWIFT SWORD: THE CENTENNIAL HISTORY OF
    THE CIVIL WAR,New York,Pocket Books.

Caute, D.,1978, THE GREAT FEAR: THE ANTI-COMMUNIST PURGE UNDER TRUMAN AND EISENHOWER,New York,Simon and Schuster.

Chadwick, J.W., J.B. Houston and J.R.W. Mason, 1972, BALLINA: A LOCAL STUDY IN REGIONAL ECONOMIC DEVELOPMENT, Dublin, Institute of Public Administration.

Channon, J.W., 1971, Probing rural change, CANADIAN FARM ECONOMICS,6(1),16-20.

Charlesworth, A. (ed), 1983, AN ATLAS OF RURAL PROTEST IN BRITAIN 1548 - 1900,London,Croom Helm.

Childs, G. and C. Minay, 1977, 'The Northern Pennines Rural Development Board', Oxford, Oxford Polytechnic,Department of Town Planning,Working Paper 30.

Chirot, D. and T.D. Hall, 1982, World-system theory,in R.H. Turner and J.F.Short (eds), ANNUAL REVIEW OF SOCIOLOGY, VOLUME EIGHT,Palo Alto,CA,Annual Reviews Inc,81-106.

Chisholm, M.,1962,Have English villages a future?,GEOGRAPHICAL MAGAZINE,35,243-52.

Chodos,R.,1973,THE CPR:A CENTURY OF CORPORATE WELFARE,Toronto, James Lewis and Samuel.

Christian, W.A., 1972, PERSON AND GOD IN A SPANISH VALLEY, New York,Seminar Press.

Clark,G.,1981,Some secular changes in land-ownership in Scotland,SCOTTISH GEOGRAPHICAL MAGAZINE,97,27-36.

Clark, G.,1982,Developments in rural geography,AREA,14,249-54.

Clark, G., 1982a, HOUSING AND PLANNING IN THE COUNTRYSIDE, Chichester,John Wiley and Sons.

Clark, G.L.,1985, JUDGES AND THE CITIES: INTERPRETING LOCAL AUTONOMY,Chicago,University of Chicago Press.

Clark, P.D.C.,1974,'Attitudes and behaviour:underlying factors in regional opportunity loss', Ithaca, NY, PhD Thesis, Cornell University.

Clark, U.E.G.,1976, The cyclical sensitivity of employment in branch and parent plants,REGIONAL STUDIES,10,293-98.

Clarkson,S.,1980,'Jobs in the countryside',Ashford,University of London, Wye College, Department of Environmental Studies and Countryside Planning Occasional Paper 2.

Claval, P.,1983, OPPOSITION PLANNING IN WALES AND APPALACHIA, Cardiff,University of Wales Press.

Cleaver, H., 1979, READING CAPITAL POLITICALLY, Brighton, Harvester.

Clement, W., 1975, THE CANADIAN CORPORATE ELITE, Toronto, McClelland and Stewart.

Clement, W.,1977, CONTINENTAL CORPORATE POWER: ECONOMIC ELITE LINKAGES BETWEEN CANADA AND THE UNITED STATES,Toronto, McClelland and Stewart.

Cloke,P.J.,1979,KEY SETTLEMENTS IN RURAL AREAS,London,Methuen.

Cloke,P.J.,1983, AN INTRODUCTION TO RURAL SETTLEMENT PLANNING, London,Methuen.

Cloke, P.J. and M.P. Laycock, 1981, Social and economic co-operation in rural communities:the case of Llanaeghaeam, Wales,SOCIOLOGIA RURALIS,21,81-93.

Cloke, P.J. and D.P. Shaw, 1983, Rural settlement policies in structure plans,TOWN PLANNING REVIEW,54,338-354.

Cloke, P.J. and G. Edwards,1986, Rurality in England and Wales 1981,REGIONAL STUDIES,20,289-306.

Clout, H., 1968, Planned and unplanned changes in French farm structures,GEOGRAPHY,53,311-5.

Clout, H.D., 1972, RURAL GEOGRAPHY,Oxford,Pergamon.

Clout, H.D., 1984, A RURAL POLICY FOR THE EEC?,London,Methuen.

Cobb, J.C.,1982, THE SELLING OF THE SOUTH:THE SOUTHERN CRUSADE FOR INDUSTRIAL DEVELOPMENT 1936-1980, Baton Rouge, Louisiana State University Press.

Cohen, A.P., 1975, THE MANAGEMENT OF MYTHS: THE POLITICS OF LEGITIMATION IN A NEWFOUNDLAND COMMUNITY, Manchester, Manchester University Press.

Cohen, B.,1975a, 'Attitudes and job perception amongst UK planners',MPhil Thesis, University College, University of London.

Cohen, R., 1979, The changing transactions economy and its spatial implications,EKISTICS,46(274),7-15.

Cohen,R.A.,1977,Small town revitalisation planning, JOURNAL OF THE AMERICAN INSTITUTE OF PLANNERS,43,3-12.

Cohen,S.B.,1973,GEOGRAPHY AND POLITICS IN A WORLD DIVIDED, New York,Oxford University Press (second edition).

Collison, P., 1963, THE CUTTESLOWE WALLS: A STUDY IN SOCIAL CLASS, London,Faber and Faber.

Commins,P.,1982, Land policies and agricultural development, in P.J.Drudy (ed), IRELAND: LAND, POLITICS AND PEOPLE, Cambridge,Cambridge University Press,217-40.

Congressional Quarterly Inc,1984, FARM POLICY, Washington D.C.

Connell,J.,1978, THE END OF TRADITION: COUNTRY LIFE IN CENTRAL SURREY,London,Routledge and Kegan Paul.

Cooke, P.,1983, THEORIES OF PLANNING AND SPATIAL DEVELOPMENT, London,Hutchinson.

Cooper,J.,1980,CLASS: A VIEW FROM MIDDLE ENGLAND,London,Corgi.

Copp, J.H.,1972, Rural sociology and rural development, RURAL SOCIOLOGY,37,515-33.

Corbridge,S.,1984,Crisis,what crisis? monetarism,Brandt II and the geopolitics of debt, POLITICAL GEOGRAPHY QUARTERLY,3, 331-45.

Cortz, D.,1978, Corporate farming: a tough row to hoe, in R.D. Rodefeld et al (eds,) CHANGE IN RURAL AMERICA, St. Louis, C.V. Mosby,144-51.

Cottrell,W.F.,1951,Death by dieselization: a case study in the reaction to technological change,AMERICAN SOCIOLOGICAL REVIEW,16,258-365.

Coughenour,C.M. and J.A.Christenson,1983,Farm structure,social class and farmers' policy perspectives, in D.E. Brewster et al (eds), FARMS IN TRANSITION, Ames, Iowa State University Press, 67-86.

Coughenour, C.M. and L. Swanson,1983,Work statuses and occupations of men and women in farm families and the structure of farms,RURAL SOCIOLOGY,48,23–43.

Couloumbis, T.A. and J.H. Wolfe,1982,INTRODUCTION TO INTERNATIONAL RELATIONS,Englewood Cliffs,NJ,Prentice-Hall.

Countryside Review Committee,1977,LEISURE AND THE COUNTRYSIDE, London,HMSO.

Cox, G. and P. Lowe,1983,A battle not the war: the politics of the Wildlife and Countryside Act, in A.W. Gilg (ed), COUNTRYSIDE PLANNING YEARBOOK 4,Norwich,Geo Books,48–76.

Cox, K.R. and G.J. Demko,1969, Agrarian structure and peasant discontent in the Russian revolution of 1905,EAST LAKES GEOGRAPHER,3,3–20.

Crabb, P., 1973, Churchill Falls – the cost and benefits of a hydro-electric development project,GEOGRAPHY,58,330–35.

Craig, R.B.,1971, THE BRACERO PROGRAM: INTEREST GROUPS AND FOREIGN POLICY,Austin,University of Texas Press.

Crenson, M.A.,1971,THE UN-POLITICS OF AIR POLLUTION:A STUDY OF NONDECISION-MAKING IN THE CITIES,Baltimore,Johns Hopkins University Press.

Crenson, M.A., 1983, NEIGHBOURHOOD POLITICS, Cambridge, MA, Harvard University Press.

Crosby, T.L.,1977, ENGLISH FARMERS AND THE POLITICS OF PROTECTION 1815-1852,Brighton,Harvester Press.

Cross,J.,1976, THE SUPERMARKET TRAP: THE CONSUMER AND THE FOOD INDUSTRY,Bloomington,Indiana University Press.

Cross, K.W. and R.D.Turner,1974,Factors affecting the visiting pattern of geriatric patients in a rural area,BRITISH JOURNAL OF PREVENTATIVE AND SOCIAL MEDICINE,28,133–39.

Cuddy, M., 1982, European agricultural policy: a regional dimension,BUILT ENVIRONMENT,7(314),200–10.

Cullingford, D. and S.Openshaw,1982,Identifying areas of rural deprivation using social area analysis,REGIONAL STUDIES, 16,409–18.

Cusack,T.R. and M.D.Ward,1980,'Military spending in the United States, Soviet Union and People's Republic of China', Berlin,International Institute for Comparative Social Research Discussion Paper,80–118.

Cutileiro,J.,1971,A PORTUGUESE RURAL SOCIETY,Oxford,Clarendon.

Dahms, F.A.,1980, The evolving spatial organisation of small settlements in the countryside, TIJDSCHRIFT VOOR ECONOMISCHE EN SOCIALE GEOGRAFIE, 71,295–306.

Dahrendorf, R.,1959, CLASS AND CONFLICT IN INDUSTRIAL SOCIETY, London,Routledge and Kegan Paul.

Dailey, G.H. and R.R.Campbell,1980,The Ozark-Ouachita uplands: growth and consequences, in D.L. Brown and J.M. Wardwell (eds), NEW DIRECTIONS IN URBAN-RURAL MIGRATION, New York, Academic Press,233–65.

Dalton, R.T.,1971, Peas for freezing: a recent development in Lincolnshire agriculture,EAST MIDLANDS GEOGRAPHER,5,133–41

Danbon,D.B.,1979,THE RESISTED REVOLUTION:URBAN AMERICA AND THE INDUSTRIALISATION OF AGRICULTURE 1900–1930, Ames, Iowa State University Press.

Davidoff, L., J. l'Esperance and H. Newby,1976, Landscape with figures: home and community in English society, in J. Mitchell and A. Oakley (eds), THE RIGHTS AND WRONGS OF WOMEN,Harmondsworth,Penguin,139–75.

Davies, T.M.,1978, Capital, state and sparse populations, in H. Newby (ed), INTERNATIONAL PERSPECTIVES IN RURAL SOCIOLOGY,Chichester,Wiley,87–104.

Davis, J., 1973, LAND AND FAMILY IN PISTICCI, London, Athlone Press.

Dawson, J.A., 1976, The country shop in Britain, in P. Jones and R.Oliphant (eds), LOCAL SHOPS,Reading,Unit for Retail Planning Information Report U2,63–71.

Dearlove, J.,1974, The control of change and regulation of community action, in D.Jones and M.Mayo (eds), COMMUNITY WORK ONE,London,Routledge and Kegan Paul,22–43.

Delacroix, J.,1979, The exports of raw materials and economic growth, in J.W. Meyer and M.T. Hannan (eds), NATIONAL DEVELOPMENT AND THE WORLD SYSTEM, Chicago, University of Chicago Press,152–67.

Demko, G.J. and R.J. Fuchs, 1984, Urban policy and settlement system change in the USSR, in G.J. Demko and R.J. Fuchs (eds), GEOGRAPHICAL PERSPECTIVES ON THE SOVIET UNION, Chicago, University of Chicago, Department of Geography Research Paper 211,53–69.

Denoon, D.,1983, SETTLER CAPITALISM: THE DYNAMICS OF DEPENDENT DEVELOPMENT IN THE SOUTHERN HEMISPHERE,Oxford,Clarendon.

Deseran, F.A., W.W. Falk and P. Jenkins,1984, Determinants of earnings of farm families in the US, RURAL SOCIOLOGY,49, 210–29.

Development Commission, 1985, THE FORTY SECOND REPORT OF THE DEVELOPMENT COMMISSIONERS,London.

Dewey, R.,1960, The rural–urban continuum: real but relatively unimportant,AMERICAN JOURNAL OF SOCIOLOGY,66,60–6.

Dicken,P.,1983,Japanese manufacturing investment in the United Kingdom,AREA,15,273–84.

Dickens,P.,S.Duncan,M.Goodwin and F.Gray, 1985, HOUSING,STATES AND LOCALITIES,London,Methuen.

Dilamarter,D.F.,1971,'The spatial impact of a marketing board on winter wheat production in Ontario',Toronto, MA Thesis,York University.

Dillman,B.L.,1982, Rural development in an austere environment: the challenge of the eighties,SOUTHERN JOURNAL OF AGRICUL-TURAL ECONOMICS,14(1),55–62.

Dix,K.,1973,Appalachia: third world pillage, ANTIPODE,5(1), 25–30.

Djilas, M., 1957, THE NEW CLASS: AN ANALYSIS OF THE COMMUNIST SYSTEM,London,Thames and Hudson.

Dobriner, W.M., 1963, CLASS IN SUBURBIA, Englewood Cliffs, NJ, Prentice-Hall.

Dorner, P., 1983, Technology and US agriculture, in G.F. Summers (ed),TECHNOLOGY AND SOCIAL CHANGE IN RURAL AREAS, Boulder,CO,Westview,73-86.

Douglass, W.A., 1971, Rural exodus in two Spanish Basque villages,AMERICAN ANTHROPOLOGIST,73,1100-14.

Douglass,W.A.,1976,Serving girls and sheepherders:emigration and continuity in a Spanish Basque village,in J.B.Aceves and W.A.Douglass (eds),THE CHANGING FACES OF RURAL SPAIN, Cambridge,MA,Schenkman,45-61.

Douglass, W.A., 1984, EMIGRATION IN A SOUTH ITALIAN TOWN, New Brunswick,NJ,Rutgers University Press.

Drudy, P.J., 1978, Depopulation in a prosperous agricultural region,REGIONAL STUDIES,12,49-60.

Duchêne,F.,E.Szczepanik and W.Legg,1985,NEW LIMITS ON EUROPEAN AGRICULTURE,Totowa,NJ,Rowman and Allanheld.

Duhl, L.J.,1963, The human measure: man and family in megalo-polis, in L.Wingo (ed), CITIES AND SPACE, Baltimore, John Hopkins University Press.

Duncan, J.S.,1973, Landscape taste as a symbol of group identity,GEOGRAPHICAL REVIEW,63,334-55.

Dunkle, R., D. Brown and S. Lovejoy,1983,Adaptation strategies of main-street merchants,RURAL SOCIOLOGIST,3,102-6.

Dunleavy, P.,1985, Bureaucrats, budgets and the growth of the state,BRITISH JOURNAL OF POLITICAL SCIENCE,15,299-328.

Dunlop,G.M.,R.J.C.Harper and S.Hunka,1957,The influence of the time spent in school buses upon achievement and atten-dance of pupils of Alberta consolidated schools,ALBERTA JOURNAL OF EDUCATIONAL RESEARCH,3,170-9.

Dunn,M.,M.Rawson and A.Rogers,1981, RURAL HOUSING: COMPETITION AND CHOICE,London,Allen and Unwin.

Dunne, F.,1980,Occupational sex-stereotyping among rural young women and men,RURAL SOCIOLOGY,45,396-415.

Dye, T.R., 1976, WHO'S RULING AMERICA, Englewood Cliffs, NJ, Prentice-Hall.

Dyer, M.C., 1978, Leadership in a rural Scottish county, in G.W. Jones and A. Norton (eds), POLITICAL LEADERSHIP IN LOCAL GOVERNMENT, Birmingham, University of Birmingham, Institute of Local Government Studies,30-50.

Eccles, J. and G.W. Fuller,1970, Tomato production in Portugal promoted by the Heinz Company,in A.H.Bunting (ed), CHANGE IN AGRICULTURE,London,Duckworth,257-62.

Economic Council of Canada,1975, LOOKING OUTWARD: A NEW TRADE STRATEGY FOR CANADA,Ottawa,Queen's Printer.

Edelstein,M.,1982, OVERSEAS INVESTMENT IN THE AGE OF IMPERIAL-ISM: THE UNITED KINGDOM 1850-1914,London,Methuen.

Effrat,M.P.,1974, Approaches to community, in M.P.Effrat (ed), THE COMMUNITY,New York,Free Press,1-32.

Elliott,B., F.Bechhofer,D.McCrone and S.Black, 1982, Bourgeois social movements in Britain,SOCIOLOGICAL REVIEW,30,71-96.

Elson,M.J.,1985,Containment in Hertfordshire, in S.Barrett and P.Healey (eds), LAND POLICY: PROBLEMS AND ALTERNATIVES, Aldershot,Gower,127-52.

Elson,M.J.,1986,GREEN BELTS,London,Heinemann.

Elson,M.J., H.Buller and P.Stanley, 1986, 'Providing for motor sports',London,The Sports Council,Study 28.

Emmett,I.,1964,A NORTH WALES VILLAGE: A SOCIAL ANTHROPOLOGICAL STUDY,London,Routledge and Kegan Paul.

Encel,S.,1970,EQUALITY AND AUTHORITY: A STUDY OF CLASS, STATUS AND POWER IN AUSTRALIA,Melbourne,Cheshire.

England, J.L. and S.L. Albrecht,1984, Boomtowns and social disruption,RURAL SOCIOLOGY,49,230-46.

Erickson, R.A.,1980, Corporate organisation and manufacturing branch plant closures in nonmetropolitan areas,REGIONAL STUDIES,14,491-501.

Erickson, R.A. and T.R.Leinbach,1979,Characteristics of branch plants attracted to nonmetropolitan areas,in R.E.Lonsdale and H.L.Seyler (eds), NONMETROPOLITAN INDUSTRIALISATION, New York:John Wiley and Sons, 57-78.

Evans, P.,1979, DEPENDENT DEVELOPMENT: THE ALLIANCE OF MULTI-NATIONAL,STATE AND LOCAL CAPITAL IN BRAZIL,Princeton,NJ, Princeton University Press.

Eversley, P.,1974,Conservation for the minority,BUILT ENVIRON-MENT QUARTERLY,3,14-5.

Ewart-Evans, G.,1956, ASK THE FELLOWS WHO CUT THE HAY,London, Faber and Faber.

Farcy,H.de,1976,Rural development - the creation of a collective will, in Y.H. Landau et al (eds), RURAL COMMUNITIES, New York,Praeger,21-8.

Fassinger,P.and H.K.Schwarzweller,1984,The work of farm women, in H. K. Schwarzweller (ed), RESEARCH IN RURAL SOCIOLOGY AND DEVELOPMENT VOLUME ONE,Greenwich,CT,JAI Press,37-60.

Ferejohn, J.A.,1974,PORK BARREL POLITICS,Stanford,CA,Stanford University Press.

First-Dilic,R.,1978,The productive roles of farm women in Yugoslavia,SOCIOLOGIA RURALIS,17,125-39.

Fischer, C.S.,1975, Toward a subcultural theory of urbanism, AMERICAN JOURNAL OF SOCIOLOGY,80,1319-41.

Fitchen, J.M., 1981, POVERTY IN RURAL AMERICA, Boulder, CO, Westview.

Fitzgerald, R.,1977, Abraham Maslow's hierarchy of needs - an exposition and evaluation, in R. Fitzgerald (ed), HUMAN NEEDS AND POLITICS,Oxford,Pergamon,36-51.

Fletcher, P.,1969,The agricultural housing problem, SOCIAL AND ECONOMIC ADMINISTRATION,3,155-66.

Fliegel,F.C.,A.J.Sofranko and N.Glasgow,1981,Population growth in rural areas and sentiments of the new migrants toward further growth,RURAL SOCIOLOGY,46,411-29.

Flinn,W.L.,1982,Communities and their relationships to agrarian values,in W.P.Browne and D.F.Hadwiger (eds), RURAL POLICY PROBLEMS,Lexington,MA,D.C.Heath,19-32.

Flint, D.,1975,THE HUTTERITES,Toronto,Oxford University Press.

Flora,C.B. and S.Johnson,1978,Discarding the distaff:new roles for rural women, in T.R. Ford (ed), RURAL USA, Ames, Iowa State Union Press,168-81.

Flynt, J.W.,1979, DIXIE'S FORGOTTEN PEOPLE: THE SOUTH'S POOR WHITES,Bloomington,Indiana University Press.

Foraie, J. and M. Dear,1978, The politics of discontent among Canadian Indians,ANTIPODE,10(1),34-45.

Forsyth, D.J.C., 1972, US INVESTMENT IN SCOTLAND, New York, Praeger.

Forsythe,D.E.,1980,Urban incomers and rural change:the impact of migrants from the city on life in an Orkney community, SOCIOLOGIA RURALIS,20,287-305.

Forsythe,D.E.,1983,'Planning implications of urban-rural migration',Gloucester,Gloucestershire College of Arts and Technology,Papers in Local and Rural Planning 21.

Forsythe,D.E.,1984,The social effects of primary school closures, in T. Bradley and P. Lowe (eds), LOCALITY AND RURALITY,Norwich,Geo Books,209-24

Fothergill, S. and G. Gudgin,1982, UNEQUAL GROWTH: URBAN AND REGIONAL EMPLOYMENT CHANGE IN THE UK,London,Heinemann.

Francis, D.R.,1982,'Community initiatives and voluntary action in rural England', Ashford, PhD Thesis, Wye College, University of London.

Frankenberg, R.,1957, VILLAGE ON THE BORDER: A SOCIAL STUDY OF RELIGION, POLITICS AND FOOTBALL IN A NORTH WALES COMMUNITY,London,Cohen and West.

Franklin, S.H.,1969,THE EUROPEAN PEASANTRY,London,Methuen.

Freudenberg, W.R.,1982, The impacts of rapid growth on the social and personal well-being of local community residents, in B.A. Weber and R.E. Howell (eds), COPING WITH RAPID GROWTH IN RURAL COMMUNITIES, Boulder, CO, Westview Press,137-70.

Frieden, B.J., 1979, THE ENVIRONMENTAL PROTECTION HUSTLE, Cambridge,MA,M.I.T. Press.

Friedland, W.H.,1980,Technology in agriculture: labour and the rate of accumulation, in F.H. Buttel and H. Newby (eds), THE RURAL SOCIOLOGY OF THE ADVANCED SOCIETIES, London, Croom Helm,201-14.

Friedland, W.H. and D. Nelkin, 1971, MIGRANT: AGRICULTURAL WORKERS IN AMERICA'S NORTHEAST, New York, Holt, Rinehart and Winston.

Fröbell,F.,J.Heinrichs and O.Kreye,1980,THE NEW INTERNATIONAL DIVISION OF LABOUR,Cambridge,Cambridge University Press.

Frundt,H.J.,1975,'American agribusiness and US foreign agri-cultural policy', New Brunswick, NJ, PhD Thesis, Rutgers University.

Fry, E.H., 1970, FINANCIAL INVASION OF THE USA, New York, McGraw-Hill.

Fuguitt, G.V.,1958,Urban influence and the extent of part-time farming,RURAL SOCIOLOGY,23,392-7.

Fuguitt, G.V.,1959, Part-time farming and the push-pull hypothesis,AMERICAN JOURNAL OF SOCIOLOGY,64,375-79.

Fuguitt, G.V., P.Voss and J. Doherty,1979,GROWTH AND CHANGE IN RURAL AMERICA,Washington DC,Urban Land Institute.

Fujishin, B.,1975, 'Styles of advocacy: the roles of voluntary associations in the Birmingham planning process', Birmingham, University of Birmingham,Centre for Urban and Regional Studies Research Memorandum 43.

Fullerton,J.,1984,THE SOVIET OCCUPATION IN AFGHANISTAN,London, Methuen.

Fulton,P.N.,1975,Setting of social contact and status advancement through marriage: a study of rural women, RURAL SOCIOLOGY,40,45-54.

Gagnon, D.,1976,Geographical and social mobility in nineteenth century Ontario, CANADIAN REVIEW OF SOCIOLOGY AND ANTHROPOLOGY,13,152-64.

Galbraith, J.K.,1967,THE NEW INDUSTRIAL STATE, Boston,Houghton Mifflin.

Gale, A.B., 1977, 'Cooperative behaviour in agriculture', Aberystwyth,PhD Thesis,University of Wales.

Gallagher, A.,1961, PLAINVILLE FIFTEEN YEARS LATER, New York, Columbia University Press.

Gamston, D.,1975, 'The designation of conservation areas', University of York, Institute of Advanced Architectural Studies monograph.

Garovich, L.,1982, Land use planning as a response to rapid population growth and community change, RURAL SOCIOLOGY, 47,47-67.

Garside, P., 1984, West End, East End: London 1890-1940, in A. Sutcliffe (ed), METROPOLIS 1890-1940, London, Mansell, 221-58.

Gasson,R.M.,1966,'The influence of urbanisation on farm ownership and practice', Ashford, University of London, Wye College,Studies in Rural Land Use 7.

Gasson, R.M., 1969, Occupational immobility of small farmers, JOURNAL OF AGRICULTURAL ECONOMICS,20,279-88.

Gasson, R.M.,1975, 'Provision of tied cottages', Cambridge, University of Cambridge, Department of Land Economy Occasional Paper 4.

Gasson, R.M.,1980, Roles of farm women in England, SOCIOLOGIA RURALIS,20(3),165-80.

Gasson, R.M.,1981, 'Opportunities for women in agriculture', Ashford, University of London, Wye College, Department of Environmental Studies and Countryside Planning Occasional Paper 5.

Gasson,R.M.,1984,Farm women in Europe; their need for off-farm employment,SOCIOLOGIA RURALIS,24,216-27.

Gaventa, J.,1980,POWER AND POWERLESSNESS:QUIESCENCE AND REBEL-
LION IN AN APPALACHIAN VALLEY,Oxford,Clarendon Press.

George, S.,1985,POLITICS AND POLICY IN THE EUROPEAN COMMUNITY,
Oxford,Clarendon Press.

Gertler, L.O., and R.W. Crowley,1977,CHANGING CANADIAN CITIES,
Toronto,McClelland and Stewart.

Giddens, A.,1971, CAPITALISM AND MODERN SOCIAL THEORY: AN
ANALYSIS OF THE WRITINGS OF MARX,DURKHEIM AND MAX WEBER,
Cambridge,Cambridge University Press.

Gilder, I.,1984, State planning and local needs, in T. Bradley
and P. Lowe (eds), LOCALITY AND RURALITY, Norwich, Geo
Books,243-57,

Gilg, A.W.,1980, COUNTRYSIDE PLANNING YEARBOOK 1, Norwich,
Geo Books.

Gilg, A.W.,1982, COUNTRYSIDE PLANNING YEARBOOK 3, Norwich,
Geo Books.

Gilg, A.W.,1984, COUNTRYSIDE PLANNING YEARBOOK 5, Norwich,
Geo Books.

Gilg,A.W.,1985, AN INTRODUCTION TO RURAL GEOGRAPHY, London,
Edward Arnold.

Gilmore,R.,1982, A POOR HARVEST: THE CLASH OF POLICIES AND
INTERESTS IN THE GRAIN TRADE,New York,Longman.

Girt,J.L.,1973,Distance to general medical practice and its
effect on revealed ill-health in a rural environment,
CANADIAN GEOGRAPHER,17,154-66.

Glenn, N.D. and L. Hill,1977,Rural-urban differences in attitudes and behaviour in the United States, ANNALS OF THE
AMERICAN ACADEMY OF POLITICAL AND SOCIAL SCIENCE,429, 36-
50.

Gmelch,G.,1986,The readjustment of return migrants in western
Ireland, in R. King (ed), RETURN MIGRATION AND REGIONAL
ECONOMIC PROBLEMS,London,Croom Helm,152-70.

Goddard, J.B. and I.J. Smith,1978,Changes in corporate control
in the British urban system, 1972-77, ENVIRONMENT AND
PLANNING,A10,1073-84.

Gold,G.L.,1975, SAINT PASCAL: CHANGING LEADERSHIP AND SOCIAL
ORGANISATION IN A QUEBEC TOWN, Toronto, Holt,Rinehart and
Winston.

Golde,G.,1975, CATHOLICS AND PROTESTANTS: AGRICULTURAL MODERN-
ISATION IN TWO GERMAN VILLAGES,New York,Academic Press.

Goldfarb,R.L.,1981, A CASTE OF DESPAIR: MIGRANT FARM WORKERS,
Ames,Iowa State University Press.

Goldman,R. and D.R.Dickens,1983, The selling of rural America,
RURAL SOCIOLOGY,48,585-606.

Goldschmidt, W.,1978, AS YOU SOW: THREE STUDIES IN THE SOCIAL
CONSEQUENCES OF AGRIBUSINESS, Montclair, NJ, Allenheld
Osmun.

Goldschmidt, W.,1978a,Large-scale farming and the rural social
structure,RURAL SOCIOLOGY,43,362-66.

Gonick, C.W.,1970, Foreign ownership and political decay, in
    I. Lumsden (ed), CLOSE TO THE 49TH PARALLEL ETC.,Toronto,
    University of Toronto Press,43-73.
Goodman,R.,1972,AFTER THE PLANNERS,Harmondsworth,Penguin.
Goss, K.F., R.D. Rodefeld and F.H. Buttel, 1980, The
    political economy of class structure in US agriculture,
    in F.H. Buttel and H. Newby (eds), THE RURAL SOCIOLOGY OF
    THE ADVANCED SOCIETIES,London,Croom Helm,83-132.
Gottdiener,M.,1977,PLANNED SPRAWL:PRIVATE AND PUBLIC INTERESTS
    IN SUBURBIA,Beverley Hills,CA,Sage.
Gould, A. and D. Keeble,1984,New firms and rural industrialis-
    ation in East Anglia,REGIONAL STUDIES,18,189-201.
Goulet,D.A.,1971,THE CRUEL CHOICE: A NEW CONCEPT IN THE THEORY
    OF DEVELOPMENT,New York,Atheneum.
Grafton,D.J.,1980,'Planning for remote rural areas: the Swiss
    experience', Southampton, University of Southampton,
    Department of Geography Discussion Paper 5.
Grafton,D.J.,1982, Net migration,outmigration and remote rural
    areas:a cautionary note,AREA,14,313-18.
Grant, G.,1965, LAMENT FOR A NATION: THE DEFEAT OF CANADIAN
    NATIONALISM,Toronto,McClelland and Stewart.
Grant,W.P.,1977, INDEPENDENT LOCAL POLITICS IN ENGLAND AND
    WALES,Farnborough,Saxon House.
Graziano,L.,1978, Centre-periphery relations and the Italian
    crisis: the problem of clientelism, in S.Tarrow,
    P.J.Katzenstein and L.Graziano (eds),TERRITORIAL POLITICS
    IN INDUSTRIAL NATIONS,New York,Praeger,290-326.
Green,G.P.,1984, Credit and agriculture: some consequences of
    the centralisation of the banking system,RURAL SOCIOLOGY,
    49,568-79.
Green, M.B. and R.G. Cromley,1982, The horizontal merger: its
    motives and spatial employment impacts, ECONOMIC
    GEOGRAPHY,58,358-70.
Green,P.,1964, Drymen: village growth and community problems,
    SOCIOLOGIA RURALIS,4,52-62.
Green,P.M.,1982, The impact of rural in-migration on local
    government, in W.P. Browne and D.F. Hadwiger (eds), RURAL
    POLICY PROBLEMS,Lexington,MA,D.C.Heath,83-97.
Green,R.J.,1971,COUNTRYSIDE PLANNING: THE FUTURE OF THE RURAL
    REGIONS,Manchester,Manchester University Press.
Greenberg,E.S.,1974,SERVING THE FEW: CORPORATE CAPITALISM AND
    THE BIAS OF GOVERNMENT POLICY,New York,Wiley.
Greenberg,M.R.,1984,Changing cancer mortality patterns in the
    rural United States,RURAL SOCIOLOGY,49,143-53.
Greenberg, S.B.,1981, Race and business enterprise in Alabama,
    in M. Zeitlin (ed) POLITICAL POWER AND SOCIAL THEORY 2,
    Greenwich,CT,JAI Press,203-38.
Greenwood,D.J.,1976, UNREWARDING WEALTH: THE COMMERCIALISATION
    AND COLLAPSE OF AGRICULTURE IN A SPANISH BASQUE TOWN,
    Cambridge,Cambridge University Press.

Gregory, D. and J. Urry(eds),1985,SOCIAL RELATIONS AND SPATIAL STRUCTURES,London,Macmillan.

Gregory,R.,1976,The voluntary amenity movement, in M. MacEwen (ed), FUTURE LANDSCAPES,London,Chatto and Windus,199-217.

Griffith, J.A.G., 1974, PARLIAMENTARY SCRUTINY OF GOVERNMENT BILLS,London,Allen and Unwin.

Griffith-Jones, S., 1984, INTERNATIONAL FINANCE AND LATIN AMERICA,London,Croom Helm.

Groop, R.E.,1978, Size and location as factors influencing village population change in Kansas, EAST LAKES GEOGRAPHER,13,34-44.

Guither, H.D., 1963, Factors influencing farm operators' decisions to leave farming,JOURNAL OF FARM ECONOMICS, 45, 567-76.

Guterbock,T.M.,1980,MACHINE POLITICS IN TRANSITION: PARTY AND COMMUNITY IN CHICAGO,Chicago,University of Chicago Press.

Hadwiger, D.F.,1976, Farmers in politics, in V.Wiser (ed) TWO CENTURIES OF AMERICAN AGRICULTURE,Washington DC,Agricultural History Society,156-70.

Hall,P., H.Gracey, R.Drewett and R.Thomas,1973,THE CONTAINMENT OF URBAN ENGLAND,London,Allen and Unwin.

Handy, C. and M. Pfeff,1975, 'Consumer satisfaction with food products and marketing services',Washington DC,US Department of Agriculture, Agricultural Economics Research Report 281.

Haney, W.G,1983, Farm family and the role of women, in G.F. Summers (ed), TECHNOLOGY AND SOCIAL CHANGE IN RURAL AREAS,Boulder,CO,Westview,179-93.

Harloe,M.,1977, Introduction, in M.Harloe (ed),CAPTIVE CITIES, Chichester,Wiley,1-47.

Harper,E.,F.Fliegel and J.van Es,1980,Growing numbers of small farms in the north central states, RURAL SOCIOLOGY, 45, 608-20.

Harris, C.K. and J. Gilbert,1982, Large scale farming, rural income and Goldschmidt's agrarian thesis,RURAL SOCIOLOGY, 47,449-58.

Harrison,A., R.B. Tranter and R.S. Gibbs,1977,'Landownership by public and semi-public institutions in the UK', Reading, University of Reading, Centre for Agricultural Strategy Paper 3.

Hart,J.A.,1983,THE NEW INTERNATIONAL ECONOMIC ORDER: CONFLICT AND COMPETITION IN NORTH-SOUTH ECONOMIC RELATIONS 1974-77,London,Macmillan.

Hart,J.F.,N.E.Salisbury and E.G.Smith, 1968, The dying village and some notions about urban growth, ECONOMIC GEOGRAPHY, 44,343-49.

Hart, P.W.E.,1978, Geographical aspects of contract farming, TIJDSCHRIFT VOOR ECONOMISCHE EN SOCIALE GEOGRAFIE, 69, 205-15.

Harvey, D.W., 1985, THE URBANISATION OF CAPITAL, Oxford, Blackwell.

Hayes, M.N. and A.L. Olmstead,1984, Farm size and community quality, AMERICAN JOURNAL OF AGRICULTURAL ECONOMICS, 66, 430–36.

Haynes,R.M. and C.G.Bentham,1979, Accessibility and the use of hospitals in rural areas,AREA,11,186–91.

Heady, E.O. and S.T. Sonka,1974, Farm size, rural community income, and consumer welfare,AMERICAN JOURNAL OF AGRICULTURAL ECONOMICS,56,534–42.

Healy, D.,1970, US EXPANSIONISM: THE IMPERIALIST URGE IN THE 1890s,Madison,University of Wisconsin Press.

Healy,D.,1976,GUNBOAT DIPLOMACY IN THE WILSON ERA: THE US NAVY IN HAITI,1915–16,Madison,University of Wisconsin Press.

Healy,J.,1968,THE DEATH OF AN IRISH TOWN,Cork,Mercier.

Healy,R.G.,1980,Landscape and landowner: issues of land tenure in rural America, in A.M.Woodruff (ed), THE FARM AND THE CITY,Englewood Cliffs,NJ,Prentice-Hall,90–108.

Hechter,M.,1975,INTERNATIONAL COLONIALISM:THE CELTIC FRINGE IN BRITISH NATIONAL DEVELOPMENT 1536–1966, London, Routledge and Kegan Paul.

Heffernan,W.D.,1972,Sociological dimensions of agricultural structures in the United States, SOCIOLOGIA RURALIS,12, 481–99.

Hefford,R.K.,1985,FARM POLICY IN AUSTRALIA,St.Lucia,University of Queensland Press.

Helleiner,G.K. and R.Laverge,1979, Intra-firm trade and industrial exports to the United States, OXFORD BULLETIN OF ECONOMICS AND STATISTICS,41,297–311.

Heller,T.,1979, Rural health and health services, in J.M. Shaw (ed), RURAL DEPRIVATION AND PLANNING, Norwich, Geo Abstracts,81–92.

Hennigh,L.,1978,The good life and the taxpayers' revolt, RURAL SOCIOLOGY,43,178–90.

Herington,J.,1984,THE OUTER CITY,London,Harper and Row.

Hertfordshire County Council,1975,'Public transport for rural communities',Hertford.

Hightower, J.,1973, HARD TOMATOES, HARD TIMES, Cambridge, MA, Schenkman.

Hightower, J.,1975, EAT YOUR HEART OUT: FOOD PROFITEERING IN AMERICA,New York,Crown.

Hill,B.E.,1984,THE COMMON AGRICULTURAL POLICY,London,Methuen.

Hill, C.,1961, THE CENTURY OF REVOLUTION 1603–1714, London, Nelson.

Hill, C.M.,1980, 'Leisure behaviour in Norfolk rural communities',Norwich,PhD Thesis,University of East Anglia.

Hillier,J.,1982,The role of CoSIRA factories in the provision of employment in rural eastern England,in M.Moseley (ed), POWER, PLANNING AND PEOPLE IN RURAL EAST ANGLIA, Norwich, Geo Books,177–98.

Hobbs,D.J.,1982, Public education in rural America: a third wave?,in W.P. Browne and D.F. Hadwiger (eds),RURAL POLICY PROBLEMS,Lexington,MA,D.C.Heath,183–201.

Hodge,G. and M.A.Qadeer,1980,The persistence of Canadian towns and villages:small is viable,URBAN GEOGRAPHY,1,335-49.

Hoffman,G.,1982, Nineteenth century roots of American world power relations,POLITICAL GEOGRAPHY QUARTERLY,1,279-92.

Hoggart,K.,1979,'The determinants of farm linkage patterns: an investigation of three townships in Huron county, Ontario', London, University of London, King's College, Department of Geography Occasional Paper 10.

Hoggart,K., 1979a, Resettlement in Newfoundland, GEOGRAPHY, 64 215-18.

Hoggart,K.,1984, Community power and local state: Britain and the United States, in D.T.Herbert and R.J.Johnston (eds), GEOGRAPHY AND THE URBAN ENVIRONMENT 6, Chichester, Wiley, 145-211.

Holland, D.W. and J.L. Beritelle,1975, School consolidation in sparsely populated rural areas, AMERICAN JOURNAL OF AGRI-CULTURAL ECONOMICS,57,567-75.

Honey,R.,1983,The social costs of space:providing services in rural areas,GEOGRAPHICAL PERSPECTIVES,51,24-37.

Horan, P.M. and C.M. Tolbert,1984, THE ORGANISATION OF WORK IN RURAL AND URBAN LABOUR MARKETS,Boulder,CO,Westview.

House,J.W.,1982,FRONTIER ON THE RIO GRANDE,Oxford,Clarendon.

Howarth,R.W.,1969, The political strength of British agricul-ture,POLITICAL STUDIES,17,458-69.

Huddleston, M.W. and M.L. Palley,1981,Shortchanging nonmetro-politan America: small communities and federal aid,PUBLIC BUDGETING AND FINANCE,1(3),35-45.

Hudson,J.C.,1985, PLAINS COUNTRY TOWNS, Minneapolis,University of Minnesota Press.

Hunter,J.,1979,The crofter, the laird and the agrarian social-ist: the Highland land question in the 1970s, SCOTTISH GOVERNMENT YEARBOOK,48-60.

Hymer,S.,1972, The multinational corporation and the law of uneven development, in J.Radice (ed), INTERNATIONAL FIRMS AND MODERN IMPERIALISM,Harmondsworth,Penguin,37-62.

Inglehart,R.,1977,THE SILENT REVOLUTION,Princeton,NJ,Princeton University Press.

Ironside, R.G.,1969, Rural regeneration in Les Landes, south west France,CAHIERS DE GEOGRAPHIE DE QUEBEC,12,365-81.

Jenkins,J.C.,1982, Why do peasants revolt?, AMERICAN JOURNAL OF SOCIOLOGY,88,487-514.

Jennings,P.,1980, Class and national division in south Texas: the farmworker strike in Raymondville, HUMANITY AND SOCIETY,4,52-69.

Johansen, H.E. and G.V. Fuguitt,1979, Population growth and retail decline: conflicting effect of urban accessibility in American villages,RURAL SOCIOLOGY,44,24-38.

Johansen,H.E. and G.V.Fuguitt,1984, THE CHANGING RURAL VILLAGE IN AMERICA,Cambridge,MA,Ballinger.

John,B.,1981,A plea from the Welsh wilderness,TOWN AND COUNTRY
PLANNING,50,256-60.

Johnson,R.W.,1972, The nationalisation of English rural
politics:Norfolk South West,1944-70,PARLIAMENTARY AFFAIRS
26,8-55.

Johnston, R.J.,1979, The spatial impact of fiscal changes in
Britain: regional policy in reverse?, ENVIRONMENT AND
PLANNING,11A,1439-44.

Johnston,R.J.,1980,THE GEOGRAPHY OF GOVERNMENT SPENDING IN THE
UNITED STATES,Chichester,Wiley.

Johnston,R.J.,1983,Politics and the geography of social well-
being, in M. Busteed (ed), DEVELOPMENTS IN POLITICAL GEO-
GRAPHY,London,Academic Press,189-250.

Johnson,S.,1974,THE POLITICS OF THE ENVIRONMENT,London,Stacey.

Joint Unit for Research on the Urban Environment, 1983, 'An
evaluation of Development Commission activities in
selected areas',Birmingham,University of Aston.

Jones,G.S.,1972,The history of US imperialism,in R. Blackburn
(ed), IDEOLOGY IN SOCIAL SCIENCE,London,Fontana,207-37.

Joseph, A., B. Smit and K. Beesley, 1982, 'Public input into
planning for municipal service provision',Guelph, University of Guelph, Department of Geography, Studies in Rural
Adjustment Report 14.

Josling, T.E. and D. Hamway, 1976, Income transfer effects of
the Common Agricultural Policy, in B. Davey et al (eds),
AGRICULTURE AND THE STATE,London,Macmillan,180-205.

Kavanagh, D.,1974, Beyond autonomy? The problems of corporations,GOVERNMENT AND OPPOSITION,9,42-60.

Keeble,D.,1984, The urban-rural manufacturing shift,GEOGRAPHY,
69,163-66.

Keith,W.J.,1974,THE RURAL TRADITION:A STUDY OF THE NON-FICTION
PROSE WRITERS OF THE ENGLISH COUNTRYSIDE,Toronto,
University of Toronto Press.

Kendall, D.,1963, Portrait of a disappearing English village,
SOCIOLOGIA RURALIS,3,157-65.

Kennett,W.,1972,PRESERVATION,London,Temple Smith.

Kenny,M.,1961,A SPANISH TAPESTRY: TOWN AND COUNTRY IN CASTILE,
New York,Harper and Row.

Key,V.O.,1964, POLITICS,PARTIES AND PRESSURE GROUPS, New York,
T.Y.Crowell (fifth edition).

Kiljunen,M-L.,1980,Regional disparities and policy in the EEC,
in D. Seers and C. Vaitsos (eds), INTEGRATION AND UNEQUAL
DEVELOPMENT,London,Macmillan,199-222.

King, R., 1973, LAND REFORM - THE ITALIAN EXPERIENCE, London,
Butterworth.

Kitching,G.,1982, DEVELOPMENT AND UNDERDEVELOPMENT IN HISTORI-
CAL PERSPECTIVE,London,Methuen.

Klein,L.,1982, The European Community's regional policy, BUILT
ENVIRONMENT,7(314),182-89.

Knoke, D. and D.E. Long,1975, The economic sensitivity of the
American farm vote,RURAL SOCIOLOGY,40,7-17.

Korsching,P.F.,1984, Farm structural characteristics and prox-
imity of purchase location of goods and services, in
H.K. Schwarzweller (ed), RESEARCH IN RURAL SOCIOLOGY AND
DEVELOPMENT VOLUME ONE,Greenwich,CT,JAI Press,261-87.

Krakover,S.,1984,Trends of spatial reorganisation of growth in
urban fields in the eastern United States,1962-78,
ENVIRONMENT AND PLANNING,A16,1361-73.

Krannich, R.S. and C.R. Humphrey,1983, Local mobilisation and
community growth: toward an assessment of the 'growth
machine',RURAL SOCIOLOGY,48,60-81.

Krumme,G.,1981,Corporate organisation and regional development
in the American federal system, in G.W. Hoffman (ed),
FEDERALISM AND REGIONAL DEVELOPMENT,Austin,University of
Texas Press,154-92.

La Follette,C.,1982,More growth and less federal aid mean new
problems for rural areas,NATIONAL JOURNAL,9(4),155-17.

Lall,S.,1979,The international allocation of research activity
by US multinationals, OXFORD BULLETIN OF ECONOMICS AND
STATISTICS,41,313-31.

Lantz, H.R.,1972, A COMMUNITY IN SEARCH OF ITSELF, Carbondale,
Southern Illinois University Press.

Lapping, M.B. and H.A. Clemenson,1984, Recent developments in
North American rural planning, in A.W. Gilg (ed) COUNTRY-
SIDE PLANNING YEARBOOK 5,Norwich,Geo Books,42-61.

Larkin,A.,1979, Rural housing and housing needs, in J.M. Shaw
(ed), RURAL DEPRIVATION AND PLANNING, Norwich,Geo Books,
71-80.

Lauter,G.P. and P.M.Dickie,1975,MULTINATIONAL CORPORATIONS AND
EAST EUROPEAN SOCIALIST ECONOMIES,New York,Praeger.

Layton,R.L.,1978, The operational structure of the hobby farm,
AREA,10,242-6.

Lee,D.R. and P.G.Helmberger,1982,'Structure changes in acreage
supply responses:an econometric analysis of US feed grain
programmes 1948-80',Ithaca,NY,Cornell University Agricultural Experiment Station AER 82-47.

Lee,T.R.,1957, On the relation between the school journey and
social and emotional adjustment in rural infant children,
BRITISH JOURNAL OF EDUCATIONAL PSYCHOLOGY,27,101-14.

Leigh, R. and D.J. North, 1978, Acquisitions in British indus-
tries: implications for regional development, in F.E.
Hamilton (ed), CONTEMPORARY INDUSTRIALISATION, London,
Longman,158-81.

Lenski,G.E.,1966,POWER AND PRIVILEGE: A THEORY OF SOCIAL STRA-
TIFICATION,New York,McGraw-Hill.

Levinson,C.,1978,VODKA COLA,London,Gordon and Gremmesi.

Lewis,D.1972,LOUDER VOICES:THE CORPORATE WELFARE BUMS,Toronto,
James Lewis and Samuel.

Lewis, J.N.,1971, Government policy and the location of agricultural industries, in G.J.R. Linge and P.J. Rimmer

(eds), GOVERNMENT INFLUENCE AND THE LOCATION OF ECONOMIC ACTIVITY,Canberra, Australian National University,Department of Human Geography Publication HG/5,161-74.

Lichter, D.T., G.V. Fuguitt and T.B. Heaton,1985,Components of nonmetropolitan population change: the contribution of rural areas,RURAL SOCIOLOGY,50,88-98.

Lijfering,J.H.W.,1974, Socio-structural changes in relation to rural out-migration,SOCIOLOGIA RURALIS,14,3-14.

Lipset, S.M.,1950, AGRARIAN SOCIALISM: THE COOPERATIVE COMMON-WEALTH FEDERATION IN SASKATCHEWAN,Berkeley,University of California Press.

Lipsky,M.,1968,Protest as a political resource, AMERICAN POLITICAL SCIENCE REVIEW,62,1144-58.

Little,J.K.,1984, 'Social change in rural areas: a planning perspective',Reading,PhD Thesis,University of Reading.

Little,J.K.,1986, Feminist perspectives in rural geography: An introduction,JOURNAL OF RURAL STUDIES,2,1-8.

Littlejohn,J.,1963,WESTRIGG:THE SOCIOLOGY OF A CHEVIOT PARISH, London,Routledge and Kegan Paul.

Littlejohn, J.,1972, SOCIAL STRATIFICATION, London, Allen and Unwin.

Lloyd, R.C. and K.P. Wilkinson,1985,Community factors in rural manufacturing development,RURAL SOCIOLOGY,50,27-37.

Lopreato,J.,1967,PEASANTS NO MORE,San Francisco,Chandler.

Lopreato, J. and L. E. Hazelrigg, 1972, CLASS, CONFLICT AND MOBILITY, San Francisco,Chandler.

Lord,G.,1979, The rise of separatist nationalism in Quebec, in L.J.Sharpe (ed), DECENTRALIST TRENDS IN WESTERN DEMOCRACIES,London,Sage,259-78.

Lowe,P.D.,1977,Amenity and equity: a review of local environmental pressure groups in Britain, ENVIRONMENT AND PLANNING A 9,35-58.

Lowe, P.D. and J.Goyder,1983,ENVIRONMENTAL GROUPS IN POLITICS, London,Allen and Unwin.

Lowe, P.D., J. Clifford and S. Buchanan,1980,The environmental movement in focus,VOLE(January),26-8.

Lowe, P.D., G.Cox, M.MacEwen, T.O'Riordan and M.Winter, 1986, COUNTRYSIDE CONFLICTS: THE POLITICS OF FARMING, FORESTRY AND CONSERVATION,Aldershot,Gower.

Lu, Y-C., and L. Tweeten,1973,The impact of busing on student achievement,GROWTH AND CHANGE,4(4),44-6.

Lukes,S.,1974,POWER:A RADICAL VIEW,London,Macmillan.

Luloff, A.E. and W.H. Chittenden,1984,Rural industrialisation, RURAL SOCIOLOGY,49,67-88.

Luloff, A.E. and K.P. Wilkinson, 1979, Participation in the national flood insurance program, RURAL SOCIOLOGY, 44, 137-52.

Lyson, T.A.,1984, Pathways into production agriculture: the structuring of farm recruitment in the United States, in H.K. Schwarzweller (ed), RESEARCH IN RURAL SOCIOLOGY AND DEVELOPMENT, VOLUME ONE,Greenwich,CT,JAI Press,79-103.

McConnell,G.,1959,THE DECLINE OF AGRARIAN DEMOCRACY,Berkeley, University of California Press.

McConnell,J.E.,1980,Foreign direct involvement in the United States,ANNALS OF THE ASSOCIATION OF AMERICAN GEOGRAPHERS, 70,259-70.

McDonald,B.,1969,'Public involvement in the planning process: the role of the local amenity society',MSc Thesis, University of Edinburgh.

McDonald,J.F.,1979,'The lack of political identity in English regions:evidence from MPs',Glasgow, University of Strathclyde,Studies in Public Policy 33.

MacDonald,J.S., 1963, Agricultural organisation, migration and labour militancy in rural Italy,ECONOMIC HISTORY REVIEW, 16,61-75.

McDonald, M.J. and J. Muldowny,1982,TVA AND THE DISPOSSESSED: THE RESETTLEMENT OF POPULATION IN THE NORRIS DAM AREA, Knoxville,University of Tennessee Press.

McDonald, O., 1977, Multinationals, spatial inequalities and workers' control, in D.B. Massey and P.W.J. Batey (eds), ALTERNATIVE FRAMEWORKS FOR ANALYSIS,London,Pion,68-85.

McEwen,J.,1977, WHO OWNS SCOTLAND? A STUDY IN LAND OWNERSHIP, Edinburgh,EUSPB.

McGranahan,D.A.,1984, Local growth and the outside contacts of influentials,RURAL SOCIOLOGY,49,530-40.

MacGregor,M.,1972,The rural culture,NEW SOCIETY,9 March,486-9.

McIntosh,F.,1969, A survey of workers leaving Scottish farms, SCOTTISH AGRICULTURAL ECONOMICS,19,191-97.

McIntyre, R. and D. Wilhelm,1985,'Corporate taxpayers and corporate freeloaders: four years of continuing, legalised tax avoidance by America's largest corporations 1981-84', Washington DC, American Federation of State, County and Municipal Employees.

Mackay, G.A. and G. Laing, 1982, 'Consumer problems in rural areas',Glasgow,Scottish Consumer Council.

McLaughlin,B.P.,1976,Rural settlement planning:a new approach, TOWN AND COUNTRY PLANNING,44,156-60.

McLuhan,M.,1973,UNDERSTANDING MEDIA,London,Abacus.

McNab, A.,1984, 'Integrated rural development in Britain', Gloucester,Gloucestershire College of Arts and Technology Papers in Local and Rural Planning 22.

Macpherson,C.B.,1977, Needs and wants: an ontological or historical problem?,in R.Fitzgerald (ed), HUMAN NEEDS AND POLITICS,Oxford,Pergamon,26-35.

McRoberts,K.,1979, Internal colonialism: the case of Quebec, ETHNIC AND RACIAL STUDIES,2,293-318.

McWilliams,C.,1939,FACTORIES IN THE FIELD: THE STORY OF MIGRATORY FARM LABOUR IN CALIFORNIA, Boston,Little,Brown.

Maddock, R.T.,1978, The economic and political characteristics of food as a diplomatic weapon,JOURNAL OF AGRICULTURAL ECONOMICS,29,31-41.

Madgwick,P.J., N. Griffiths and V. Walker,1973,THE POLITICS OF RURAL WALES,London,Hutchinson.

Mage,J.A.,1974,'Part-time farming in southern Ontario', Waterloo,PhD Thesis,University of Waterloo.

Mage,J.A.,1976,A typology of part-time farming,in A.M. Fuller and J.A. Mage (eds), PART-TIME FARMING, Norwich, Geo Books,6-37.

Majka, L.C. and T.J. Majka,1982,FARM WORKERS,AGRIBUSINESS AND THE STATE,Philadelphia,Temple University Press.

Mann,S.A.,1984,Sharecropping in the cotton South: a case of uneven development in agriculture, RURAL SOCIOLOGY, 49, 412-29.

Mann,S.A. and J.M.Dickinson,1978, Obstacles to the development of a capitalist agriculture, JOURNAL OF PEASANT STUDIES, 5,466-81.

Mann,S.A. and J.M.Dickinson,1980, State and agriculture in two eras of American capitalism, in F.H. Buttel and H. Newby (eds), THE RURAL SOCIOLOGY OF THE ADVANCED SOCIETIES, London,Croom Helm,283-325.

Manzer ,R., 1974, CANADA: A SOCIO-POLITICAL REPORT, Toronto, McClelland and Stewart.

Marantz,J.K.,K.Case and H.Leopard,1976,DISCRIMINATION IN RURAL HOUSING,Lexington,MA,D.C.Heath.

Markusen, A.R., 1985, PROFIT CYCLES, OLIGOPOLY AND REGIONAL DEVELOPMENT,Cambridge,MA,MIT Press.

Martindale, D. and R.G. Hanson,1969,SMALL TOWN AND THE NATION: THE CONFLICT OF LOCAL AND TRANSLOCAL FORCES,Westport, CT,Greenwood.

Martins, M.R. and J. Mawson,1982,The revision of the European Regional Development Fund regulations, BUILT ENVIRONMENT 7,(3/4),190-9.

Maslow,A.H.,1970, MOTIVATION AND PERSONALITY, New York, Harper and Row (second edition).

Massey,D.,1984,SPATIAL DIVISIONS OF LABOUR,London,Macmillan.

Massey, D. and R.A. Meegan, 1979, The geography of industrial re-organisation,PROGRESS IN PLANNING,10,(3),155-237.

Mathias, P., 1971, FORCED GROWTH: FIVE STUDIES OF GOVERNMENT INVOLVEMENT IN THE DEVELOPMENT OF CANADA, Toronto, James Lewis and Samuel.

Matthews,D.R.1976,'THERE'S NO BETTER PLACE THAN HERE': SOCIAL CHANGE IN THREE NEWFOUNDLAND COMMUNITIES, Toronto, Peter Martin Associates.

Matthews,E.M.,1965,NEIGHBOUR AND KIN;LIFE IN A TENNESSEE RIDGE COMMUNITY,Nashville,TN,Vanderbilt University Press.

Medler, J. and A. Mushkatel,1979,Urban-rural class conflict in Oregon land-use planning, WESTERN POLITICAL QUARTERLY,32, 338-49.

Meier, K.J. and W.P.Browne, 1983, Interest groups and farm structure,in D.E. Brewster,et al (eds), FARMS IN TRANSITION,Ames,Iowa State University Press,47-56.

Mellor, I. and R.G. Ironside,1974,'An evaluation of the multiplier effect at Slave Lake and its tributary area since the establishment of the Lesser Slave Lake Special Incentive Area', Edmonton, Alberta Office of Programme Coordination.

Memmott,F.W.,1963, The substitutability of communications for transportation,TRAFFIC ENGINEERING,33,5:20-5.

Merton,R.K.,1957, SOCIAL THEORY AND SOCIAL STRUCTURE, Glencoe, IL,Free Press (revised and enlarged edition).

Mewett,P.G.,1979,The emergence and persistence of peripheral areas in relationship to the processes of development and underdevelopment, in J.Sewel(ed), THE PROMISE AND THE REALITY - LARGE-SCALE DEVELOPMENTS IN MARGINAL REGIONS, Aberdeen, University of Aberdeen, Institute for the Study of Sparsely Populated Areas,142-56.

Mighels,R.L.,1955,AMERICAN AGRICULTURE,New York,Wiley.

Miller, M.K. and A.E. Luloff,1981, Who is rural? RURAL SOCIOLOGY, 46,608-625.

Minay,C.,1985,'The Development Commission's rural industrial development programme', Oxford, Oxford Polytechnic, Department of Town Planning Working Paper 87.

Mingione,E.,1974,Territorial division of labour and capitalist development,CURRENT SOCIOLOGY,22,223-78.

Mitchell,G.D.,1950, Depopulation and rural social structure, SOCIOLOGICAL REVIEW,42,69-85.

Mizruchi,E.H.,1969, Romanticism, urbanism and small town in mass society, in P. Meadows and E.H. Mizruchi (eds), URBANISM, URBANISATION AND CHANGE, Reading,MA, Addison-Wesley.

Mollenkopf, J.H., 1983, THE CONTESTED CITY, Princeton, NJ, Princeton University Press.

Molnar, J.J. and M.D. Lawson,1984,Perceptions of barriers to black political and economic progress in rural areas, RURAL SOCIOLOGY,49,261-83.

Molotch,H.,1976,The city as a growth machine, AMERICAN JOURNAL OF SOCIOLOGY,82,309-32.

Molyneux,J.K.,1975,'Changing service and amenity patterns in an area of south Lincolnshire 1951-71',Nottingham,PhD Thesis,University of Nottingham.

Momsen,J.H.,1983, 'Settlement change in Alberta: the urbanisation of the countryside',London, Canada House Occasional Paper 3.

Moore,B., J.Rhodes and P.Tyler, 1986, THE EFFECTS OF GOVERNMENT REGIONAL ECONOMIC POLICY, London, HMSO.

Moore, K.,1976, Modernisation in a Canary island village, in J.B. Aceves and W.A. Douglas (eds), THE CHANGING FACES OF RURAL SPAIN,Cambridge,MA,Schenkman.

Moran,M.,1981,Finance capital and pressure group politics in Britain,BRITISH JOURNAL OF POLITICAL SCIENCE,11,381-404.

Morin,E.,1970, THE RED AND THE WHITE: REPORT FROM A FRENCH VILLAGE,New York,Random House.

Morkeberg,H.,1978,Working conditions of women married to self-employed farmers,SOCIOLOGIA RURALIS,18,95-105.

Morrill,R.L.,1973,On the size and spacing of growth centres, GROWTH AND CHANGE,4(2),21-4.

Moseley,M.J.,1973, The impact of growth centres in rural regions I: An analysis of spatial patterns in Brittany, REGIONAL STUDIES,7,57-75.

Moseley,M.J.,1973a, The impact of growth centres in rural regions II: an analysis of spatial flows in East Anglia, REGIONAL STUDIES,7,77-94.

Moseley, M.J.,1973b, Growth centres - a shibboleth?, AREA, 5, 143-50.

Moseley,M.J.,1979, ACCESSIBILITY: THE RURAL CHALLENGE, London, Methuen.

Moseley, M.J. and J. Darby,1978, The determinants of female activity rates in rural areas, REGIONAL STUDIES, 12, 297-309.

Mosley,P.,1981, Models of the aid allocation process,POLITICAL STUDIES,29,245-53.

Moss, L. and S.R. Parker,1967,THE LOCAL GOVERNMENT COUNCILLOR, London,HMSO (Committee on the Management of Local Government,Volume Two).

Mughan,A. and I.McAllister,1981,The mobilisation of the ethnic vote,ETHNIC AND RACIAL STUDIES,4,189-204.

Munro,J.M.,1975, Highways in British Columbia: economics and politics,CANADIAN JOURNAL OF ECONOMICS,8,192-204.

Munton,R.J.C.,1977, Financial institutions: their ownership of agricultural land in Great Britain,AREA,9,29-37.

Myers, R., 1976, The National Sharecroppers Fund and the Farm Co-op Movement in the South, in R. Merrill (ed), RADICAL AGRICULTURE,New York,Harper and Row,129-42.

Myrdal,G.,1957,ECONOMIC THEORY AND UNDERDEVELOPED REGIONS, London,Duckworth.

Nagle,J.C.,1976, AGRICULTURAL TRADE POLICIES,Farnborough,Saxon House.

Napier, T.L. and R.C. Maurer,1978,'Correlates of commitment to community development efforts',Columbus,Ohio State University, Department of Agricultural Economics and Rural Sociology Report ESS 555.

Nash,G.D.,1985,THE AMERICAN WEST TRANSFORMED:THE IMPACT OF THE SECOND WORLD WAR,Bloomington,Indiana University Press.

Nash,R.,1967, WILDERNESS AND THE AMERICAN MIND, New Haven, CT,Yale University Press.

Nash,R.,1980,SCHOOLING IN RURAL SOCIETIES,London,Methuen.

Naylor, R.T.,1975, THE HISTORY OF CANADIAN BUSINESS, Toronto, Lorimer (two volumes).

Nelson,L.,1957,Rural life in a mass industrial society, RURAL SOCIOLOGY,22,20-30.

Nelson,W.E.,1979,Black rural land decline and political power, in L. McGee and R. Boone (eds), THE BLACK RURAL LANDOWNER - ENDANGERED SPECIES,Westport,CT,Greenwood Press,83-96.

Newby,H.,1977,THE DEFERENTIAL WORKER,Harmondsworth,Penguin.

Newby,H.,1978,The rural sociology of advanced capitalist societies, in H.Newby (ed), INTERNATIONAL PERSPECTIVES IN RURAL SOCIOLOGY,Chichester,Wiley,3-30.

Newby,H.,1980,Rural sociology,CURRENT SOCIOLOGY,28(1),3-141.

Newby,H.,1980a,Urbanisation and the rural class structure, in F. Buttel and H. Newby (eds), THE RURAL SOCIOLOGY OF ADVANCED SOCIETIES,London,Croom Helm,255-79.

Newby,H.,1983, European social theory and the agrarian question: towards a sociology of agriculture, in G.F. Summers (ed),TECHNOLOGY AND SOCIAL CHANGE IN RURAL AREAS, Boulder,CO,Westview Press,109-23.

Newby, H., C. Bell, D. Rose and P. Saunders, 1978, PROPERTY, PATERNALISM AND POWER,London,Hutchinson.

Newman, J. and C. McCauley,1977,Eye contact with strangers in city, suburb and small town, ENVIRONMENT AND BEHAVIOUR,9, 547-58.

Nicholls,D.C.,1976,Agencies for rural development in Scotland, in P.J.Drudy (ed),REGIONAL AND RURAL DEVELOPMENT,Chalfont St.Giles,Alpha Academic,69-89.

Nicholls,W.H.,1969,The US South as an underdeveloped region, in U. Pappi and C. Nunn (eds), ECONOMIC PROBLEMS OF AGRI-CULTURE IN INDUSTRIAL SOCIETIES,London,Macmillan.

Nooij,A.T.J.,1969, Political radicalism among Dutch farmers, SOCIOLOGIA RURALIS,9,43-61.

Norton,R.D. and J.Rees,1979, The product cycle and the spatial decentralisation of American manufacturing, REGIONAL STUDIES,13,141-51.

Norton-Taylor, R.,1982, WHOSE LAND IS IT ANYWAY? AGRICULTURE, PLANNING AND LAND-USE IN THE BRITISH COUNTRYSIDE, Wellingborough,Turnstone Press.

Nove,A.,1982,Is there a ruling class in the USSR,in A. Giddens and D.Held (eds), CLASSES,POWER AND CONFLICT,Basingstoke, Macmillan,588-605.

O'Connor,J.,1973,THE FISCAL CRISIS OF THE STATE, New York, St. Martin's Press.

Offe,C. and V.Ronge,1982, Theses on the theory of the state,in A. Giddens and D.Held (eds), CLASSES, POWER AND CONFLICT, Basingstoke,Macmillan,249-56.

Olgyay,G.,1982,Environment, land-use and development: the case of Vermont, in G.Enyedi and I. Volgyes (eds), THE EFFECT OF MODERN AGRICULTURE ON RURAL DEVELOPMENT, New York, Pergamon.

Olson, M., 1965, THE LOGIC OF COLLECTIVE ACTION, Cambrige, MA, Harvard University Press.

Oster,G.,1979,A factor analytic test of the theory of the dual economy,REVIEW OF ECONOMICS AND STATISTICS,61,33-9.

O'Sullivan,P.,1986,GEOPOLITICS,London,Croom Helm.

Overton,J.,1979,A critical examination of the establishment of national parks and tourism in underdeveloped areas, Gros Morne National Park in Newfoundland,ANTIPODE,11(2),34-47.

Owen,N.,1983, ECONOMIES OF SCALE, COMPETITIVENESS AND TRADE PATTERNS WITHIN THE EUROPEAN COMMUNITY,Oxford,Clarendon.

Owens,S.E.,1984,Energy and spatial structure: a rural example, ENVIRONMENT AND PLANNING,A16,1319-37.

Oxley,H.G.,1974, MATESHIP IN LOCAL ORGANISATION: A STUDY OF EGALITARIANISM, STRATIFICATION, LEADERSHIP, AND AMENITIES PROJECTS IN A SEMI-INDUSTRIAL COMMUNITY OF INLAND NEW SOUTH WALES,St.Lucia,University of Queensland Press.

Pacione,M.,1984,RURAL GEOGRAPHY,London,Harper and Row.

Pahl,R.E.,1965, URBS IN RURE: THE METROPOLITAN FRINGE IN HERT-FORDSHIRE,London, London School of Economics Geographical Paper 2.

Pahl,R.E.,1966,The rural-urban continuum,SOCIOLOGIA RURALIS,6, 299-329.

Pahl,R.E.,1968, Is the mobile society a myth? NEW SOCIETY,11 January,46-8.

Pareto, V.,1901, THE RISE AND FALL OF THE ELITES, Totowa, NJ, Bedminster Press (1968 Edition).

Pareto,V.,1935, MIND AND SOCIETY: A TREATISE ON GENERAL SOCIO-LOGY,London,Jonathan Cape.

Parker,G.,1983,A POLITICAL GEOGRAPHY OF COMMUNITY EUROPE, London,Butterworth.

Parker,R.C. and J.M.Connor,1979,Estimates of consumer loss due to monopoly in the US food manufacturing industries, AMERICAN JOURNAL OF AGRICULTURAL ECONOMICS,61,626-39.

Parkin,F.,1971,CLASS INEQUALITY AND POLITICAL ORDER,St.Albans, Paladin.

Parsons,D.J.,1980,'Rural gentrification:the influence of rural settlement planning policies', Brighton, University of Sussex,Research Paper in Geography.

Payne,S.,1971, Catalan and Basque nationalism, JOURNAL OF CON-TEMPORARY HISTORY,6(1),15-51.

Pellenberg, P.H. and J.A.A.M. Kok,1985,Small and medium-sized innovative firms in the Netherlands' urban and rural regions, TIJDSCHRIFT VOOR ECONOMISCHE EN SOCIALE GEOGRAFIE,76,242-52.

Penfold,S.F.,1974, 'Housing problems of local people in rural pressure areas', Sheffield, University of Sheffield, Department of Town and Regional Planning TRP 7.

Perelman,M.,1977,FARMING FOR PROFIT IN A HUNGRY WORLD: CAPITAL AND THE CRISIS IN AGRICULTURE,Montclair,NJ,Allenheld, Osmun.

Perkins,B.B.,1972,'Multiple job holdings among farm operators' Guelph, Ontario, University of Guelph, School of Agricul-tural Economics and Extension Education Publication AE/72/5.

Perlin,G.,1971,Patronage and paternalism: politics in Newfoun-dland, in D.I.Davies and K.Herman (eds), SOCIAL SPACE: CANADIAN PERSPECTIVES,Toronto,New Press,190-6.

Perroux, F., 1983, A NEW CONCEPT OF DEVELOPMENT,London,Croom Helm.

302

Perry,R.L.,1971,GALT,USA:THE 'AMERICAN PRESENCE' IN A CANADIAN CITY,Toronto,Maclean-Hunter.

Pfeffer,M.J.,1983,Social origins of three systems of farm production in the United States,RURAL SOCIOLOGY,48,540-62.

Phelan, J. and R. Pozen, 1973, THE COMPANY STATE, New York, Grossman.

Phillips, D.R. and A.M. Williams,1982, RURAL HOUSING AND THE PUBLIC SECTOR,Farnborough,Gower.

Phillips, D.R. and A.M. Williams,1983, The social implications of rural housing policy, in A.W. Gilg (ed), COUNTRYSIDE PLANNING YEARBOOK 4,Norwich,Geo Books,77-102.

Pinder,D.,1983,REGIONAL ECONOMIC DEVELOPMENT AND POLICY:THEORY AND PRACTICE IN THE EUROPEAN COMMUNITY, London, Allen and Unwin.

Pitt-Rivers,J.,1960, Social class in a French village, ANTHROPOLOGICAL QUARTERLY,33,1-13.

Ploch, L.A.,1978, The reversal in migration patterns - some rural development consequences, RURAL SOCIOLOGY, 43, 293-303.

Ploch,L.A.,1980, Effects of turnaround migration on community structure in Maine, in D.L.Brown and J.M.Wardwell (eds), NEW DIRECTIONS IN URBAN-RURAL MIGRATION,New York,Academic Press,291-311.

Popplestone,G.,1967, Conflict and mediating roles in expanding settlements,SOCIOLOGICAL REVIEW,15,339-55.

Prebble,J.,1963,THE HIGHLAND CLEARANCES,Harmondsworth,Penguin.

Preston,P.W.,1982,THEORIES OF DEVELOPMENT,London,Routledge and Kegan Paul.

Price,B.L.,1983,THE POLITICAL ECONOMY OF MECHANISATION IN US AGRICULTURE,Boulder,CO,Westview Press.

Price,M.L. and D.C.Clay,1980, Structural disturbances in rural communities: some repercussions of the migration turnaround in Michigan,RURAL SOCIOLOGY,45,591-607.

Ragin, C.,1977, Class, status and 'reactive ethnic cleavages', AMERICAN SOCIOLOGICAL REVIEW,42,438-50.

Reece,J.E.,1979, Internal colonialism: the case of Brittany, ETHNIC AND REGIONAL STUDIES,2,275-93.

Rees,A.D.,1950,LIFE IN A WELSH COUNTRYSIDE,Cardiff, University of Wales Press.

Reichenbach,H.,1980,A politico-economic overview, in D. Seers and C.Vaitsos (eds), INTEGRATION AND UNEQUAL DEVELOPMENT, London,Macmillan,75-99.

Richards,R.O.,1984, Decision-making in small towns: vertical integration revisited,RURAL SOCIOLOGIST,4(1),51-60.

Richling,B.,1985, You'd never starve here: return migration to rural Newfoundland, CANADIAN REVIEW OF SOCIOLOGY AND ANTHROPOLOGY,22,236-49.

Robson,B.,1982, The Bodley barricade: social space and social conflict, in K.R. Cox and R.J. Johnston (eds), CONFLICT, POLITICS AND THE URBAN SCENE,Harlow,Longman,45-61.

Rodefeld,R.,1974,'The changing organisational and occupational structure of farming and the implications for farm work-force individuals, families and communities',Madison,PhD Thesis,University of Wisconsin.

Rokkan,S.,1970,CITIZENS,ELECTIONS AND PARTIES,Oslo,Universit-etsforlaget.

Rose, D., P. Saunders, H. Newby and C. Bell,1976,Ideologies of property: a case study,SOCIOLOGICAL REVIEW,24,699-730.

Rose-Ackerman, S. and R. Evenson,1985,The political economy of agricultural research and extension, AMERICAN JOURNAL OF AGRICULTURAL ECONOMICS,67,1-14.

Rossiter,C.,1985,THE BUREAUCRATIC STRUGGLE FOR CONTROL OF US FOREIGN AID,Boulder,CO,Westview.

Rotstein,A.,1976,Canada: the new nationalism, FOREIGN AFFAIRS, 55,97-118.

Runciman, W.G., 1968, Class, status and power?,in J.A. Jackson (ed), SOCIAL STRATIFICATION, Cambridge, Cambridge University Press,25-61.

Ryan,B.,1970,The criteria for selecting growth centres in Appalachia, PROCEEDINGS OF THE ASSOCIATION OF AMERICAN GEOGRAPHERS,2,118-123.

Sachs,C.E.,1983,THE INVISIBLE FARMERS: WOMEN IN AGRICULTURAL PRODUCTION,Totowa,NJ,Rowman and Allanheld.

Sacks,P.M.,1976,THE DONEGAL MAFIA: AN IRISH POLITICAL MACHINE, New Haven,CT,Yale University Press.

Salamon, L.M. and S. van Evera,1973,Fear,apathy and discrimin-ation,AMERICAN POLITICAL SCIENCE REVIEW,67,1288-1306.

Salamon,S. and A.M.Keim,1979, Land ownership and women's power in a midwestern farming community,JOURNAL OF MARRIAGE AND THE FAMILY,41,109-19.

Sale,K.,1975,POWER SHIFT: THE RISE OF THE SOUTHERN RIM AND ITS CHALLENGE TO THE EASTERN ESTABLISHMENT,New York,Vintage.

Sampson,A.,1981,THE MONEY LENDERS:BANKERS IN A DANGEROUS WORLD London,Hodder and Stoughton.

Sampson,A.,1982,THE CHANGING ANATOMY OF BRITAIN,London,Hodder and Stoughton.

Saunders,P.,1982,The relevance of Weberian sociology for urban political analysis, in A.Kirby and S.Pinch (eds), PUBLIC PROVISION AND URBAN POLITICS, Reading, University of Reading,Geographical Paper 80,1-24.

Saunders,P.,1984,'We can't afford democracy too much: findings from a study of regional state institutions in South East England',Brighton,University of Sussex Urban and Regional Studies Working Paper 43.

Saunders,P., H. Newby, C. Bell and D.Rose,1978,Rural Community and rural community power, in H.Newby (ed), INTERNATIONAL PERSPECTIVES IN RURAL SOCIOLOGY,Chichester,Wiley,55-85.

Saville,J.,1957,RURAL DEPOPULATION IN ENGLAND AND WALES, 1851-1951,London,Routledge and Kegan Paul.

304

Schaffer,J.H.,1976,'Poor town, proud people:the socio-economic structure of an Irish community',Boulder, PhD Thesis, University of Colorado.

Schneider,P.,J.Schneider and E.Hansen, 1972, Modernisation and development: the role of regional elites and noncorporate groups in the European Mediterranean, COMPARATIVE STUDIES IN SOCIETY AND HISTORY,14,328-50.

Schulman, M.D., 1981, Ownership and control in agribusiness corporations,RURAL SOCIOLOGY,46,652-68.

Schumacher,E.F.,1973,SMALL IS BEAUTIFUL:ECONOMICS AS IF PEOPLE MATTERED,New York,Harper and Row.

Schumpeter, J.A., 1943, CAPITALISM, SOCIALISM AND DEMOCRACY, London,Allen and Unwin.

Schwartz,M.,1976, RADICAL PROTEST AND SOCIAL STRUCTURE: THE SOUTHERN FARMERS' ALLIANCE AND COTTON TENDENCY 1880-1890, New York,Academic Press.

Schwarzweller, H.K., J. Brown and J.J. Mangalam,1971, MOUNTAIN FAMILIES IN TRANSITION: A CASE STUDY OF APPALACHIAN MIGRATION, University Park, Pennsylvania State University Press.

Scott,J.,1982, THE UPPER CLASSES: PROPERTY AND PRIVILEGE IN BRITAIN,London,Macmillan.

Selznick,P.,1949,TVA AND THE GRASS ROOTS,New York,Harper and Row.

Shaffer,R.E.,1974, Rural industrialisation: a local income analysis,SOUTHERN JOURNAL OF AGRICULTURAL ECONOMICS,6(1), 97-102.

Sharp,C., D.Gwynn and O.E.Thompson,1986,Farm size and the role of farm women,RURAL SOCIOLOGIST,6,259-64.

Shaw,M.,1976, Can we afford villages?, BUILT ENVIRONMENT QUARTERLY,2 (June),135-7.

Shaw,T., and K.Willis,1982, 'Green Belts',unpublished paper to the Rural Study Group of the Institute of British Geographers,Annual Conference,Southampton.

Sheail, J.,1975, The concept of national parks in Britain, TRANSACTIONS OF THE INSTITUTE OF BRITISH GEOGRAPHERS,66, 41-56.

Sheail, J.,1981, RURAL CONSERVATION IN THE INTER-WAR YEARS, Oxford,Oxford University Press.

Sher, J.P. and R.B. Tompkins,1977, Economy, efficiency and equality: the myths of rural school and district consolidation, in J.P. Sher (ed), EDUCATION IN RURAL AMERICA, Boulder,CO,Westview,43-77.

Shoard, M., 1980, THE THEFT OF THE COUNTRYSIDE, London, Temple Smith.

Shover,J.L.,1965,CORNBELT REBELLION:THE FARMERS' HOLIDAY ASSO-CIATION,Urbana,University of Illinois Press.

Shucksmith,D.M.,1980,Petrol prices and rural recreation in the 1980s, NATIONAL WESTMINSTER BANK QUARTERLY REVIEW, February,52-9.

Shucksmith,D.M.,1981,NO HOMES FOR LOCALS,Aldershot,Gower.

Shucksmith, D.M., and G. Lloyd,1982, The Highlands and Islands Development Board regional policy and the Invergordon closure, NATIONAL WESTMINSTER BANK QUARTERLY REVIEW,May, 14–24.

Sigelman,L.,1983, Politics, economics and the American farmer, RURAL SOCIOLOGY,48,367–85.

Silk,C.P.and J.A.Silk,1985,Racism,nationalism and the creation of a regional myth:the southern states after the American civil war, in J.Burgess and J.R.Gold (eds), GEOGRAPHY,THE MEDIA AND POPULAR CULTURE,London,Croom Helm,165–91.

Silverman,S.F.,1965, Patronage and community-nations relationships in central Italy,ETHNOLOGY,4,172–89.

Simmons,J.W.,1976,Short-term income growth in the Canadian urban system,CANADIAN GEOGRAPHER,20,419–31.

Simon,W. and J.H.Gagnon,1967,The decline and fall of the small town,TRANS-ACTION,4(5),42–51.

Sinclair, P.R. and K. Westhues,1974,VILLAGE IN CRISIS,Toronto, Holt,Rinehart and Winston.

Sjoberg, G., R.A. Brymer and B. Farris,1966,Bureacracy and the lower class, SOCIOLOGY AND SOCIAL RESEARCH,50,325–37.

Smith,E.G.,1980, America's richest farms and ranches,ANNALS OF THE ASSOCIATION OF AMERICAN GEOGRAPHERS,70,528–41.

Smith,I.J.,1978,'Ownership status and employment changes in northern region manufacturing industry 1963–73', Newcastle-upon-Tyne, University of Newcastle-upon-Tyne, Centre for Urban and Regional Development Studies Discussion Paper 7.

Smith,I.J.,1982, The role of acquisition in the spatial distribution of the foreign manufacturing sector in the United Kingdom, in M.J. Taylor and N.J. Thrift (eds), THE GEOGRAPHY OF MULTINATIONALS,London,Croom Helm,221–51.

Smith,J.P.,1983,Depression political patterns and the current farm crisis,RURAL SOCIOLOGIST,3,111–4.

Solomon,L.,1980,ENERGY SHOCK: AFTER THE OIL RUNS OUT,Toronto, Doubleday Canada.

Solstad,K.J.,1971,Education in a Highland region: geographical factors in school progress, SCOTTISH EDUCATIONAL STUDIES, 3,3–9.

Spindler,G.D.,1973, BURGBACH: URBANISATION AND IDENTITY IN A GERMAN VILLAGE,New York,Hold,Rinehart and Winston.

Spooner,D.J.,1972,Industrial movement and the rural periphery, REGIONAL STUDIES, 6,197–215.

Springate,D.,1973, REGIONAL INCENTIVES AND PRIVATE INVESTMENT, Montreal,C.D. Howe Research Institute.

Stacey,M.,1969, The myth of community studies, BRITISH JOURNAL OF SOCIOLOGY,20,134–7.

Standing Conference of Rural Community Councils,1978,'The decline of rural services',London, National Council of Social Services.

Stebbing,S.R.,1982,'Some aspects of the relationship between rural social structure and the female sex role', Ashford, PhD Thesis,Wye College,University of London.

Stein, M.R., 1972, THE ECLIPSE OF COMMUNITY, Princeton, NJ, Princeton University Press (expanded version).

Stephens, J.D.,1982, Metropolitan areas as decision-making centres, in C.M. Christian and R.A. Harper (eds), MODERN METROPOLITAN SYSTEMS, Columbus, OH, Charles E.Merrill, 111–46.

Stevens,J.B.,1980,The demand for public goods as a factor in the nonmetropolitan migration turnaround,in D.L.Brown and J.M. Wardwell (eds), NEW DIRECTIONS IN URBAN–RURAL MIGRATION,New York,Academic Press,115–35.

Stewart,W.,1969, 'Whatever you write about me, don't call me a goddam fugitive from justice': J.Doyle's gamble with millions, heads he wins – tails you lose, MACLEAN'S MAGAZINE,82(2),30–6.

Stillwell,F.J.B.,1982, Capital accumulation and regional economic performance, AUSTRALIAN GEOGRAPHICAL STUDIES,20, 131–43.

Stinchcombe,A.L.,1961, Agricultural enterprise and rural class relations,AMERICAN JOURNAL OF SOCIOLOGY,67,165–76.

Stokes,B.,1985,The divided farm lobby, NATIONAL JOURNAL,17, 632–38.

Strathern,M.,1982, The village as an idea, in A.P. Cohen (ed), BELONGING: IDENTITY AND SOCIAL ORGANISATION IN BRITISH RURAL CULTURES,Manchester,Manchester University Press.

Strathern,M.,1984,The social meaning of localism,in T. Bradley and P.D. Lowe (eds), LOCALITY AND RURALITY,Norwich, Geo Books,181–98.

Strong, A.L.,1975, PRIVATE PROPERTY AND THE PUBLIC INTEREST, Baltimore,Johns Hopkins University Press.

Summers,G.F.,S.D.Evans,F.Clemente,E.M.Beck and J.Minkoff,1976, INDUSTRIAL INVASION OF NONMETROPOLITAN AMERICA, New York, Praeger.

Suttles, G.D.,1972, THE SOCIAL CONSTRUCTION OF COMMUNITIES, Chicago,University of Chicago Press.

Swann, D., 1984, COMPETITION AND INDUSTRIAL POLICY IN THE EUROPEAN COMMUNITY,London,Methuen.

Taaffe, E.J., H.L. Gauthier and T.A. Maraffa, 1980, Extended commuting and the inter metropolitan periphery,ANNALS OF THE ASSOCIATION OF AMERCIAN GEOGRAPHERS,70,313–29.

Tarrant,J.R.,1980, FOOD POLICIES,Chichester,Wiley.

Tarrant,J.R.,1981,Food as a weapon? The embargo on grain trade between USA and USSR,APPLIED GEOGRAPHY,1,273–86.

Tarrant,J.R.,1982,EEC food aid,APPLIED GEOGRAPHY,2,127–41.

Taylor, M.J. and N.J. Thrift,1981, British capital overseas: direct investment and corporate development in Australia, REGIONAL STUDIES,15,183–212.

Taylor, M.J. and N.J. Thrift,1982,Models of corporate develop-
    ment and the multinational corporation,in M.J. Taylor and
    N.J.Thrift (eds), THE GEOGRAPHY OF MULTINATIONALS,London,
    Croom Helm,14-32.
Taylor,M.J. and N.J.Thrift,1986,New theories of multinational
    corporations,in M.J. Taylor and N.J. Thrift (eds), MULTI-
    NATIONALS AND THE RESTRUCTURING OF THE WORLD ECONOMY,
    London,Croom Helm,1-20.
Taylor,P.J.,1980,'A materialist framework for political geo-
    graphy',Newcastle-upon-Tyne,University of Newcastle-upon-
    Tyne,Department of Geography Seminar Paper 37.
Taylor,P.J.,1983,The question of theory in political geography
    in N.Kliot and S.Waterman (eds), PLURALISM AND POLITICAL
    GEOGRAPHY,London,Croom Helm,9-18.
Thoman,G.R.,1973, FOREIGN INVESTMENT AND REGIONAL DEVELOPMENT,
    New York,Praeger.
Thomas,B.,1972,MIGRATION AND URBAN DEVELOPMENT,London,Methuen.
Thompson, J.D., 1967, ORGANISATIONS IN ACTION, New York,
    McGraw-Hill.
Thorpe,D.,1972, Village, town or settlement system?, in W.D.C.
    Wright and D.H. Stewart (eds), THE EXPLODING CITY,
    Edinburgh,Edinburgh University Press,106-18.
Tickamyer,A.R.,1983, Rural-urban influences in legislative
    power and decision-making,RURAL SOCIOLOGY,48,133-47.
Tilly,C.,1974, Do communities act?, in M.P. Effrat (ed), THE
    COMMUNITY,New York,Free Press,209-40.
Todd,D.,1983, Observations on the relevance of the industrial-
    urban hypothesis for rural development, REGIONAL STUDIES,
    14,45-54.
Tolbert, C., P.M. Horan and E.M. Beck,1980, The structure of
    economic segmentation: a dual economy approach, AMERICAN
    JOURNAL OF SOCIOLOGY,85,1095-116.
Took,L.,1986, Land tenure,return migration and rural change in
    the Italian province of Chieti, in R. King (ed) RETURN
    MIGRATION AND REGIONAL ECONOMIC PROBLEMS,London, Croom
    Helm,79-99.
Tornedon,R.L.,1975, FOREIGN DISINVESTMENT BY US MULTINATIONAL
    CORPORATIONS,New York,Praeger.
Townsend, A.R. and C.C. Taylor, 1975, Regional culture and
    identity in industrialised societies: the case of North
    East England,REGIONAL STUDIES,9,379-93.
Trelease, A.W., 1981, Southern violence: the Ku Klux Klan, in
    M.Black and J.S.Reed (eds), PERSPECTIVES ON THE AMERICAN
    SOUTH, VOLUME ONE,New York,Gordon and Breach,23-33.
Tricker,M.,1984,Rural education services:the social effects of
    reorganisation, in G. Clark et al (eds), THE CHANGING
    COUNTRYSIDE,Norwich,Geo Books,111-9.
Tweeton, L. and G.L. Brinkman, 1976, MICROPOLITAN DEVELOPMENT,
    Ames,Iowa State University Press.

Twight,B.W.,1983,ORGANISATIONAL VALUES AND POLITICAL POWER:THE
    FORESTRY SERVICE VERSUS THE OLYMPIC NATIONAL PARK,Univer-
    sity Park,Pennsylvania State University Press.
Tylecote,A.B.,1982, German ascent and British decline, 1870-
    1980: the role of upper-class structure and values, in
    E. Friedman (ed), ASCENT AND DECLINE IN THE WORLD SYSTEM,
    Beverley Hills,Sage,41-67.
Urry,J.,1981,Localities,regions and social class,INTERNATIONAL
    JOURNAL OF URBAN AND REGIONAL RESEARCH,5,455-73.
Urry,J.,1984,Capitalist restructuring,recomposition and the
    regions, in T. Bradley and P. Lowe (eds), LOCALITY AND
    RURALITY,Norwich,Geo Books,45-64.
US Department of Commerce,Economic Development Administration
    (1972),PROGRAMME EVALUATION: THE ECONOMIC DEVELOPMENT AD-
    MINISTRATION GROWTH CENTRE STRATEGY,Washington DC,Govern-
    ment Printing Office.
Useem, M. and A. McCormack,1981, The dominant segment of the
    British business elite,SOCIOLOGY,15,381-406.
Van Den Ban,A.W.,1960,Locality group differences in the adop-
    tion of new farm practices,RURAL SOCIOLOGY,25,308-320.
Vanhove, N., and L. Klaassen,1980, REGIONAL POLICY: A EUROPEAN
    APPROACH,London,Saxon House.
Veblen,T.,1923, ABSENTEE OWNERSHIP AND BUSINESS ENTERPRISE IN
    RECENT YEARS:THE CASE OF AMERICA,New York,B.W.Huebsch.
Vernon,J.,1981,'Protecting the environment: grass roots action
    in Surrey and Newham', Guildford, PhD Thesis, University
    Surrey.
Vernon,R.,1979,The product cycle hypothesis in a new indus-
    trial environment,OXFORD BULLETIN OF ECONOMICS AND STATI-
    STICS,41,255-67.
Vidich, A.J., and J. Bensman,1968,SMALL TOWN IN MASS SOCIETY:
    CLASS,POWER AND RELIGION IN A RURAL COMMUNITY,Princeton,
    NJ,Princeton University Press (second edition).
Vincent,J.,1978,'The political economy of Alpine development:
    tourism or agriculture in St.Maurice',Exeter,University
    of Exeter,Exeter Research Discussion Paper 3.
Vogeler,I.,1981,THE MYTH OF THE FAMILY FARM:AGRIBUSINESS DOMI-
    NANCE OF US AGRICULTURE,Boulder,CO,Westview Press.
Wade,R.,1971,Political behaviour and world view in a central
    Italian village,in F.G.Bailey (ed), GIFTS AND POISON: THE
    POLITICS OF REPUTATION,Oxford,Blackwell,252-80.
Wallman,S.,1977,The shifting sense of 'us': boundaries against
    development in the western Alps, in S. Wallman (ed),
    PERCEPTIONS OF DEVELOPMENT, Cambridge, Cambridge
    University Press,17-29.
Walmsley,M.,1982,Non-participation in planning:the case of
    Mundford in Norfolk,in M.Moseley (ed), POWER,PLANNING AND
    PEOPLE IN RURAL EAST ANGLIA,Norwich,Geo Book,43-74.

Walters, P.,1975,Farming politics at the grass roots: the
    Cheshire County Branch of the National Farmers' Union, in
    DECISION-MAKING IN BRITAIN III:AGRICULTURE,Milton Keynes,
    Open University Press,251-74.
Warde,A.,1986,Space,class and voting in Britain,in K. Hoggart
    and E.Kofman (eds), POLITICS,GEOGRAPHY AND SOCIAL STRATI-
    FICATION,London,Croom Helm,33-61.
Warner,P.W.,1973,Aspects of school transport in a rural scene,
    ASPECTS OF EDUCATION,17,19-37.
Warner, W.K. and W.D. Heffernan,1967,The benefit-participation
    contingency in voluntary farm organisations, RURAL SOCIO-
    LOGY,32,139-53.
Warren,R.L.,1978,THE COMMUNITY IN AMERICA,Chicago,Rand McNally
    (third edition).
Watkin,D.G.,1978,'The problems of small independent grocery
    and provisions shops in the market towns of mid-Wales',
    Lampeter, PhD Thesis, St. David's University College,
    University of Wales.
Watts,H.D.,1981,THE BRANCH PLANT ECONOMY,London,Longman.
Weaver,T.,1977, Class conflict in rural education, in J.P.Sher
    (ed), EDUCATION IN RURAL AMERICA, Boulder, CO, Westview,
    159-203.
Weber,M.,1948, FROM MAX WEBER, London,Routledge and Kegan Paul
    (H.H. Gerth and C.W. Mills eds).
Weekley,I.G.,1977, Lateral dependence as an aspect of rural
    service provision,EAST MIDLAND GEOGRAPHER,6,361-74.
Welch,R.V.,1984,The meaning of development,NEW ZEALAND JOURNAL
    OF GEOGRAPHY,76,2-4.
Wenger,G.C.,1980,Learning a living in mid-Wales,CAMBRIA,7(1),
    86-95.
Wenger,G.C.,1981,'The elderly in the community: mobility and
    access to services', Bangor, University College of North
    Wales,Department of Social Theory and Institutions,Social
    Services in Rural Areas Working Paper 16.
West, C.,1981, 'Rural female economic activity', Gloucester,
    Gloucestershire College of Arts and Technology, Papers in
    Local and Rural Planning 13.
West,J.,1945, PLAINVILLE, USA, New York, Columbia University
    Press.
Westergaard, J. and H. Resler,1975, CLASS IN A   CAPITALIST
    SOCIETY:A STUDY OF CONTEMPORARY BRITAIN,London,Heinemann.
White, D., 1974, The village life, NEW SOCIETY, 26 September,
    790-4.
White, F. and L. Tweeten,1973,Internal economics of   rural
    elementary and   secondary   schooling, SOCIO-ECONOMIC
    PLANNING SCIENCES,7,353-69.
White, R.L. and H.D. Watts,1977, The spatial evolution of an
    industry: the example of broiler production, TRANSACTIONS
    OF THE INSTITUTE OF BRITISH GEOGRAPHERS,2,175-91.

White,S.,1979,BRITAIN AND THE BOLSHEVIK REVOLUTION: A STUDY IN THE POLITICS OF DIPLOMACY 1920-1924,London,Methuen.

White, S.E.,1983, Return migration to Appalachian Kentucky, RURAL SOCIOLOGY,48,471-91.

Whitt,J.A.,1984, Structural fetishism in the new urban theory, in M.P. Smith (ed), CITIES IN TRANSFORMATION, Beverley Hills,Sage Urban Affairs Annual Review 26,75-89.

Wilczynski,J.,1976,THE MULTINATIONALS AND EAST-WEST RELATIONS, London,Macmillan.

Wild,R.A.,1974,BRADSTOW:A STUDY OF STATUS,CLASS AND POWER IN A SMALL AUSTRALIAN TOWN,Sydney,Angus and Robertson.

Wild,T.,1983,The residential dimension to rural change, in T. Wild (ed), URBAN AND RURAL CHANGE IN WEST GERMANY, London,Croom Helm,161-99.

Wildavsky,A.,1971,THE REVOLT AGAINST THE MASSES,New York, Basic.

Wilkinson,K.P.,1978,Rural community change,in T.R.Ford (ed), RURAL USA,Ames,Iowa State University Press,115-25.

Wilkinson, K. P., 1984, Rurality and patterns of social disruption,RURAL SOCIOLOGY,49,23-36.

Wilkinson,K.P.,1985,Rural community development: a deceptively controversial theme in rural sociology,RURAL SOCIOLOGIST, 5,119-24.

Willard,M.,1923, HISTORY OF THE WHITE AUSTRALIA POLICY, Melbourne,Melbourne University Press.

Williams,C.H.,1980, Ethnic separatism in western Europe, TIJDSCHRIFT VOOR ECONOMISCHE EN SOCIALE GEOGRAFIE,71,142-58.

Williams,G.,1984, Development agencies and the promotion of rural community development,in A.W.Gilg (ed), COUNTRYSIDE PLANNING YEARBOOK 5,Norwich,Geo Books,62-86.

Williams,G.,1985,The achievement of specialist agencies in rural development, in S.Barrett and P.Healey (eds), LAND POLICY:PROBLEMS AND ALTERNATIVES,Aldershot,Gower,249-72.

Williams,W.M.,1956,THE SOCIOLOGY OF AN ENGLISH VILLAGE: GOSFORTH,London,Routledge and Kegan Paul.

Williams,W.M.,1963, A WEST COUNTRY VILLAGE, ASHWORTH: FAMILY, KINSHIP AND LAND,London,Routledge and Kegan Paul.

Wilson,G.K.,1977,SPECIAL INTERESTS IN POLICY-MAKING: AGRICULTURAL POLICIES AND POLITICS IN BRITAIN AND THE UNITED STATES OF AMERICA,1956-1970,Chichester,Wiley.

Wilson,G.,1978,Farmers' organisations in advanced societies,in H. Newby (ed), INTERNATIONAL PERSPECTIVES IN RURAL SOCIOLOGY,Chichester,Wiley,31-53.

Wittkopf,E.R.,1974,The concentration and concordance of foreign aid allocations, in K.R. Cox et al (eds), LOCATIONAL APPROACHES TO POWER AND CONFLICT,New York, Wiley,301-39.

Wood,L.A.,1924, A HISTORY OF FARMERS' MOVEMENTS IN CANADA, Toronto,Ryerson Press.

Woollett,S.,1982,'Alternative rural services:a community init-
    atives manual,London,Bedford Square Press.
World Food Institute,1985,WORLD FOOD TRADE AND US AGRICULTURE
    1960-1984,Ames,Iowa State University Press.
Worster,D.,1979,DUST BOWL:THE SOUTHERN PLAINS IN THE 1930s,
    New York,Oxford University Press.
Wright,G.,1955,Peasant politics in the third French republic,
    POLITICAL SCIENCE QUARTERLY,70,75-86.
Wylie, L.,1966, CHANZEAUX: A VILLAGE IN ANJOU, Cambridge, MA
    Harvard University Press.
Wylie,L.,1974, VILLAGE IN THE VAUCLUSE, Cambridge, MA, Harvard
    University Press (third edition).
Yeats,A.J.,1981, TRADE AND DEVELOPMENT POLICIES, London,Mac-
    millan.
Youngs,M.J.,1985,The English television landscape documentary:
    a look at Granada,in J.Burgess and J.R.Gold (eds), GEO-
    GRAPHY,THE MEDIA AND POPULAR CULTURE,London,Croom Helm,
    144-64.

INDEX

AGRICULTURE: agribusiness 67-70,142,146-7,148,150,156,237,239;
   cost-price squeeze 70,136,140,147,149,232,233; family
   farming 3,13,15,65,66,147,156,237; farming as an
   occupation 15,136,154,196; government subsidies 87,90,
   157,162,211; industrialised 70,111,139,140-52,233,238;
   plantations 65,66; product prices 60,86,87,97,100,102,
   113,138,139; sharecropping 13,66,144-5.
BANKS: 136,137,138,142,169; credit 64,97,137-9,257.
CLASS STRUCTURES: 2,4,12,15,33-40,60; local 12,13,44,49-50,
   193,210,214,230-40; in structuring development 40,50,126,
   272.
COMMUNICATIONS: 133,183,255,266.
DEVELOPMENT: as a process 25-7; definition 18-27; for geographical areas 50,51; Maslow's schema 21-3,218; time framework 25,27; value differences 19,20,24,237.
DEVELOPMENT INSTITUTIONS: 54,145,176,177; Development Board
   for Rural Wales 173,185; Development Commission 171,174,
   175-6; Highlands and Islands Development Board 170,173,
   174-5,176-7; Tennessee Valley Authority 38,168,170,171,
   176,261.
EMPLOYMENT CONDITIONS: absentee-ownership 67-81,123,128,134-5,
   164; job creation 119,133,187-8,214,225; job stability
   76,116,128,134-5; segmented labour markets 80,131-2,259;
   trade unions 127,128,129,130,170,187,233-5; worker exploitation 14,19,45,120,233,235,258.
ENVIRONMENTAL ISSUES: 94,125,128,129,159-60,217-8; landscape
   23,221,223,226; local lobbies 218-26.
EUROPEAN COMMUNITY: 50,81-94,112,113; Common Agricultural
   Policy 81-91,94,95; industrial effects 81,88,91; regional
   policy 88,90-3.
FARM STRUCTURE: bifurcation 14-5,64-5,138,151,161,162,163;
   sizes 86,145,169,235,257,258; social effects 3,137,236-9,
   244-5.
FOOD: 146; aid 105,107,146; prices 60,68,86-7,107,142,160,205.
GENDER: 253-61.
GOVERNMENT: agricultural policy 68,81,86,140,145,152,154,156-

PLANNING: land-use 19,129,171,175,188-9,220-5; settlement policy 119,178,212-7,259.
POPULATION: in-migration 10,13,27,117-8,190-1,220,227,228,248, 251; international migration 58,127,139,232,235; out-migration 13,14,19,22,189-90,193,212,215,243,255; turn-over 12,13.
POWER STRUCTURES: agrarian-industrial interactions 48,58,59, 60,124,125,136,138,139,170; community power structures 4, 46,49,121,154,187-9,222,234,240-52; conflictual 51,125, 248-52; elitist 36-7,123-4,156-7,242-5; structured 31,49, 56,59,65,253.
PROTEST MOVEMENTS: agrarian 32,59,97,105,119,124,125,136-40, 170,241; civil rights 26,119,252; regional 43,44,59,122.
RESEARCH: romanticism 1,4,10,147,182; weaknesses 2,3,4,5,231, 238,253.
RURAL: area types 17,114-5,117,133; causal importance 9,16,17, 56; definition 2,8-18; ecological interpretation 16-8; imagery and idealism 10,11,137,147,155-6,197-8,230; occupational interpretation 14-6; rural-urban continuum 2,10,11,13,116,231,240,261; socio-cultural interpretation 9-14,19,27,117,208,231,240-1.
RURAL GEOGRAPHY: 3,4,5,6,39,197,265-8.
RURAL SOCIOLOGY: 4,8,197.
SERVICES: centralisation 3,165,203-12,216,237-8,259,260; col-lective 168,193,205-10,213,248,251,259; provision stan-dards 206-8,216; retailing 146,147,165-6,189,204-5,237.
SETTLEMENT STRUCTURES: 200-3; central place systems 202,213, 216,238; commuter villages 12,23,49,177,183,201,231; counterurbanisation 10,13,119,120,121,249,269; metropoli-tan regions 4,16,47,194,201,204,225; rapid growth 46,184, 227; restructuring 178,214-6; socio-economic specialis-ation 49,216-7.
THEORY: 1,3,4,9,29,34,76,115,268,270-2; conflict theory 31-40, 49,126,166; diffusionism 2,54,116-21,198-9,201-3; geo-graphically-based development ideas 9,37,40-50,54,55,114-5; integration theory 30-2,166,266; internal colonialism 121-6; marxist 33-5,37,39,77,126; uneven development 126-35; world economy 53-5,56.
TRANSNATIONAL CORPORATIONS: 9,48,50,58,67-81,95,100,106,167, 266; spatial patterns 71,75,77-8; truncation 58,73-4,76, 78,81.
TRANSPORT: 169,204,211,214,220,258; friction of distance 9, 265-6.
US CIVIL WAR: 58,59,62-4,66,120.

As the main empirical reference points for this book are Britain and the United States, these nations have only been included in the geographical index when reference was made to a region (i.e. not to the nation as a whole).

THE AUTHORS

KEITH HOGGART obtained BSc and MSc degrees from the University of Salford and has a PhD from the University of London. Currently Lecturer in Geography at King's College, University of London, he has also held a Commonwealth Scholarship at the University of Toronto and a Fulbright Scholarship at the University of Maryland and Temple University. Editor, with Eleonore Kofman, of Politics, Geography and Social Stratification (Croom Helm, 1986), his current research interests are in the fields of rural development and urban public finance.

HENRY BULLER was educated at the University of London, where he obtained a BA degree from University College and a PhD from King's College. Currently working in Paris and Reims, he is an Honorary Research Fellow at the Department of Geography, King's College, University of London and a Research Associate in the Department of Town Planning, Oxford Polytechnic. Having worked on a number of research projects concerning recreational conflict in the countryside, his current research interests are land policy and development processes in rural France.